Cases on Strategic Information Systems

Mehdi Khosrow-Pour, D.B.A.
Editor-in-Chief, Journal of Cases on Information Technology

IDEA GROUP PUBLISHING
Hershey • London • Melbourne • Singapore

Acquisitions Editor:	Michelle Potter
Development Editor:	Kristin Roth
Senior Managing Editor:	Amanda Appicello
Managing Editor:	Jennifer Neidig
Typesetter:	Jennifer Neidig
Cover Design:	Lisa Tosheff
Printed at:	Integrated Book Technology

Published in the United States of America by
Idea Group Publishing (an imprint of Idea Group Inc.)
701 E. Chocolate Avenue, Suite 200
Hershey PA 17033
Tel: 717-533-8845
Fax: 717-533-8661
E-mail: cust@idea-group.com
Web site: http://www.idea-group.com

and in the United Kingdom by
Idea Group Publishing (an imprint of Idea Group Inc.)
3 Henrietta Street
Covent Garden
London WC2E 8LU
Tel: 44 20 7240 0856
Fax: 44 20 7379 0609
Web site: http://www.eurospanonline.com

Library of Congress Cataloging-in-Publication Data

Cases on strategic information systems / Mehdi Khosrow-Pour, editor.
    p. cm.
  Summary: "This book provides practitioners, educators, and students with examples of the successes and failures in the implementation of strategic information systems in organizations"--Provided by publisher.
  Includes bibliographical references and index.
  ISBN 1-59904-414-5 (hardcover) -- ISBN 1-59904-415-3 (softcover) -- ISBN 1-59904-416-1 (ebook)
  1. Management information systems--Case studies. 2. Strategic planning--Case studies. I. Khosrowpour, Mehdi, 1951-
  HD30.213.C37 2006
  658.4'038011--dc22
                                2006003569

British Cataloguing in Publication Data
A Cataloguing in Publication record for this book is available from the British Library.

The views expressed in this book are those of the authors, but not necessarily of the publisher.

Cases on Information Technology Series

ISSN: 1537-9337

Series Editor
Mehdi Khosrow-Pour, D.B.A.
Editor-in-Chief, *Journal of Cases on Information Technology*

- Cases on Database Technologies and Applications
  Mehdi Khosrow-Pour, Information Resources Management Association, USA
- Cases on Electronic Commerce Technologies and Applications
  Mehdi Khosrow-Pour, Information Resources Management Association, USA
- Cases on Global IT Applications and Management: Success and Pitfalls
  Felix B. Tan, University of Auckland, New Zealand
- Cases on Information Technology and Business Process Reengineering
  Mehdi Khosrow-Pour, Information Resources Management Association, USA
- Cases on Information Technology and Organizational Politics and Culture
  Mehdi Khosrow-Pour, Information Resources Management Association, USA
- Cases on Information Technology Management In Modern Organizations
  Mehdi Khosrow-Pour, Information Resources Management Association, USA & Jay Liebowitz,
  George Washington University, USA
- Cases on Information Technology Planning, Design and Implementation
  Mehdi Khosrow-Pour, Information Resources Management Association, USA
- Cases on Information Technology, Volume 7
  Mehdi Khosrow-Pour, Information Resources Management Association, USA
- Cases on Strategic Information Systems
  Mehdi Khosrow-Pour, Information Resources Management Association, USA
- Cases on Telecommunications and Networking
  Mehdi Khosrow-Pour, Information Resources Management Association, USA
- Cases on the Human Side of Information Technology
  Mehdi Khosrow-Pour, Information Resources Management Association, USA
- Cases on Worldwide E-Commerce: Theory in Action
  Mahesh S. Raisinghani, Texas Woman's University, USA
- Case Studies in Knowledge Management
  Murray E. Jennex, San Diego State University, USA
- Case Studies on Information Technology in Higher Education: Implications for Policy and
  Practice
  Lisa Ann Petrides, Columbia University, USA
- Success and Pitfalls of IT Management (Annals of Cases in Information Technology, Volume 1)
  Mehdi Khosrow-Pour, Information Resources Management Association, USA
- Organizational Achievement and Failure in Information Technology Management
  (Annals of Cases in Information Technology, Volume 2)
  Mehdi Khosrow-Pour, Information Resources Management Association, USA
- Pitfalls and Triumphs of Information Technology Management
  (Annals of Cases in Information Technology, Volume 3)
  Mehdi Khosrow-Pour, Information Resources Management Association, USA
- Annals of Cases in Information Technology, Volume 4 - 6
  Mehdi Khosrow-Pour, Information Resources Management Association, USA

IDEA GROUP INC.

Cases on Strategic Information Systems

Detailed Table of Contents

Infosys Technologies Ltd., one of the world's most profitable IT services company, implemented a customer relationship management (CRM) system called CIMBA (Customer Information Management By All). This customer-focused system was conceived and designed to improve communication and collaboration between the company and its customers. By seamlessly integrating the front-end sales system with the back-end delivery system, CIMBA was expected to further enhance the company's IT solutions delivery capability. This case provides insights into the factors that triggered the need for developing such an integrated CRM solution and how the company went about developing and launching this system. It also brings to light the various challenges associated with the implementation of this IS solution.

This case will focus on Keane's approach to Project Management and how they provide this service to their clients. This includes not only how Keane is hired for Project Management but how they train their clients on how they too can implement the Keane philosophy of Productivity Management. The goal of this case is to provide the student with an example of business-technology strategy in action and allow them to explore future paths that Keane may take based on how they use technology today and in the decade to come.

This case study examines an often overlooked context of information system failures, that of pre-implementation failure. It focuses on an Information Systems Development (ISD) project at a large public university that failed even before implementation could get under way. Specifically, it describes the vendor selection process of a proposed computerized maintenance management system.

Celerity Enterprises competes in the semiconductor manufacturing industry. At the start of the case, business conditions are favorable for them to launch a new production facility to manufacture flash memory. The new facility must achieve exceptionally ambitious productivity and cost goals. A facility-level strategic planning process reveals opportunities to substitute information for other more-expensive resources. By the end of the case, just a few months later, worldwide economic conditions change radically and the future of the new facility is in jeopardy. The case describes the participants, the planning process and findings. It provides a rich setting to discuss aligning information and business planning, realities of the volatile industry, outsourcing for IS planning leadership, and using a combination of top-down and bottom-up planning.

This case gives a detailed description of the adoption of an e-business initiative by Miracle Industries Limited (MIL), a fast-moving consumer goods (FMCG) organization in India. The initiative involved linking up with key distributors so as to get important sales-related data on a real-time basis. The case describes how the decision to adopt the project was taken after a comprehensive analysis involving a detailed cost-benefits study, and an examination of the roles of various stakeholders — the distributors and the Territory Sales Officers.

the design and development of the Lotus Notes™ workflow management system. The design description includes process maps for the as-is and the new system. In addition, descriptions of the testing phase, the pilot, and the roll-out are included. The case concludes with a discussion of project success factors and planned future enhancements.

The implementation of the payroll and human resources modules of an integrated software product in a large manufacturing organization is described in this case. Some of the problems encountered include conflicts between the accounting and human resources departments, technical difficulties in building interfaces to existing systems, inadequate staffing of the project team, the IT director who left during the project, and a poorly functioning steering committee.

This case describes the experiences of Vicro Communications, who sought to reengineer its basic business processes with the aid of data-centric enterprise software. Vicro management however made the mistake of relying completely on the software to improve the performance of its business processes. It was hoped that the software would increase information sharing, process efficiency, standardization of IT platforms, and data mining/warehousing capabilities. Unfortunately for Vicro, the reengineering effort failed miserably even after investing hundreds of millions of dollars in software implementation. As a result, performance was not improved and the software is currently being phased out.

This case study highlights the concept that the "management" of the technology is usually the limiting factor causing the demise of a project rather than the "technology" itself. This real case study involves creating an awareness of a new technology within the company and trying to start a much-needed project using this technology.

This case discusses how a small specialist medical clinic (named ECS) tries to stay competitive by applying innovations through information technology and the Web. The barriers to success include the lack of financial and human resources but through forming strategic alliances, it managed to implement some systems prototypes.

The subject of this case is a computer-based cost-benefit forecasting model (CBFM), developed to investigate possible long-term effects of improved productivity from the use of modern software engineering. The primary purpose of the model was to generate comparative data to answer "what-if" questions posed by senior corporate management attempting to understand possible overall effects of introducing the new software development methodologies.

The case study describes the process of planning and implementation of integrated software designed to automate the sales schedule process. The application was released with numerous software, hardware and network problems. The effects on the customer community, the information systems department, and other stakeholders were sharp and far reaching.

This case presents data concerning the choices among information system development strategies, tools, systems which could be selected for upgrade or development, and implementation decisions for an insurance company facing a dynamic business environment.

The development of an information system at the Missouri-based General Chemicals Inc., which is one of the largest pharmaceuticals in the world, is described in this case. The system efficiently utilizes company's research resources by archiving, organizing them in a searchable manner, and allowing researchers to reuse and modify them as needed. This resulted in saving a considerable amount of money and time.

This case study describes the gradual evolution of the use of information technology, first to support basic transaction processing, and ultimately to support the strategic issues that such an operation faces The Service Employees International Union, Local 36 Benefits Office, provides service to over 3,500 union members in Pennsylvania's Delaware Valley area by the Service Employees International Union, Local 36 Benefits Office, which provides service to over 3,500 union members.

This case study examines Nzmilk, a small, successful, fresh milk supplier. The company has lost some of their market to competitors, but also gained a significant contract. Its general manager was convinced that a new IS strategy was needed, but did not know how to proceed.

This case describes how a majority of users of an Electronic Medical Record (EMR) at a family medicine clinic located in a small city in the western United States are currently quite dissatisfied with the system. The practice experienced a disastrous implementation of the EMR in 1994 and has not recovered.

The process of planning, analysis, design and implementation of an integrated voice interactive device (VID) for the Navy is presented in this case. The goal of this research is to enhance Force Health Protection and to improve medical readiness by applying voice interactive technology to environmental and clinical surveillance activities aboard U.S. Navy ships.

In Bulgaria, mineral resources are property of the State. The case specifies the functions of "Geochemia," information systems and technologies used to support the execution of its activities in the time of transition to market economy.

This case describes how a not-for-profit agency, the Appalachian Center for Economic Networks (ACEnet), facilitates the use of e-commerce by rural small businesses as a part of an overall strategy for spurring economic development through small businesses.

J. Martin Santana, ESAN, Peru
Jaime Serida-Nishimura, ESAN, Peru
Eddie Morris-Abarca, ESAN, Peru
Ricardo Diaz-Baron, ESAN, Peru

The case describes the implementation process of an ERP (Enterprise Resource Planning) system at Alimentos Peru, one of the largest foods manufacturing companies in Peru. The case explains the criteria used to evaluate and select the system, as well as the main issues and problems that arose during the implementation process. More specifically, the case focuses upon a set of implementation factors, such as top management support, user participation, and project management.

Preface

Strategic information systems have certainly transformed the way that modern organizations manage their resources and compete within today's global market. During the past two decades, many successful organizations have discovered that through the use of information technology, they can utilize their resources in support of developing very successful strategies for managing their organizations effectively. *Cases on Strategic Information Systems*, part of Idea Group Inc.'s *Cases on Information Technology Series*, presents a wide range of the most current issues, challenges, problems, opportunities, and solutions related to the development and management of strategic information systems. Real-life cases included in this publication demonstrate successes and pitfalls of organizations throughout of the world related to the planning and utilization of strategic information systems.

The cases included in this volume cover a wide variety of topics focusing on strategic information systems, such as customer relationship management system implementation, project management companies, information systems pre-implementation failure, information and business planning within the semiconductor manufacturing industry, an e-business initiative, accounting software adoption for a municipal social services agency, the rise of a dot-com firm, information system implementation, workflow automation production part testing process experiences, the implementation of an integrated software product at a large manufacturing organization, BPR and ERP system relations, the introduction of expert systems at a corporation, a business network for a small medical practice, software engineering in product development environments, a company's new sales management application, information system upgrades at an insurance company, IT for the utilization of enterprise information resources, organizational growth through IT, information systems for a milk supplier, a clinic's experience with electronic medical records, military applications of natural language processing and software, geological research information systems, small business electronic commerce, and ERP systems implementation.

Strategic information systems have emerged as one of the most effective applications of information technology in support of the strategic management of organizations. *Cases on Strategic Information Systems* will provide practitioners, educators and students with important examples of the successes and failures in the implementation of strategic information systems, and stand as an outstanding collection of current, real-life situations associated with the effective utilization of strategic information systems. Lessons that can be learned from the cases included in this publication will be instrumental for those learning about the issues and challenges related to the development and management of strategic information systems.

Note to Professors: Teaching notes for cases included in this publication are available to those professors who decide to adopt the book for their college course. Contact cases@idea-group.com for additional information regarding teaching notes and to learn about other volumes of case books in the IGI *Cases on Information Technology Series.*

ACKNOWLEDGMENTS

Putting together a publication of this magnitude requires the cooperation and assistance of many professionals with much expertise. I would like to take this opportunity to express my gratitude to all the authors of cases included in this volume. Many thanks also to all the editorial assistance provided by the Idea Group Inc. editors during the development of these books, particularly all the valuable and timely efforts of Mr. Andrew Bundy and Ms. Michelle Potter. Finally, I would like to dedicate this book to all my colleagues and former students who taught me a lot during my years in academia.

A special thank you to the Editorial Advisory Board: Annie Becker, Florida Institute of Technology, USA; Stephen Burgess, Victoria University, Australia; Juergen Seitz, University of Cooperative Education, Germany; Subhasish Dasgupta, George Washington University, USA; and Barbara Klein, University of Michigan, Dearborn, USA.

Mehdi Khosrow-Pour, D.B.A.
Editor-in-Chief
Cases on Information Technology Series
http://www.idea-group.com/bookseries/details.asp?id=18

Chapter I

Infosys Technologies Limited:
Unleashing CIMBA

Debabroto Chatterjee, The University of Georgia, USA

Rick Watson, The University of Georgia, USA

EXECUTIVE SUMMARY

Infosys Technologies Ltd., one of the world's most profitable IT services company, implemented a customer relationship management (CRM) system called CIMBA — Customer Information Management By All. This customer-focused system was conceived and designed to improve communication and collaboration between the company and its customers. By seamlessly integrating the front-end sales system with the back-end delivery system, CIMBA was expected to further enhance the company's IT solutions delivery capability. This case provides insights into the factors that triggered the need for developing such an integrated CRM solution and how the company went about developing and launching this system. It also brings to light the various challenges associated with the implementation of this IS solution.

BACKGROUND

There was a lot on Nitin's mind as he came out of yet another marathon meeting, a meeting to discuss problems the Infosys sales team was facing with the current customer relationship management (CRM) solution. As the head of the Sales & Marketing Group in the Information Systems Department, Nitin Gupta was responsible for the software support needs of Infosys' sales teams.

Only two years ago, he along with a team of four, had deployed a CRM package called CRMX,[1] for meeting the contact management and opportunity tracking needs of the sales team. The disappointing performance of CRMX had been the reason for this and numerous earlier meetings.

At the time it was implemented, CRMX, a sales force automation (SFA) tool, marked the first automation attempt by Infosys in the area of customer relationship management. One of the primary objectives was to establish a centralized repository for the contact database of the enterprise and help the sales teams target potential customers. Another major objective was to improve responsiveness to customer needs by enabling seamless sharing of information between the onsite sales team and offshore delivery teams. Thus, the system was expected to provide synergistic benefits by facilitating communication and knowledge sharing during every stage of the order generation and fulfillment process. The system was also expected to be easily scalable to meet the growing information needs of the company.

Though CRMX was a good and robust tool, it was not meeting Infosys' expectations, especially those of the sales people. The business development managers (BDMs), who were supposed to be the key users of CRMX, found it difficult and cumbersome to use. Moreover, it did not enable real-time sharing of information between the onsite (at customer locations) sales personnel and offshore delivery personnel. Phone, fax, and e-mail continued to be the primary means of communication and information exchange for the personnel located at client offices and sales offices in different parts of the world. As a result, these sales personnel operating from remote locations felt disconnected and isolated and often called themselves the *lone warriors*.

Nitin, a graduate of one of India's premier business schools, was a smart young man, who fully understood the changing business needs of his company in a rapidly growing software consultancy market. A multitude of ideas ran through his mind. He knew something more scalable and dynamic was needed and that Infosys had to do away with the current tool, once and for all. Infosys had done enough performance tuning of the system to realize the futility of any further performance enhancement exercise. Sivashankar, the IS Head at Infosys and Basab Pradhan, the Regional Manager for the Chicago office, both key players in the technology initiatives for the sales team, were also very keen on a new system. Like Nitin, they were convinced that CRMX had outlived its usefulness.

Company History and Growth

Infosys Technologies Ltd. was founded in 1981 by seven engineers working for Patni Computers, a small reseller of U.S.-based Data General. Its aim was to be a key player in the software solutions market. Narayana NR Murthy, the CEO and founder of the company, was a visionary, who could see the potential in the still infantile software solutions market — a market in which Infosys built its reputation by not only providing high-quality solutions at low cost, but also by reducing customer risk through effective execution of fixed-time and fixed-price contracts.

The first employees were hired in 1982 and the company started by offering onsite services to foreign customers. It quickly built a reputation as a provider of quality turnkey software development and maintenance services (Appendix 1). The client list grew and included many major firms across the globe. To expand its customer base in the U.S., Infosys also entered into a strategic marketing alliance with Kurt Salmon and Associates, a management consulting firm, and this move helped the company gain valuable name recognition.

Through its initial public offering (IPO) in February 1993, Infosys raised much-needed cash to expand operations at its headquarters in Bangalore. By 2002, Infosys was employing more than 10,000 software professionals. Its revenues had grown significantly — from a mere $3 million in 1990 to $545 million in 2002. Its compounded annual growth rate (CAGR) from 1996-2001 was a staggering 50%. *Business Week* (June 24, 2002) ranked it as the third most profitable IT company in the world, ranking higher than IBM and Dell.

This India-based company has come a long way. Starting from a one-room office in 1982, Infosys today has sales offices in 17 countries, and several software development and training centers in India. It also has many notable firsts to its credit. For instance, it was the first Indian software solutions provider to be listed on NASDAQ (INFY) and also receive the Capability Maturity Model (CMM) Level 5 certification from the Software Engineering Institute at Carnegie Mellon University (www.infosys.com; Trivedi & Singh, 1999). The CMM certification process assesses the level of maturity and capability of software development processes used by IT services companies; Level 5 is the highest achievable.

Its growing reputation as a world-class provider of high-quality solutions attracted the brightest talents from India's top business and engineering schools. For the last several years, it had been consistently voted as India's most admired company and best employer in a variety of business publications (Trivedi & Singh, 1999).

The Sales Organization

Infosys has sales offices in several countries. The U.S. headquarters, which also serves as the global sales headquarters, is in Fremont, California. In the U.S., it has sales offices in Phoenix (AZ), Fremont (CA), Lake Forest (CA), Atlanta (GA), Lisle (IL), Quincy (MA), Troy (MI), Berkeley Heights (NJ), Dallas (TX), and Bellevue (WA).

At the core of the Sales Team are the Business Development Managers (BDMs). They are responsible for sales prospecting and generating clients and business for Infosys. They are assigned to one particular territory and are responsible for business growth in that area. They report to a Regional Manager, who is responsible for a very large geographical area, termed a region in Infosys terminology.

Infosys has its delivery team divided into Strategic Business Units (SBUs). These units are organized in terms of geographic or domain specializations. For instance, there is an SBU called West and North America (WENA) that handles projects in most of North America, and there is an Enterprise Solutions (ES) SBU that handles projects in SAP, Baan, and other ERP technologies.

Other than the sales people, there are Account Managers (AMs) for large accounts. AMs are located at clients' sites and are the faces of the Offshore Delivery teams for the client. They are responsible for maintaining the client relationship, ensuring project deliveries, and generating new business from existing clients.

The IS Organization

The IS organization is a 150-member group and has the challenge of handling the technology needs of a technology-savvy company. The demands on the group are high, as system design requirements keep changing all the time — "our customers sit next door, so it's inevitable" is how the MIS group describes the situation. It has played a key role in making Infosys a truly paperless workplace. As Sivashankar often said, "The IS Department at Infosys is a key enabler and inherent part of all company systems and processes."

The IS Department is divided into smaller groups, each of which specializes in serving the needs of the various departments in the company. For instance, there is a Finance group that fulfills the software requirements of the Finance Department, an HR group for Human Resources, and the Sales & Marketing group that develops software for the Sales Team. Thus, each business function is mirrored within the MIS Department.

The Industry

The birth of the IT professional services industry (in which Infosys operates) can be traced back to the mid-1960s when companies such as Electronic Data Systems (www.eds.com) started providing data processing services. The industry continued to grow and the service offerings extended beyond data processing. For instance, the service offerings ranged from custom software development to facilities management, system design, software consulting, and training and documentation (Trivedi & Singh, 1999).

While the demand for software solutions continued to grow, companies in developed economies, like the U.S., were finding it prohibitively expensive and time consuming to develop solutions internally. Moreover, as technology cycles shortened and the complexity of computer systems grew, companies were finding outsourcing to be a more viable IT management strategy (Nalven & Tate, 1989; Clabum, 2003). Under these circumstances, Infosys found itself ideally positioned to make the most of this opportunity. This India-based company had a talented pool of programmers to draw from, and an efficient and mature software development and delivery process (Trivedi & Singh, 1999).

SETTING THE STAGE

The Infosys Global Delivery Model

Infosys's ability to deliver low-cost and high-quality solutions was primarily due to its Global Delivery Model (GDM). This GDM model (Appendix 2) relies on geographically dispersed teams seamlessly working at the lowest work breakdown level, and in multiple time zones to deliver significant customer value. GDM leverages key company strengths like a wide global presence and fast-acting offshore development teams.

While GDM was proving to be quite effective, its optimal utilization greatly depended on the effective coordination, communication, and collaboration between the onsite customer-facing sales teams and the offshore delivery teams. However, the support systems, to facilitate such interactions and information sharing, were either inaccessible or not integrated. For instance, the onsite personnel at client locations had to rely on fax and e-mail to communicate with their offshore counterparts. These onsite personnel had very limited access to the different corporate systems; many a times, they had literally no access to the data/information generated from these systems. To make matters worse, the existing customer relationship management system (CRMX) was not integrated with the back-end delivery systems. Such lack of integration greatly impeded the ability of the sales force to effectively respond to queries from current and prospective clients, promptly relay customer feedback to the delivery teams, and identify cross-selling opportunities.

Thus, CRMX had to be replaced with a more powerful and integrated CRM platform. Such empowering of the electronic infrastructure was essential for the future success of Infosys's GDM.

CASE DESCRIPTION

CRMX: The Current CRM Platform

CRMX, once an innovative product, was launched as an off-the-shelf customizable CRM package for mid-sized enterprises. It was easy to deploy and had a good contact management framework. Another major advantage with CRMX was that it had an easy-to-understand user interface. Infosys in the mid-1990s was looking for just such a package — easy to understand and deploy, and relatively inexpensive.

When Infosys implemented CRMX, it marked a breakthrough in sales force automation; but over a period of time, for many reasons, it started losing its usefulness. Primarily, it created another data island in the company, and the IS Department found integration with other existing systems extremely complex. Secondly, access to all the stakeholders (essential for effective visibility across the service chain) was proving to be prohibitively expensive because access was controlled through a custom security set-up in the system and required individual license for each named user. Each user also had to have the system installed on her/his machine. Another major reason for wanting to replace CRMX was that it was only a sales-facing system. Infosys at that time was a very rapidly evolving company, but with few systems that linked sales with the delivery group. "We just don't know what you guys do out there in the field," was a common refrain among the Offshore Delivery people.

Most of the CRMX users were highly mobile and based out of small offices. Since CRMX did not offer Web-based access, it was becoming a problem for the mobile users to effectively use the system. The overall system performance was also a major irritant to the sales force. For instance, uploading client data by synchronizing the client PC with a "satellite server" was a perennial problem. Some Business Development Managers (BDMs) were often heard complaining, "I dread Mondays when I'm back in office and have to sit in front of the stupid server and endlessly sync my data." Comments like, "why can't it (CRMX) improve my productivity for a change?" were also quite common.

Needs Analysis

Careful analysis of existing systems before considering the adoption and implementing of a new system is standard practice at Infosys. It was a CMM level 5 company and the change management process was clearly documented. The change initiative in this case was code-named "foCus." The team formed to oversee this system replacement initiative had broad representation from all the involved groups. It included Basab (the Regional Manager, Chicago), Nitin, Jith (SBU head for the Asia-Pacific Delivery group), and many other Delivery Unit heads and other key people in the Sales Team. In fact, Nitin went the extra distance to ensure that each and every user group was represented in the steering team.

The needs analysis exercise led to some interesting findings. For instance, one of the key requirements was the need for a very different looking CRM system. This was surprising as there was an existing belief (among the "foCus" team members) that the users were happy with the look and feel of the previous CRM solution.

In addition to expressing the need for a new user interface, the respondents also emphasized the need for an integrated, fast, and cost-effective system. Since Infosys had a high level of expertise in a broad spectrum of application development and business analysis, senior management wanted an integrated system that would enable the various departments to share their respective knowledge and expertise and thereby realize synergistic benefits. The current environment of disparate applications serving disconnected user groups was no longer acceptable. Every system, it was felt, should be integrated with other systems for complete elimination of data duplication and data redundancy.

The needs analysis process led to the following guiding principles for developing the new system.

1. The customer relationship management platform will facilitate and support an integrated service — from lead generation to opportunity identification, proposal generation and submission, contract finalization, project set up, software development, and delivery.
2. From each stage in the project life cycle, relevant data will be handed over to the next stage. Duplication of data entry will be minimized.
3. Access will be provided to all stakeholders.
4. The system will be intuitive and easy to use.

Evaluating Development Options

The options essentially boiled down to (a) buying a new package and customizing it or (b) building a customized system from scratch. Nitin argued that buying an off-the-shelf system could once again take the company down the CRMX road. Rather than adjust to a company's unique business model and process competences, an outside system often forces the company to modify its processes to adapt to the system's needs (Davenport, 1998; Roy & Juneja, 2003). Moreover, the vision was to build a CRM platform that would ultimately integrate all aspects of the business — from marketing to product development and delivery and customer support.

But there were others who felt that the company should not divert its resources toward a project that was going to be fairly long drawn, and there was no guarantee that the final product would live up to expectations. They did not buy into the vision of an integrated CRM platform that would give a significant boost to Infosys's Global Delivery Model. They further suggested that Infosys send out a request for proposal from leading CRM vendors.

An intense and long drawn meeting followed to discuss the buy-versus-make issue. Compelling arguments were made in support of each of the alternatives. The committee failed to reach a consensus and a vote was taken to make the decision. By a small margin of two votes, the motion in support of developing a CRM platform was passed.

Nitin was pleased, especially because he strongly believed that the company possessed the requisite software development experience and expertise to deliver a leading-edge and long-term solution. He and his committee members then started brainstorming on how best to build a dynamic and scalable CRM system for the company. After numerous brainstorming sessions, the committee finally put together a proposal for building an integrated CRM platform.

A name was needed for the system. It had to be something with which users could identify and in Sivashankar's words "catchy." An internal e-mail was sent out seeking suggestions for a suitable name. Almost everyone made a suggestion, and finally "CIMBA" was selected. CIMBA stood for: Customer Information Management By All. Overnight, the team became the CIMBA Team. The proposal was called the CIMBA (Customer Information Management By All) program.

Thus, the intent of the CIMBA program was to build an integrated CRM platform that would effectively link the front-end customer support systems with the back-end delivery systems (Maddox, 2003; Hill, 2003). It was envisioned to be a full-cycle automation and process deployment program that would, both directly and through spin-off gains, change forever the level of integration between the onsite and offshore teams at Infosys.

Appendix 3A provides a comparative depiction of the functionalities of CIMBA and CRMX. CIMBA was designed to offer a more comprehensive set of functionalities. For instance, CIMBA would offer additional capabilities in the areas of knowledge management, campaign management, and sales revenue forecasting. CIMBA was also expected to be a more integrated CRM platform that would effectively link the front-end customer support systems with the back-end delivery systems. Such integration would facilitate the handing down of data from one stage of the project life cycle to the next. Moreover, its superior data synchronization capabilities would provide (both the sales and delivery teams) access to real-time data. CIMBA would also support unique business rules and access controls, to insure, among other things, accuracy of revenue projections and avoidance of losses from unauthorized projects. It was also expected to (a) be more scalable, (b) provide greater information visibility and access to all stakeholders, and (c) be intuitive and easy to use.

Potential Payoffs

The CIMBA program had revenue generation, risk reduction, and streamlining processes in client interaction as major goals.

For Infosys, the most important metric for measuring the impact of CIMBA was *average relationship size* — "relationship size" (it is an Infosys term) refers to the amount of revenue being generated from each client. It was expected that the roll-out of CIMBA will enable the company to identify more cross-selling opportunities and thereby increase revenue.

Infosys had lost money on projects that were not properly authorized. By significantly improving process controls, CIMBA could cut down such losses. Moreover, CIMBA-enabled process controls would facilitate new levels of process compliance in the workings of the client-facing personnel, thereby greatly improving sales and revenue predictability.

The CIMBA application would also offer (management) granular visibility to field activity and the forecasting process. Finally, CIMBA would greatly enhance connectivity and visibility across the customer service chain. Superior connectivity and visibility would result in:

- Better targeting and follow-up of prospective clients.
- Improved quality and speed of response to customer queries.
- Offshore managers having a much better sense of the clients' and the market's pulse.
- Greater communication and coordination between sales and delivery teams.

Establishing a Development Team

Nitin set out to form the team that would develop the new system. He chose Sunil Thakur to lead the team. Sunil had recently joined the IS Department and had just finished a project as a Team Leader. He was very good at mapping user requirements to system functionalities. The other members in the team were Krishna, who was the Technical lead, and four developers who had recently joined the department. It was a young team and the developers had little previous experience in building a large-scale Web-based system. The database administrator and graphics designer were sourced from a common pool in the MIS Department. Development of the system started in November 2000.

Development and Implementation of CIMBA

Sunil decided on following an iterative approach for developing the system prototype. The next six months were spent developing the system from scratch. While building an integrated CRM platform seemed like a powerful idea, translating that idea into reality entailed overcoming several types of implementation hurdles. These hurdles ranged from technical to procedural and organizational (Corner & Hinton, 2002; Kenyon & Vakola, 2003; Schmerken, 2003).

The CIMBA program called for building a class of Web applications that were never attempted before in the organization. It also called for seven other applications to be modified on synchronized timelines for seamless integration. For the first time, an application was being built for internal use that was not only to be used extensively from locations outside India on a true global scale, but also used in a sustained manner. Users were expected to stay connected for sessions as long as a full working day. To deliver acceptable performance in a sustained usage scenario to globally located users, the team had to find new technologies and approaches.

The team selected XML as the technology of choice for developing a system that was scalable and could be seamlessly integrated with all types of devices and all other systems (Hagel & Brown, 2001; Marvich, 2003). Since none of the team members had prior experience with XML, a four-day training session was organized for the developers; the developers were given another week to become comfortable with the new technology.

The CIMBA initiative also involved a large-scale process definition and mapping exercise that required organization-wide consensus building. Key organizational entities had to agree on the complete set of life cycle processes — from prospecting to opportunity management, forecasting, engagement initiation and execution.

A listserv called CIMBA TEAM was set up, and all stakeholders and future users were made members of this CIMBA TEAM distribution list. This listserv was used to share ideas, seek suggestions, discuss problems, post progress reports, and make announcements.

The CIMBA program also called for one of the most complex multi-phased data migration and process integration roll-outs. It called for migrating data from disparate sources into a centralized repository. It was clear that there were a lot of data inconsistencies, and hence the migration exercise would involve a large data cleanup exercise. The cleanup effort was not a simple technical cleanup; it required continuous liaison with business users and data owners across the organization to determine the correct mappings in the destination data set.

The development team also had to deal with scope creep challenges. Requests for changes kept coming even after the steering committee had come to an agreement on design and functionality specifications. Many of these requests led to changes in the look and feel of the system.

By June 2001, the first phase of the project was complete. CIMBA provided customer and contact management functionalities and a means to record opportunities (see Appendix 3B for a CIMBA homepage screen shot). It was rolled out to select "pilot" users, most of whom belonged to the WENA business unit. The initial user response was one of excitement. The pilot users found the system easy to use and intuitive to navigate. Some bugs and small user issues did crop up, but the CIMBA team resolved these problems fairly quickly.

Phase 2 involved integrating CIMBA with all other relevant systems. These were systems that were already in use and automated processes like business proposal generation, letter of engagement, and project set up. It was here that the steering committee hit the first significant roadblock. The stakeholders from Sales and Delivery had different ideas on how the integration should be carried out, and they found it difficult to reach consensus. There were numerous conference calls, meetings, and brainstorming sessions, but no decision was reached for a full four months.

Meanwhile, Sunil and the CIMBA team kept on working to improve the system capabilities. Since the application was hosted on servers at Bangalore (a city in India) and the pilot users were in North America, system response was at times slow. To improve response time, programming code was optimized, cache settings were adjusted, and database tables were clustered. As Sunil once said, "Since we have the time, let us ensure that not a single piece of code is heavy. Let's ensure that when we send out data over the network, we send it in such a way that it's lean, mean, and quick."

As October 2001 dawned, there was intense pressure (from the Board of Directors) on the steering committee to reach consensus on the CIMBA integration process. Finally, at the end of October, consensus was reached on how best to integrate CIMBA with the other systems. With these new specifications in place, the CIMBA team resumed work in earnest. Sunil and Krishna sat down with the teams that were maintaining the other systems and determined the changes required to integrate with CIMBA. Once these technical details were identified, it took just a month to finish Phase 2. By the first week of December, the integrated version of CIMBA was ready for delivery. In Nitin's words, "The lion king is ready to roar," referring to Simba, the lion cub character in the movie *The Lion King*.

CIMBA is Rolled Out

In mid-December the integrated CIMBA suite was rolled out for one customer account, Insureco,[2] a U.S.-based insurance provider and one of Infosys' prime clients. The reasons for choosing this particular company as the pilot account were many — a large account and a very tech-savvy AM, Michael Hudson, who was asked to be the key driver of CIMBA adoption in the Insureco team. Before the roll-out, many training sessions were held for the Insureco team, both in India and the U.S., to make it comfortable with the system.

Enabling CIMBA for an account can be quite complex as it involves synchronizing multiple databases and enforcing "gates" across multiple applications. But, the development team came up with a very elegant solution to make this process fairly simple. A combined database "switch" was made, which was a roll up of all the downstream switches. Once the data were migrated to the CIMBA database, all that was left was to turn the "CIMBA Switch" on and the account was automatically migrated to the CIMBA platform. This one solution must have saved the CIMBA team many hours of error-prone work.

Even though there were some initial hitches with the use of the CIMBA system for the INSURECO account, its implementation was deemed a success. Visibility of account-related information greatly increased across the entire management chain — from business development managers to account managers, delivery managers, and the entire top management team.

Bolstered by the success, it was decided to roll-out CIMBA, one by one, to the various sales regions (Roberts, 2004). The roll-out involved certain preparatory activities such as data migration from the CRMX database to the new CIMBA database, extensive training for both onsite and offshore users, and a whole list of other checks. Alongside the roll-out came carefully planned training and, as one Regional Manager put it, "impeccable user support." By the time the roll-out was completed, 356 users across the globe had been trained; total training time was about 2,026 hours, out of which 896 hours were done outside India.

However, to the surprise of the development team, within the first few days of the roll-out, the support calls would almost quadruple. There were BDMs requesting access to accounts, harried financial analysts complaining that they couldn't access their region's accounts, and marketing analysts claiming "access denied" for creation of Letter of Engagements. As one developer remarked, "It has become a madhouse here at the CIMBA support hotline." Sunil's response was prompt, "No amount of training and pre-roll-out can prepare us for the post-roll-out period. These calls will come no matter how smooth the roll-out is. So folks, it makes no sense to wish them away!"

As Nitin and Basab looked forward to 2002, and the upheavals in the software services market, they were thankful that they had CIMBA in place. While CIMBA was off to a good start, Nitin and Basab remained hopeful that this CRM platform would survive the test of time and prove to be a very good investment for the company. A lot of time, money, and effort had gone into building this platform. Some of the skeptics in the organization were still of the opinion that it would have been more prudent to have bought or leased a CRM solution from a leading vendor and thereby reduced some of the development, maintenance, and upgrade costs.

CURRENT CHALLENGES/PROBLEMS FACING THE ORGANIZATION

While the CIMBA platform has greatly improved communication and collaboration between the sales and delivery people, it has yet to realize its ultimate goal of electronically integrating the entire value chain. While it does integrate the front-end customer support systems with many of the back-end delivery systems, the extent of integration is far from being total and seamless. The current version of the system is yet to enable the seamless sharing of data with several other corporate systems that are responsible for functions like project execution and operations, finance, accounting, and human resource management. For instance, the following systems are yet to be integrated with CIMBA.

- **PSWEB:** a manpower allocation (to projects) system
- **A/R Tracking:** used to track the recovery of amount billed to customers
- **IPM:** a tool used to manage an active project

Mobilizing organization-wide support and commitment to bring about that level of integration has been difficult. Moreover, the relative newness of CIMBA made it difficult to convince those who still doubt the stability and scalability of the platform. Several of the key decision makers want to watch the performance of the current version of the system for a couple of years before expanding its scope. They are also keen on realizing some quick and significant returns from the time and money invested in developing CIMBA.

In addition to the challenge of achieving greater process and functional integration, there is also the challenge of developing and/or adopting a suitable set of metrics to measure the effectiveness of CIMBA. While the CIMBA proponents have claimed victory by suggesting that this new system has realized its three major goals — revenue generation, risk reduction, and streamlining client interaction-related processes — these claims are yet to be well substantiated with a relatively objective set of metrics. For instance, it was claimed that CIMBA resulted in an improvement in the average relationship size, that is, the average revenue earned from current clients. But several of the skeptics questioned the claim and argued that the increase in revenue was due to a boom in the outsourcing business and the company's solid reputation; one of them commented, "I can't see how the CIMBA platform has improved our capability to effectively execute the Global Delivery Model."

Computing the total cost of ownership (TCO) of CIMBA is yet another operational hurdle. Identifying and included all relevant expense categories — ranging from direct and indirect labor costs to hardware and software costs and the costs of providing on-going training, maintenance, and user support — has been a challenge. The assumptions and estimates that have to be made for computing TCO have also been grounds for disagreement and dispute. For instance, while many felt that the TCO should be computed for three-to-five years, there were others who wanted it to be more long term. This disagreement essentially stemmed from two distinct views about the potential benefits from CIMBA. According to one school of thought, CIMBA was a short to medium-term technology solution for automating marketing, service, and sales functions. They had a narrow view of CIMBA — a CRM solution that was a competitive necessity and not a source of sustained competitive edge. But then there was the other group that envisioned the CIMBA program to be much more than a technology solution for automating sales, service, and the marketing function. To them, CIMBA: (a) represented a concerted effort to improve all business processes to better meet the needs of the customer; and (b) epitomized a customer-focused, long-term approach to achieve a better alignment between the company's business and IT strategy.

The other challenge that CIMBA implementers are grappling with is to get the users, both at the client sites and offshore delivery sites, to learn to use CIMBA more effectively. Despite providing several training sessions, several users are found to use only a limited set of the functionalities. For instance, many in the sales organization still treat and use it as a tool for contact management; they have yet to learn to use its various analytical capabilities to gauge customer needs and behavior and develop effective marketing strategies.

ACKNOWLEDGMENTS

This case is intended to be the basis for class discussion rather than to illustrate either effective or ineffective handling of a management situation. While the situation discussed in this case is real, some distortions have been intentionally made for enhancing its effectiveness as a teaching tool.

We would like to acknowledge the contributions of Mr. Amit Gupta and Mr. Nitin Gupta in helping us prepare this case.

REFERENCES

Clabum, T. (2003, November 10). Bear Stearns signs up for overseas services. *InformationWeek*, *963*, 34.

Corner, I., & Hinton, M. (2002). Customer relationship management systems: Implementation risks and relationship dynamics. *Qualitative Market Research*, 5(4), 239-252.

Davenport, T. H. (1998). Putting the enterprise into the enterprise system. *Harvard Business Review*, 76(4), 121-133.

Ebner, M., Hu, A., Levitt, D., & McCrory, J. (2002). How to rescue CRM. *The McKinsey Quarterly*, (Special Edition), 49-57.

Georgiadis, M., Seshadri, R., & Yulinsky, C. (2002). Tactical CRM. *McKinsey Marketing Solutions*, 1-9.

Hagel, J., III, & Brown, J. S. (2001). Your next IT strategy. *Harvard Business Review*, 79(9), 105-113.

Hill, S., Jr. (2003). Next-generation CRM. *Customer Management*, 44-45.

Kenyon, J., & Vakola, M. (2003). Customer relationship management: A viable strategy for the retail industry. *International Journal of Organization Theory and Behavior*, 6(3), 329-353.

Maddox, K. (2003). CRM focus will be on integration in 2003. *B to B*, 88(2), 12.

Marvich, V. (2003). CRM across the enterprise: Integrating the channels. *Customer Inter@ction Solutions, 21*(9), 42-45.

Nalven, C., & Tate, P. (1989, February 1). Software: Asia — taking the offshore option. *Datamation, 35*(3), 72-74.

Roberts, B. (2004). Big bang or phased roll-out. *HRMagazine, 49*(1), 89.

Roy, A., & Juneja, A. (2003). CRM applications: Licensed or hosted — Which is better for you? *Customer Inter@ction Solutions,* 32-34.

Schmerken, I. (2003). Earning a payback on CRM: Smaller projects, more homework. *Wall Street & Technology,* 12-16.

Trivedi, B., & Singh, J. (1999). *Infosys Technologies Limited (A).* Wharton Case Study.

APPENDIX 1. INFOSYS PRODUCTS AND SERVICES

The table below provides a snapshot of the IT solutions and services provided by Infosys to a wide range of industry verticals.

Illustrative solution areas	Engagement Life-cycle Stages	Typical Engagements
• Business Process Management • Customer Relationship Management • Knowledge Management • Mobile Computing and M-Strategy • Supply Chain Management • IT Strategy • Infrastructure Management Services	• Consulting and Strategy • Architecture and Integration • Custom Systems Development • Re-engineering and migration services • Maintenance and Support	**Automotive** • Structural analysis of engine mounting bracket. • Weight optimization of EDH bracket. • Full range analysis and solutions for seating systems. **Aerospace** • Design and analysis of airframe sections for aerospace vehicles. • Software for buckling based design of rocket motor casing. • Load analysis software for aerospace application. **Financial Services** • Development of online trading systems for brokers. • Development of a transaction processing system known as the Authorization and Capture system. • Development of an application architecture. **Heathcare** • Development of an electronic health record system. • Development of a web-based integrated application to automate the authorization process. • Development of an application architecture.

Source: Infosys corporate Web site, www.infosys.com

APPENDIX 2. INFOSYS GDM IN ACTION

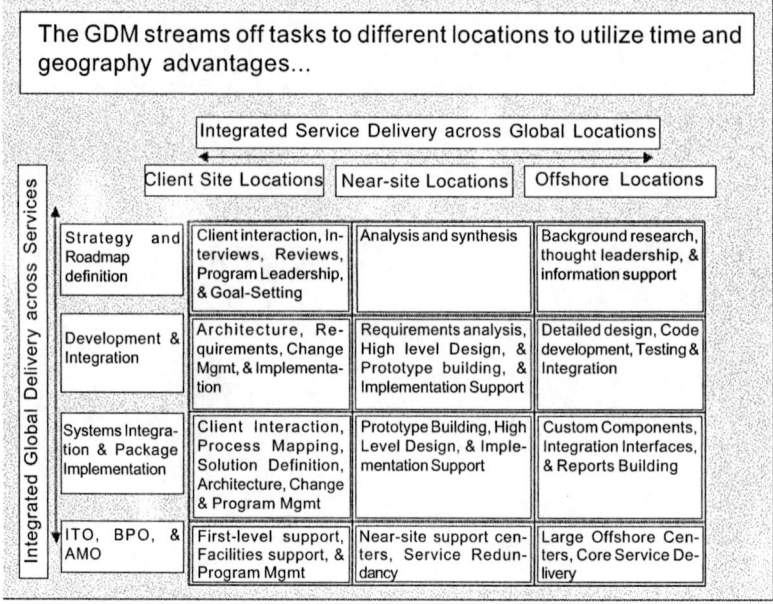

ITO: *Information Technology Outsourcing*
BPO: *Business Process Outsourcing*
AMO: *Application Maintenance/ Management Outsourcing*

APPENDIX 3A. CRMX AND CIMBA FUNCTIONALITIES

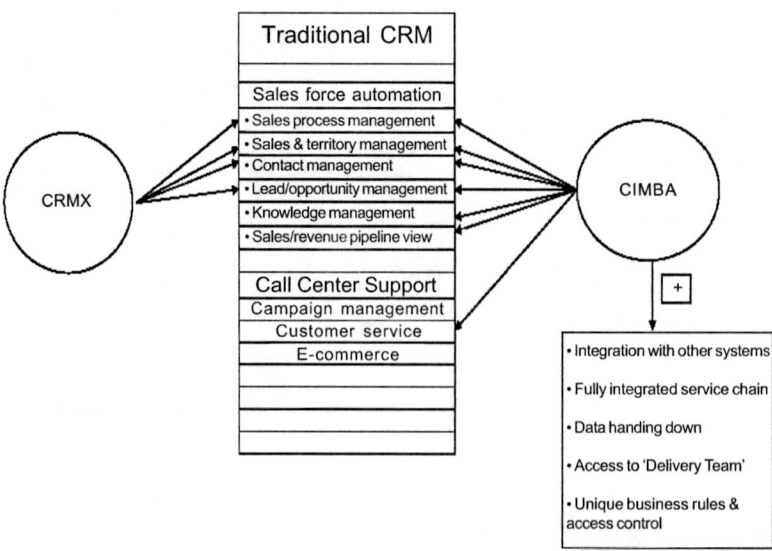

APPENDIX 3B. CIMBA HOMPAGE

Debabroto Chatterjee (Dave) is assistant professor at the Terry College of Business, The University of Georgia. His research interests lie at the interface between information technology and strategy. *His work has been published in several reputed journals like* MIS Quarterly, Journal of Management Information Systems, Communications of the ACM, Electronic Markets, Information Resource Management Journal, *and* Annals of Cases on Information Technology. *He has given invited talks in premier academic and business institutions, both in the U.S. and abroad. Other notable recognitions have come in the form of the Boeing Faculty Fellowship award in 1998, nomination to the United Nations panel of e-commerce experts in 2000, and invited talk on e-business at the SIM 2003 National Conference.*

Richard Watson is the J. Rex Fuqua distinguished chair for Internet strategy and director of the Center for Information Systems Leadership in the Terry College of Business, The University of Georgia. He has published in leading journals in several fields, as well as authored books on data management and electronic commerce. His current research focuses primarily on electronic commerce and IS leadership. He has given invited seminars in more than 20 countries for companies and universities. He is president-elect of AIS, a visiting professor at Agder University College, Norway, and a consulting editor to John Wiley & Sons.

This case was previously published in the *Journal of Cases on Information Technology*, 7(4), pp. 127-142, © 2005.

Chapter II

The Application of IT for Competitive Advantage at Keane, Inc.

Mark R. Andrews, State of Arizona, USA

Raymond Papp, Central Connecticut State University, USA

EXECUTIVE SUMMARY

The Keane Company, founded in 1965 by John F. Keane, has grown from a local software service company into a national firm which has three operating divisions and over 45 branches throughout the United States, Canada and the United Kingdom. Within these operating divisions are multitudes of consulting opportunities, ranging from supplemental staffing, project management and application outsourcing. This case will focus on Keane's approach to project management and how they provide this service to their clients. This includes not only how Keane is hired for Project Management but how they train their clients on how they too can implement the Keane philosophy of productivity management. Instead of focusing on any one client of Keane, their overall technology strategy will be highlighted, from their early days through the present to illustrate how Keane has successfully incorporated information technology and project management to become a major player in the software service and consulting field. The goal of this case is to provide the student with an example of business-technology strategy in action and allow them to explore future paths that Keane may take based on how they use technology today and in the decade to come. Several discussion questions are included which focus on Keane's IT strategies and their implementation. These questions can be used to stimulate class discussion or given as written assignments to be handed in.

BACKGROUND

In 1965, John F. Keane, having worked for IBM for several years, saw the need for a software service company that would cater to firms interested in using technology to improve their operations. It was at this time that Keane decided to venture out onto his own and established The Keane Company in a small, one room office above a donut shop in Hingham, Massachusetts. From these humble beginnings, Keane has grown over the past three decades to a national firm with over 45 branch offices throughout the United States, Canada, as well as the United Kingdom. The firm has grown to such an extent that by the end of 1980s Keane was incorporated and listed as one of *Forbes*' "Top 200 Small U.S. Companies." Revenue has correspondingly increased from $99 million in 1992 to over $1 billion in 1998, and based on the most recent earnings report, the trend will continue.[1]

Keane has not always been so successful. Problems in the early 1970s with project overruns led Keane to develop its own project management process called Productivity Management. Today, this process is the foundation of Keane's application development and outsourcing methodologies. During the 1980s, Keane continued to grow and its application management methodology (AMM) was the catalyst for it outsourcing solutions (Appendix A). 1989 saw the standardization of its software development process into its Waterfall, Rapid Application Development and Client-Server Frameworks development life cycles. These frameworks continued to evolve into the 1990s.

The late 1980s and early 1990s saw Keane acquire several consulting firms, completing 17 acquisitions over a 10-year period. These acquisitions have allowed Keane to enhance its repository of processes and methodologies and prepared them to take their next major step.

In 1996, Keane debuted its Resolve 2000 service to address the Year 2000 (Y2K) problem. The growth of its Y2K solutions has allowed it to surpass its goal of $1 billion a year in revenue, two years earlier than originally planned. It has also diversified in the past few years, expanding into European markets with its acquisition of Bricker & Associates (http://www.keane.com/about/history.shtml).

SETTING THE STAGE

Today, Keane has three operating divisions: information services, healthcare services, and the federal systems business unit. Within these operating divisions are numerous consulting opportunities, among them supplemental staffing, project management, and application outsourcing. Keane is always expanding the services of its software development life cycle to include Year 2000 compliance management, project management, information systems planning, application management outsourcing, application development, help desk outsourcing, and healthcare information systems (Appendix A).

Despite the focus on Y2K issues, Keane's non-Y2K business continues to grow, up 30-35% in 1998. The areas of operations improvement consulting, e-solutions, customer relationship management, and application management outsourcing are leading Keane's growth.[2] Project and productivity management, since its inception in the early 1970s, has been the cornerstone of Keane's application development and outsourcing methodologies. Project management will be introduced briefly, followed up Keane's approach to its implementation and directions for future use of information technologies.

What is Project Management?

Project management is a specialized branch of management that has evolved in order to coordinate and control some of the complex activities of modern industry. It is concerned with planning, scheduling, and controlling non-routine activities within certain time and resource constraints. Harold Kerzner, Professor of Systems Management at Baldwin-Wallace College, defines project management as:

Project management is the planning, organizing, directing, and controlling of company resources for a relatively short-term objective that has been established to complete specific goals and objectives. Furthermore, project management utilizes the systems approach to management by having functional personnel assigned to a specific project. (Kerzner, 1995, p. 4)

Figure 1 is a representation of project management. It is intended to show that project management is designed to manage or control company resources on a given activity, within time, within cost, and within performance/technology. The outer lines, time, cost, and performance/technology signify the constraints of the project. As a consulting firm, Keane must also be concerned with an additional constraint — good customer relations. The importance of this last constraint lies within the fact that although the project is managed within time, cost, and performance parameters, but if customer relations are poor, the company may not receive future business.

The Project Manager

Usually, the identity of the project manager is hidden behind some other organizational role. This leads to the manager having increased responsibility without having any real authority. It is understood that the project manager is responsible for coordinating and integrating activities across many functional lines. The job role itself, however, is in need of a fairly standard job title, complete with job description, although not many

Figure 1.

companies can agree on qualifications. It is hardly possible to lay down an ideal personality specification for a project manager. J. Robert Fluor has tried to do just that, though. During his keynote address to the Project Management Institute, October 1977, in Chicago, he described the new responsibilities of project managers at Fluor Corporation:

Project management continues to become more challenging and we think this trend will continue. This means we have to pay special attention to the development of project managers who are capable of coping with jobs that range from small to mega projects and with life spans of several months to ten years. At Fluor, a project manager must not only be able to manage the engineering, procurement and construction aspects of a project, he or she must also be able to manage aspects relating to finance, cost engineering, schedule, environment considerations, regulatory agency requirements, inflation and cost escalations, labor problems, public and client relations, employee relations and changing laws. That's primarily on the domestic side. On the international projects, the list of additional functions and considerations adds totally different considerations. (Kerzner, 1995, p. 11)

To be effective as a project manager, an individual must have managerial, as well as technical skills. The difficulty lies within having business people that struggle to think as business people. This is where Keane, Inc. is able to excel. They are able to employ technically skilled people that have the aptitude for business. This translates into Keane being able to provide a high level of quality customer satisfaction and technical proficiency to all jobs.

Evaluation Techniques

The management of any organization involves the efficient allocation of resources. They are continually seeking new and better control techniques to cope with the complexities, tremendous amounts of data, and tight deadlines that are mainstays in many of today's industries. By seeking better methods, they are able to present technical and cost data to their customers more effectively.

Keane's project managers use network techniques to assist them. Two such techniques have come into prominence in recent times, those being the program evaluation and review technique (PERT) and the critical path method (CPM). PERT and CPM techniques can be used on small projects, as well as large projects. The more complex the activities of the projects, and the more coordination that is needed, the greater the value of PERT/CPM systems to the project manager.

These two techniques are almost identical; the difference lies in the fact that PERT attempts to handle some of the problems of uncertainty. PERT has several distinguishing characteristics:

*   It forms the basis for all planning and predicting and provides management with the ability to plan for best possible use of resources to achieve a given goal within time and cost limitations.
*   It provides visibility and enables management control "one-of-a-kind" programs as opposed to repetitive situations.

- It helps management handle the uncertainties involved in programs by answering such questions as how time delays in certain elements influence project completion, where slack exists between elements, and what elements are crucial to meet the completion date. This provides management with a means for evaluating alternatives.
- It provides a basis for obtaining the necessary facts for decision making.
- It utilizes a so-called time network analysis as the basic method to determine manpower, material, and capital requirements as well as providing a means for checking progress.
- It provides the basic structure for reporting information.

PERT/CPM systems assist the project manager in planning, scheduling, and controlling on a regular basis. It is important that the project manager realizes that their responsibilities are not static, they are definitely dynamic. Because there will be continual changes throughout the life of any project, PERT/CPM methods require there be constant revising and updating. Figure 2 illustrates the complete life cycle of project management.

Figure 2. Weiss, Joseph, and Wysocki (1992, p. 5)

PLANNING			IMPLEMENTATION	
1	2	3	4	5
DEFINE	**PLAN**	**ORGANIZE**	**CONTROL**	**CLOSE**
State the problem	Identify Project Activities	Determine Personnel Needs	Define Management Style	Obtain Client Acceptance
Identify Project Goals	Estimate Time & Cost	Recruit Project Manager	Establish Control Tools	Install Deliverables
List the Objectives	Sequence Project Activities	Recruit Project Team	Prepare Status Reports	Document the Project
Determine Preliminary Resources	Identify Critical Activities	Organize Project Team	Review Project Schedule	Issue the Final Report
Identify Assumptions & risks	Write Project Proposal	Assign Work Packages	Issue Change Orders	Conduct Post-Implementation Audit

DELIVERABLES

• Project Overview	• WBS • Project Network • Critical Path • Project Proposal	• Recruitment Criteria • Work Package Description • Work Package Assignments	• Variance Reports • Status Reports • Staff Allocation • Reports	• Final Report • Audit Report

As has been previously stated, project management is divided into different phases — planning, scheduling, and controlling. In order to gain an understanding of how these phases are applied, it is necessary to grasp what each one involves.

Planning

Planning is the process of deciding a future course of action and is something that must be done throughout the life of the project. It involves answering the questions of what, when, and how to reach the objectives of the project. PERT/CPM systems answer these questions for the project manager by creating a network, or matrix, of the project. This requires the manager to itemize every critical task that needs to be completed before the project can be considered complete. By establishing this task list, the project manager is almost guaranteeing that no section of the project will be overlooked. In addition, it helps to provide a logical sequence to the steps of the project. The manager has a plethora of tools to assist them in devising the formal project plan such as work-breakdown structures, Gantt charts, network diagrams, resource allocation charts, resource loading charts, linear responsibility chars, cumulative cost distribution, etc. (Frame, 1987, p. 8).

Scheduling

Scheduling is the process of converting a plan into an operation timetable. PERT/CPM systems provide methods for assisting the project manager in setting up the timetable for the use of project resources. Specifically, PERT scheduling is broken down into six steps:

- Steps one and two are actually combined and start with the project manager listing the activities, their priority, and their interrelationship with other components.
- Step three is for reviewing of charts and diagrams to ensure that there is neither too many nor too few activities identified, and the relationships are indeed correct.
- In the fourth step, the functional manager converts all charts and diagrams to a PERT by identifying the time lines for each activity.
- Step five is the first step on the critical path. This is where the manager first defines the actual requirements of the project itself. If the critical path does not satisfy the chronological needs, then they must modify the critical path so that it will fit.
- Step six is a matter of having the project manager begin to plug in calendar dates onto the PERT chart.

Once these steps have been completed the project manager must continually review the process to ensure that they stay as proactive as possible in preventing mishaps.

Controlling

Controlling is the process of comparing planned performance with actual performance and taking corrective action where significant differences exist. The project manager must constantly look at what has been done and compare it to the original plan, then determine if there are any major discrepancies. If there are they must then figure out if these discrepancies are of a detrimental nature and, if so, what corrective action needs to be taken. PERT/CPM systems present the manager with a consolidation of the current and future status of the project. The systems also assist by updating and reporting expected versus actual performance and by identifying new critical tasks that may need special attention.

Although managing a project is not an easy task, many project managers have a tendency to make them more difficult than needed. It is believed that effective project management can be learned. In order for this to occur, there are two basis lessons that the effective project manager must learn. First, they must be able to identify and avoid some of the common pitfalls that are encountered while managing a project. The second is how to organize and carry out the project for success. The successful project manager should proactively guide the project forward in the best, most professional manner possible.

CASE DESCRIPTION

There are several areas that Keane confronts with every project. These involve having to answer questions about the project like "How much" and "How long," implementing systems with time constraints and a limited budget, and determining why some project managers can get results more often than others.

To answer the questions "How much" and "How long" Keane holds a seminar called "The Project Estimating and Risk Management." This seminar shows what requirements are needed to have successful project planning and implementation. There are four points that Keane touches on. They are:

- Management of the estimating process,
- Identification of the risk,
- Articulation of the risk, and
- Proactive risk management.

Productivity management seminars are held to show associates how to implement projects on time within the specified cost and to the users' expectations. The seminar also shows the importance of teamwork and individual accountability. The techniques that are practiced during the seminar can be used for any project life cycle methodology or project control software.

The Project Manager Development Program teaches people how to be competent project managers. Project managers must follow project philosophies, processes, methods, and tool sets to guarantee the success of a project's execution. The project manager job functions include planning, managing tasks, managing project team, interfacing with client and the organization.

With 35 years of experience, Keane offers a unique approach to projects. With their proprietary methodologies, they ensure consistent and reputable results for their clients. Their solutions are always sized to whatever the problem, not making solutions too big or too small for the particular client or project. Keane's strategic services involve following all the steps through during the software development life cycle.

Keane follows a project management checklist that is found in the *Principles of Productivity Management*, written by Donald H. Plummer, PMP. This text is presented

The Initial Project Review
- Begin the preparation of the statement of work
- Meet with the customers and information processing management to review objectives & responsibilities
- Produce a project organization chart to identify key individuals and their positions relative to the project
- Determine staff availability
- Review budgets, objectives, and milestones
- Review schedules
- Review standards and methodologies
- Review any project control/accounting reporting procedures and systems
- Decide on a development approach
- Clearly define all products
- Prepare a document identifying all customer-significant, deliverable products

The Preliminary Planning
- Break the job down into manageable tasks; create the work breakdown structure
- Determine assignments by skill level
- Develop time estimates with team members
- Determine and finalize individual schedules; assign in writing individual team members to each task
- Enter all data into the project management software
- Define the change procedures
- Define the acceptance criteria and procedures
- Finalize all budgets
- Prepare a project plan with a software package or according to organization guidelines
- Prepare a schedule of deliverable products
- Prepare a deliverable-vs-budget schedule indicating the hours or dollars to be expanded by week versus the number of products delivered by week

The Project Confirmation
- Break the job into acceptable short intervals (80 hours or less)
- Produce a task list in which all tasks are no longer than 80 hours or two calendar weeks
- Formalize change procedure; confirm in writing how changes will be handled for the project
- Formalize acceptance criteria; confirm in writing how acceptance will occur for the project
- Prepare a project control book
- Orient team members in a team meeting
- Develop management report formats and project reporting framework
- Identify the specific weekly or monthly reports that will be prepared and distributed for the project

Tracking
- Determine task dependencies and critical paths
- Prepare weekly timesheets for all team members to report against
- Take appropriate steps to maintain team morale and focus
- Track all progress against the plan by weekly reporting of hours and dollars
- Occasionally track projects through some form of trend analysis
- Begin the project completion process
- Complete all testing
- Turn completed project over to user (Plummer, 1995, pp. 171-172)

to all employees upon attending their initial orientation when they are hired. It serves as a means for a constant reference. The checklist includes the Initial Project Review, Preliminary Planning, Project Confirmation, and Tracking.

Keane's future objective is to strengthen and improve its application development, application outsourcing, Year 2000 compliances, and staff augmentation. John Keane identified his visions for the company in his 1997 Letter from the President:

To help clients respond to business requirements with software solutions, Keane is focused primarily on application development, application outsourcing, Year 2000 compliance, and staff augmentation. These services respond to the critical needs of our clients to successfully deploy new technology, cost-effectively manage existing software assets, and complete Year 2000 compliance efforts on time. Keane is also focused on providing a full range of integrated, open information systems to clients throughout the entire Healthcare spectrum.

Our objective is to continue to strengthen and improve our capabilities in all these areas, to be the best available solution for a larger and larger number of clients. We are positioned to do this because of our strong network of branch offices, which facilitate responsive and cost-effective service delivery, and because of our project management and outsourcing methodologies, which embody Keane's 30 years of experience. Working with Keane, our clients are efficiently managing their information technology assets, improving customer satisfaction, and building new information systems more quickly.

Keane's effort to continue to strengthen and improve in all areas to be the best solution for their growing clientele is the result of their goal to be "the leading IT solution firm focused on helping clients plan, build and manage application software to achieve business advantage" (http://www.keane.com/investor_center/ index.shtml).

CURRENT CHALLENGES AND PROBLEMS

Keane, while currently successful in their application development, application outsourcing, and Year 2000 compliance, cannot rely on past performance and success to carry it through the next decade or even the next few years. Changes in the very nature in which technology is applied and leveraged in organizations requires that Keane continually assess its vision, mission, and alignment of its business strategies with its technology strategies. To be able to do this effectively, Keane must examine where it has been, where it is at present, and where it plans to be in the future. The strategic alignment model is one approach to assessing the extent of business and information technology strategy integration. This model will be explained in the next section and used to assess Keane's level of business-IT strategy integration.

Strategic Use of IT at Keane

Strategic alignment is the appropriate use of information technology in the integration and development of business strategies and corporate goals. In order for a company to assess its current alignment, it is important to address a number of questions geared toward determining whether strategies that are presently in place are profitable and IT is being utilized appropriately. Those questions focus on the current business and IT alignment, whether the business strategies and plans are being developed appropriately, and the implications of misalignment between business and IT (Papp, 1998, p. 814).

The theory of strategic alignment was first introduced in the late 1980s (Henderson & Venkatraman, 1990). Alignment looks at the relationship between business and IT. A model is used to identify the relationship and is based on two linkages (Figure 3). The model is divided into four quadrants: Business Strategy, Organizational Infrastructure, IT Infrastructure and IT Strategy. How a company assesses its alignment is by looking at the overall strengths and weaknesses. Based on the results of the assessment it is possible to determine the strongest area (this is referred to as the anchor) — this is where any change begins — and the weakest area (referred to as the pivot area) — this is the area that will be addressed. Once these areas are identified, it is possible to determine the are that will be impacted by any change.

There are a total of 12 perspectives, which include four fusion perspectives (when there are two pivot domains of equal strength, which creates a combination of two perspectives). The four original perspectives described by Henderson and Venkatraman (1993) include strategy execution, technology potential, service level, and competitive potential. The four other non-fusion perspectives are organization IT infrastructure, IT infrastructure strategy, IT organization infrastructure and organization infrastructure strategy perspective (Papp, 1996, p. 227). Table 1 shows the breakdown of these perspectives.

Figure 3.

The Strategic Alignment Model

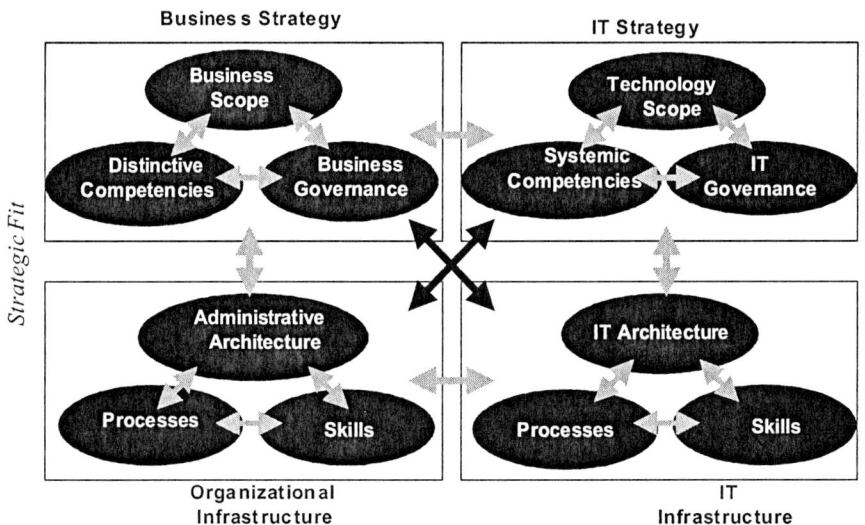

Functional Integration

Table 1. Alignment perspectives: Anchor, pivot and impacted domain (Papp, 1998, p. 7)

	Strategy Execution	Technology Potential	Competitive Potential	Service Level
Anchor	Business Strategy	Business Strategy	IT Strategy	IT Strategy
Pivot	Business Infrastructure	IT Strategy	Business Strategy	IT Infrastructure
Impact	IT Infrastructure	IT Infrastructure	Business Infrastructure	Business Infrastructure
	Organizational IT Infrastructure	IT Infrastructure Strategy	IT Organizational Infrastructure	Organizational IT Infrastructure
Anchor	Business Infrastructure	IT Infrastructure	IT Infrastructure	Business Infrastructure
Pivot	IT Infrastructure	IT Strategy	Business Infrastructure	Business Strategy
Impact	IT Strategy	Business Strategy	Business Strategy	IT Strategy

Companies that have achieved alignment can facilitate building a strategic competitive advantage that will provide them with increased visibility, efficiency, and profitability to compete in today's changing markets (Papp, 1998, p. 11).

The Keane IT Infrastructure Strategy

John Keane founded the company above a donut shop in Massachusetts with the vision of helping businesses bridge the gap he saw between the power of technology and the ability of companies to utilize that technology. With his knowledge of the technology and a vision, he was able to create an "anchor" in IT Infrastructure. What was needed was an IT strategy. As the company developed and went through growing pains, it began to develop a clear and dependable project management approach. The resulting process was Keane's means of surviving a highly competitive market.

By its 10th anniversary, Keane had continued to develop its IT strategy by recognizing the special needs of the healthcare field, which led to the establishment of another division, the healthcare services division. Again, this was a continuous process of enhancing its IT strategies which, in turn, improved the business strategy of the company. However, with the advancement of technologies came the need to refocus. They had, by the mid-1980s, achieved the IT strategy that was envisioned in the beginning.

A New Perspective

The safety and stability that Keane had grown accustomed to changed in the second quarter of 1986 when they experienced their first (and only) operating loss. This loss caused the company to refocus its efforts on its core business — software services. To capture and replicate its experience within a growing organization, Keane began to focus on internal training. Most noteworthy, an alliance with Boston University combined Keane's business capabilities with the university's technical training expertise to form an intensive training program for Keane employees called "Accelerated Software

Development Program." The idea behind this was to provide a significant competitive advantage by quickly bringing entry-level technical people up to speed in Keane's software development and project management methodologies.

In an effort to further improve its business strategy, Keane began to standardize its approach to software development, rather than relying on a handful of individual experts. This evolved into Keane's creation of numerous development life cycles which were needed for clear and disciplined guidelines for developing better business application software.

During the 1990s, Keane has made tremendous leaps in expanding market share, geographic coverage, capabilities, and healthcare products with the acquisitions of GE Consulting and PHS in 1993, AGS Computers in 1994, as well as Fourth Tier, Inc. and Bricker & Associates, Inc., in 1998. This critical mass helped the company reduce costs by spreading major expenses over a broader revenue base, which has strengthened the entire business strategy for the company. This strength has translated into a stronger organizational infrastructure that should help Keane move into the 21st century.

REFERENCES

Frame, J. D. (1987). *Managing projects in organizations.* San Francisco: Jossey-Bass.

Henderson, J., & Venkatraman, N. (1990). *Strategic alignment: A model for organizational transformation via information technology* (Working Paper 3223-90). Massachusetts Institute of Technology, Sloan School of Management.

Henderson, J., & Venkatraman, N. (1993). Strategic alignment: Leveraging information technology for transforming organizations. *IBM Systems Journal, 32*(1), 4-16.

Keane, J. (1995). *President's letter.* Retrieved from http://www.keane.com

Kerzner, H. (1995). *Project management.* New York: Van Nostrand Reinhold.

Luftman, J. (1996). *Competing in the information age: Practical applications of the strategic alignment model.* New York: Oxford University Press.

Luftman, J., & Papp, R. (1996, August). Business and IT strategic alignment: New perspectives and assessments. *Proceedings of the 2nd Americas Conference on Information Systems*, Phoenix, Arizona.

Papp, R. (1995). *Determinants of strategically aligned organizations: A multi-industry, multi-perspective analysis.* Dissertation, Stevens Institute of Technology, Hoboken, NJ.

Papp, R. (1998). *Achieving business and IT alignment: Emerging information technologies.* Hershey, PA: Idea Group Publishing.

Papp, R. (1998). Alignment of business and information technologies strategy: How and why. *Information Management, 11*(3), 4.

Plummer, D. H. (1995). *Productivity management.* Boston: Keane, Inc.

Weiss, J. W., & Wysocki, R. K. (1992). *5-phase project management.* Reading, MA: Addison-Wesley.

ENDNOTES

[1]   Revenues for the first six months ending June 30, 1999 were $565 million, up 13.8% over the first half of 1998. Net income was also up almost 25%, rising from $45.5

million in the first half of 1998 to $56.8 million for the first six months ending June 30, 1999. (http://www.keane.com/investor_center/2ndq99.shtml).

[2]   Y2K revenue was 22% of total revenue in the second quarter of 1999, down from 36% for the same period last year. Keane's outlook is that its growth will continue to lead the industry. (http://www.keane.com/investor_center/2ndq99.shtml)

APPENDIX A. SERVICES PROVIDED BY KEANE, INC.

Year 2000 Compliance Management helps to manage the complicated endeavors of reaching century compliance on time. Keane combines project management and technology migration expertise with the tool-assisted capabilities from industry leading tool vendors.

Information System Planning service identifies the clients most critical information needed for the organization, projects, and infrastructure that will provide the largest return on investment.

Application Management Outsourcing deals with maintenance and management of all or portions of the client's application. During this time Keane offers benefits such as reduced costs, improved control, and established metrics for continuous improvements.

Application Development services includes planning, design, and implementation services. All of these services can be customized to meet the clients needs and requirements.

Help Desk Outsourcing helps the client control hidden costs of end-user support and provides consistent Information System Support services for all business areas and applications.

Its *Project Management* services help to show participants how to complete projects successfully and efficiently. Keane's project managers provide seminars to their clients and employees to assist them in managing project environment, develop capable project managers, and identify and migrate project risks.

APPENDIX B. STRATEGIC ALIGNMENT MATRIX

Characteristics	Current Marketplace	Emerging Marketplace
BUSINESS		
Business Scope		
Distinctive Competencies		
Business Governance		
Administrative Structure		
Processes		
Human Resources		
INFORMATION TECHNOLOGY		
Technology Scope		
Systemic Competencies		
IT Governance		
IT Architecture		
Processes		
Human Resources		
BUSINESS		
Entry Barriers		
Executive Role		
INFORMATION TECHNOLOGY		
IT Focus		
CIO Role		
IT Value		
Alignment Perspective		
Strategic Planning Method		

(Papp, 1996)

Strategic Alignment Components

I.   BUSINESS STRATEGY

1.Business Scope – Includes the markets, products, services, groups of customers/clients, and locations where an enterprise competes as well as the buyers, competitors, suppliers and potential competitors that affect the competitive business environment.

2. Distinctive Competencies – The critical success factors and core competencies that provide a firm with a potential competitive edge. This includes brand, research, manufacturing and product development, cost and pricing structure, and sales and distribution channels.

3. Business Governance – How companies set the relationship between management stockholders and the board of directors. Also included are how the company is affected by government regulations, and how the firm manages their relationships and alliances with strategic partners.

II.   ORGANIZATION INFRASTRUCTURE & PROCESSES

4. Administrative Structure – The way the firm organizes its businesses. Examples include central, decentral, matrix, horizontal, vertical, geographic,, and functional.

5. Processes - How the firm's business activities (the work performed by employees) operate or flow. Major issues include value added activities and process improvement.

6. Skills – H/R considerations such as how to hire/fire, motivate, train/educate, and culture.

III.   IT STRATEGY

7. Technology Scope - The important information applications and technologies.

8. Systemic Competencies - Those capabilities (e.g., access to information that is important to the creation/achievement of a company's strategies) that distinguishes the IT services
.
9. IT Governance - How authority and responsibility for IT is shared between users, IT management and service providers. Project selection and prioritization issues are included here.

IV.   IT INFRASTRUCTURE AND PROCESSES

10. Architecture -The technology priorities, policies, and choices that allow applications, software, networks, hardware, and data management to be integrated into a cohesive platform.

11. Processes - Those practices and activities carried out to develop and maintain applications and manage IT infrastructure.

12. Skills - IT human resource considerations such as how to hire/fire, motivate, train/educate, and culture.

(Luftman, 1996)

12. Skills - IT human resource considerations such as how to hire/fire, motivate, train/educate

---

Mark R. Andrews has been in the information technology industry since 1996, primarily as a consultant. He has provided PC technical support for such firms as General Electric ED&C and BancOne Commercial Credit. Mr. Andrews served as an external consultant at the Government Information Technology Agency, State of Arizona, as a Web developer and application support analyst. He was a member of the Web Masters Task Group, whose goal was to develop a set of guidelines for all State of Arizona Web masters.

Raymond Papp was an assistant professor in the Department of Management Information Systems at Central Connecticut State University. Dr. Papp's research interests include strategic impacts of information technology, Internet-based learning, strategic alignment, and emerging information technologies. His research has appeared in several academic and practitioner journals, and he has presented research at professional and executive conferences. He has worked as a computer programmer and senior analyst and continues to practice as an independent consultant.

This case was previously published in the *Annals of Cases on Information Technology Applications and Management in Organizations*, Volume 2/2000, pp. 214-232, © 2000.

Chapter III

A Case of Information Systems Pre-Implementation Failure:

Pitfalls of Overlooking the Key Stakeholders' Interests

Christoph Schneider, Washington State University, USA

Suprateek Sarker, Washington State University, USA

EXECUTIVE SUMMARY

This case study examines an often overlooked context of information system failures, that of pre-implementation failure. It focuses on an Information Systems Development (ISD) project at a large public university that failed even before implementation could get under way. Specifically, it describes the vendor selection process of a proposed computerized maintenance management system. While the managers in charge of the project took great care to avoid commonly discussed types of information systems failures by emphasizing user involvement and trying to select the best possible system they could afford, non-functional requirements, procedures as outlined in the RFP, and the roles of relevant but relatively "hidden" decision makers during the pre-implementation stage of the project were overlooked. This led to the termination of the project after an appeal was lodged by a software vendor whose product had not been selected for implementation.

ORGANIZATIONAL BACKGROUND

UMaint is the maintenance department of a large public university (BigU[1]) in the northwest of the United States. Currently, about 18,000 students are enrolled at BigU, a large proportion of whom reside on-campus. This makes BigU's main campus one of the largest residential campuses in the Pacific Northwest. In addition to the student body, about 7,000 faculty and staff work on campus.

UMaint's employees are responsible for the maintenance of BigU, the campus area of which encompasses more than 400 buildings and over 1,930 acres of land. In a typical year, UMaint handles approximately 60,000 service calls, and schedules and completes 70,000 preventive maintenance projects for 69,000 pieces of equipment.

The primary departments of UMaint are Architectural, Engineering, and Construction Services, Utility Services, Custodial Services, and Maintenance Services. These departments are supported by UMaint's Administrative Services. Architectural, Engineering, and Construction Services are involved in all new construction projects as well as all modifications to existing facilities. The Utility Services Department operates the university's power plant and is responsible for providing utilities such as steam, electricity, and water. Custodial Services, UMaint's largest department, handles the custodial work for all buildings and public areas on campus. Maintenance Services is divided into environmental operations, life safety and electronics, plant maintenance and repair, and operations, and is responsible for the upkeep of the university's buildings and facilities.

The Administrative Services Department encompasses units such as operational accounting, personnel and payroll, storeroom, plant services (including motor pool, heavy equipment, trucking and waste, and incinerator operations), and management information systems. This department handles all supporting activities needed to coordinate and facilitate UMaint's primary activities. Overall, more than 450 employees work for UMaint in order to support the university's operations. Please refer to Appendix A for the organizational chart of UMaint.

SETTING THE STAGE

The major challenges faced by UMaint arose from the state's tight budget situation and increased competition from outside service suppliers. In order to deal with these challenges, UMaint had to constantly strive to reduce costs and streamline operations. One major obstacle to providing services efficiently and effectively, as is the case in many universities and even business organizations, was UMaint's outdated information systems infrastructure.

This infrastructure consisted primarily of an outdated mainframe, in which the applications were written in Natural and the databases were hosted in ADABAS. Administrative functions were conducted using form-based systems that had been developed in-house. Over the years, the systems had grown with the needs of UMaint. In the absence of an Information Systems (IS) Department that was internal (or dedicated) to the needs of UMaint, the growth of the systems in this area has been rather uncontrolled, leading to a variety of different applications, the majority of which were incompatible with one another. In order to perform accounting, inventory management, maintenance, and project development functions, the employees had to work with over

100 different databases. This situation led to a huge paper trail, the need for multiple paper copies of documents and considerable redundancy of work, which in turn resulted in the lack of data integrity, a major hindrance to efficient operations. As one manager explained "... *[UMaint has] someone three cubicles down from another person replicating the same work, unnecessarily.*"[2]

In the mid-1990s, UMaint's director and departmental managers (hereafter referred to as top management) decided to implement a computerized maintenance management system (CMMS) in order to consolidate the legacy applications into one integrative system. With this, UMaint hoped to be able to provide more efficient and higher quality service, obtain more timely and accurate information for top management, and reduce costs for the customers by eliminating the need for multiple data entry and simultaneously reducing the potential for errors associated with the maintenance business processes. Additional goals were to increase the accountability of the organization as well as to better maintain the university's facilities. Appendix B displays typical features of a CMMS.

In order to achieve these goals, UMaint formed an implementation team consisting of employees representing different levels of the organization. These were charged with initiating the CMMS project. In the first stage, the primary task was to gather information about different CMMS vendors. As the members of the implementation team conducted all project-related activities in addition to their regular tasks, many of them spent additional hours in the evenings to work on the CMMS project, often leading to dissatisfaction among the project team-members. Unfortunately, the personal sacrifices of the team-members proved to be in vain, since shortly thereafter, work on the project was halted due to lack of adequate funds. A timeline of the events is shown in Appendix C. The idea of introducing a CMMS however was never completely abandoned. In the year 2000 (after the Y2K crisis never materialized), top management again decided to make a renewed commitment to implementing a CMMS. After a new president had been hired to take on leadership of BigU, a new strategic plan was set up for the entire university; one of the goals was to "create a shared commitment to quality in all ... activities," which included to "develop strategies that foster a university culture dedicated to adopting and extending best practices that promote an ongoing commitment to continuous improvement." As this goal was closely linked to the services of UMaint, Jack, the director of UMaint, was invited to be a member of the implementation team for this goal. Since Jack had been with BigU for a long time, he therefore enjoys high credibility and good relationships with higher management, but as in most large institutions, he is constrained by his budget and the strategic goals of BigU. With the introduction of a CMMS, UMaint would clearly help BigU in achieving this goal, thus, funding seemed much more likely, and consequently, an internal IS Department responsible for providing information services and technical support for UMaint's information systems was created. The IS Department was also charged with the project of implementing the computerized maintenance management system. Currently, the Information Systems Department consists of an Information Systems Manager ("Frank") and two Computer Systems Administrators, in addition to several computer-savvy college students serving as Support Staff.

Top management was very excited about the project and had a good grasp on the project's potential implications for the organization, a factor that would help in the selection process (West & Shields, 1998). Specifically, Jack (the director of UMaint) strongly believed that "*the selection of a CMMS will affect the way [UMaint does] business for the next 10 years.*"

In addition to serving UMaint, the CMMS was supposed to serve BigU's Housing Department and Central Stores. University Housing would use the system to support all maintenance related aspects of their operations, as well as to manage its warehouse, and Central Stores would use the system for all procurement-related activities. Several other departments played a role in the process as well; for example, the university's Budget Office allocated the funds for the project and was hence involved in the purchasing process. As the total amount budgeted for the purchase of the CMMS exceeded $2,500, the acquisition had to be made through the university's Purchasing Department in the form of a bidding process; furthermore, due to the administrative, rather than academic or research related nature of the project and the fact that BigU is a state university, unsuccessful bidders had the option of filing a protest with the state's Department of Information Services (DIS) after the announcement of the final decision. The DIS had the authority to review and override any decisions made by UMaint. Furthermore, acting as an outside consultant (Piturro, 1999), the university's IS Department provided guidance to UMaint's IS Department during the selection process. Please refer to Appendix D for a diagram displaying how UMaint fits into the university's structure; Appendix E shows the major stakeholders of the proposed system.

Given that upper administration was well aware of the large impact the system would have, everyone agreed to implement and routinize the "best possible alternative" for the proposed system. Due to the limited resources, developing an integrated system in-house was not seen to be feasible. Therefore, it was decided to purchase a system from an outside vendor. Even though highly customized solutions can be problematic in the long run (Ragowsky & Stern, 1995), both off-the-shelf packages and solutions specifi-cally designed for UMaint by interested vendors were considered.

CASE DESCRIPTION

In the practitioner literature, there has been a limited number of articles on software vendor selection processes; most of these articles have offered somewhat simplistic implementation guidelines such as having sufficient human resources, securing commit-ment by top management, conducting site visits, or being aware of possible problems associated with customized software (e.g., Ciotti, 1988; Piturro, 1999; Ragowsky & Stern, 1995; West & Shields, 1998). In addition, few academic IS researchers have directed attention to critically examining the process of software vendor selection; consequently, normative rational processes (Howcroft & Light, 2002) are seen to guide systems acquisition initiatives (please refer to Appendix F for a model of the software vendor selection process). Even fewer have empirically studied the vendor selection process (e.g., Gustin, Daugherty, & Ellinger, 1997; Howcroft & Light, 2002; Lucas & Spitler, 2000; Weber, Current, & Benton, 1991). A recent meta-analysis found that in information systems development process, user participation can influence IS success (e.g., Hwang & Thorn, 1999). Consistent with these findings, noted researchers such as Land (1982) have explained that in the software vendor selection process, user input is seen as an important step in determining business needs, which is fundamental to implementing a successful system. According to Land (1982), user participation can be consultative, democratic, or responsible, where responsible participation implies the greatest influence on the part of the users. At UMaint, one of the first steps in the vendor selection process was to inquire about experiences with a CMMS implementation at similar institutions. Frank briefly described the experiences other universities had:

I talked to a lot of schools [that had] been through [this process], read a lot of implementations, and saw where the weaknesses were; the weaknesses were people not getting involved, I spent some time over at [WesternU], looking at them and their implementation, and one of the biggest regrets and problems was they didn't get people involved.

Just a few years ago, WesternU, another large university in the Northwest region of the US went through a similar process and faced severe implementation problems, partly due to lack of stakeholder involvement with the project *from the very beginning*. On talking to WesternU, Frank found out that:

...they went through this a couple of years ago, and have a vendor and have implemented it and are still in partial implementation, and they said the biggest problem that they had is when they went to train people and so many people down at the very lowest level claiming not to know what is going on, or not to be happy with the system.

Knowing about the beneficial effects of user involvement (Hwang & Thorn, 1999; Land, 1982; Schwab & Kallman, 1991), top management tried to involve their employees from the initiation stages of the project. Joan (the Assistant Director for Administrative Services) recognized the fact that it would be impossible to involve all employees directly; therefore, she strived to *inform* as many people as possible, rather than ensure *direct* participation from every employee:

We obviously selected a core group to work on it because every one of 400 people can't be involved...I'm not sure if they have the need to say vendor A is better than C or D or E, it's more the idea that, okay, it's going to change what I do, am I comfortable with that idea and then as far as which vendor we choose, they're going to have to learn something new, no matter what. So it's more trying to get them the information... them knowing about it, knowing that it's coming, and giving input where they can. But it's...it's very specific what they're going to be doing, and I think it's really important for them to know that it's happening and feel comfortable with the idea.

Therefore, committees were formed in order to involve people from all departments. The Executive Committee was responsible for making the formal decisions while the CMMS Evaluation Team was heavily involved in ranking and scoring the different vendors. In addition to these committees, process teams were created; these were charged with analyzing the workflows in different areas such as accounting, human resources, scheduling, preventive maintenance, or engineering and design. The Executive Committee consisted of UMaint's Director, his two assistants, and the IS manager. Members of Central Stores and University Housing were only represented in the Evaluation Team. In fact, the CMMS Evaluation Team included members of every department of UMaint, with members being nominated by their respective department heads. Of the 51 members, 16 were managers/directors, 18 were supervisors/leads, and 17 were line employees. The process teams consisted of line employees as well as managers. To ensure adequate say of individuals who actually get the work done, several teams were led by line employees. A requirement from the beginning was that "every supervisor, every manager has to be involved." The committee members were expected to "involve all of their people in some way." In this way, the project leadership hoped to have as many people as possible participating in the selection process.

The implementation of the CMMS was planned in three stages. The first stage involved the request for proposal (RFP) process; the second stage consisted of the decision-making process, and the third stage, the actual implementation.

During the RFP process, the different process teams had to analyze the various business functions. This phase was not completed as smoothly as expected, since some of the groups did not put in the necessary effort. This was seen as a big concern during the RFP process when:

...every area that was going to be involved was requested to go out and say what's important to them. And that's where they didn't participate as fully as maybe another group. Some groups did an excellent job of saying I need this, this, this, this, this. And this is the ranking, this is what we need, this is what we've got to have, this is what would be nice. And we may have had...I think it is mainly one area that just kind of didn't do that very well. And I guess how that will be affected as if they, after we've chosen, come and say, oh, I've got to have this...well, you didn't have it in the RFP, we, I, can't know what's important for you. So that would probably... and laziness maybe is a bad word, but I think that's...they're just not choosing their priorities. Some people realize just how important the process was and some don't realize it... and may be impacted.

As the IS group was concerned with choosing the best possible system for the organization, its members decided to start analyzing the business functions themselves. This helped the groups to get started in the RFP process. Frank described his experience with the process as follows:

[We] charged all the groups to come up with [the business functions]; that didn't work out too well, the groups really didn't do a good job putting it together, so what happened was, we put them together. Our group did all of them and then what we did, we took 'em back to the groups and said okay, here's the storeroom piece, here is what we think is your function, but we're the wireheads, we're not the workers, we don't know their business, but from working with them, and doing what we have been doing up to this point, we felt we could at least get the ball rolling.

Having started the process, the IS group saw some improvement in participation, and most of the groups seemed more actively engaged in the RFP process:

...so we gave them basically something to react to. And that helped a lot. Then people started coming up with a lot of ideas, a lot of changes, and we started getting more participation that way.... These folks really struggled with coming up with something on their own.

Participation was not seen as mandatory by top management, therefore, the CMMS Decision Team tried to motivate the employees to participate as much as possible. However, they also had the option to exercise power in order to get things done. Since UMaint's executive director was a big supporter of the CMMS project, Frank (the IS manager) mentioned that:

I also have an ace in the hole with the director, who's willing to come at any point and lay the hammer down if somebody don't do what we need them to. We don't do that unless we have to, but if we get into that situation where there's a work group or a person who's simply just not going to pull their share. We're trying to work with them and we're trying to motivate them, and get 'em excited and get their involvement, but otherwise, we just have to take it to another level if it is too critical. It's a big project.

Finally, the mandatory and desirable requirements were put together and the RFP was deemed acceptable by the state's Department of Information Services after a few minor revisions. According to the RFP (please refer to Appendix G for the required structure of the proposals), three vendors were supposed to be invited to conduct onsite product demonstrations. In regards to the requirements, the RFP stated that the proposed CMMS system had to be able to integrate the large number of databases currently used to manage accounting, maintenance, inventory, and project development needs. Furthermore, Microsoft Project Server 2002, Sharepoint Portal and Team Services, as well as several Access Data Projects needed to be integrated into the proposed system. As UMaint already had an existing information systems infrastructure consisting of more than 130 workstations running a combination of Windows NT 4.0, Windows 2000 professional and Windows XP operating systems, no substantial changes in client hardware were desired; however, if additional hardware such as servers would be needed for the implementation, this would have to be covered by the budget initially allocated to the project. Furthermore, it was expected that the vendors' systems would be compatible with the operating systems of the existing workstations. Finally, it was not anticipated to allocate human resources in addition to the IS Department's current staff to the support of the CMMS system.

While certain non-functional requirements,[3] such as security of the transactions or scalability of the database, were considered to be important factors, a proven track record of successful implementations was considered more important than a system using the newest technologies.

Using the mandatory functional requirements, such as the ability to track the status of each shop assignment, request materials for a work order from a "Materials Management" module, trigger notices to responsible persons for potential problems with contract processes, or track equipment maintenance labor history, the executive committee was able to screen out a large number of vendors that initially had responded to the RFP, bringing down the number of potential vendors to six. Following recommendations provided in the practitioner literature (e.g., Raouf, Ali, & Duffuaa, 1993; Weber et al., 1991), these vendors were scored by the CMMS Evaluation Team according to the CMMS requirements, technical requirements and capabilities, and vendor qualifications. Initially, only one bidder provided the required cost proposal (which did not comply with the RFP), hence it was decided to eliminate this criterion in this stage of the process and to reassign the weights of the remaining criteria. Based on a comparison of the summary scores, four vendors were invited to present their products to the CMMS Evaluation Team according to a scripted scenario. The attendees at the demonstrations had the opportunity to score each vendor's product using a standardized scoring package. The scores given were to be used during the final decision-making; at this point, the vendors that were invited to present their products were asked to re-submit their cost proposals.

In addition to helping with the selection, giving the attendants of the vendor demonstrations the opportunity to score the products was seen as a good way to involve people and to get their "buy in." Most members of the executive committee believed that many employees "just want to be involved, just want to be important," no matter "if things go their way or not." According to Joan, the scoring could be compared to voting, where everyone's vote counts. However, in contrast to management's beliefs, many members of the process teams felt that their input did not count at all, which led to dissatisfaction with the process.

Furthermore, some areas felt "shortchanged" during the vendor demonstrations, having felt that their area has not been considered enough during the demonstrations. Another factor adding to the dissatisfaction with the decision process was a lack of feedback from top management. One administrative employee complained:

...they had us do all kinds of stuff. You know, questions we needed to ask, how everything flows, how it needs to be done.... I'm just a little nervous because ... we put a lot of input in, but we didn't get much information back, or how it would affect us.

Even though the process team members were intended to funnel down the information, the information flow did not take place as expected; indeed, many employees did not receive much information about the proposed system. This problem was of great concern to upper management, as information sharing was seen as critical to the success of the process. Frank stated:

And one of the biggest challenges this place has is the information distribution. The information does not go from the top to the bottom. The manager hears it at a Tuesday manager meeting; it may be six months until the employee hears it. It's really embarrassing, but it happens a lot around here. With a project of this magnitude and complexity, you can't do that. The information has to reach everybody at every time. So one of the tasks for us was, how do we get information out. So we developed the project information site....

The project Web site was developed to provide access to announcements about the status of the vendor selection, vendor scores and rankings, background information about the vendors, and the like. Even though the Web site was seen as pretty effective by the IS group, it did not help to effectively spread the information throughout the entire organization, as the Web site was only accessible to CMMS Evaluation Team members and Executive Committee members.

Since the IS team was in charge of the CMMS project, meetings with line employees were set up in order to provide them with information and solicit feedback. Nevertheless, many employees were unaware of the proposed system's impact on their work. Generally, they saw "other" departments as being impacted to a greater extent, and did not anticipate their own day-to-day responsibilities to change much. Many employees regarded the

CMMS as a tool for the higher echelons of administration. This led to a lack of interest in participating in the decision process, which thereafter translated to the perception that their departments had been left out of the process. For example, an employee mentioned:

*...I don't know how it will really affect me, unless they would get a couple of modules that would really help out **back here**. It's my understanding that they're just doing administration modules.... I do know that administration, **at the other end of this building**, had a lot of input.... I don't know that it will probably help **us** at this point.* (emphasis added)

Antagonism between departments added to such perceptions. As the criteria established for the vendor selection process were viewed as relatively inflexible and determined a priori , many employees saw their area as "covered," and did not see the point of providing additional input into the decision process, many hoping that the IS Department, consisting of acknowledged experts, would select the right system for them. One employee, for example, stated that *"our info tech group... they do have the knowledge to make something run right and I don't."*

The Information Systems Department was well aware of its power arising from the myths of their expertise and magic representing systems professionals as "high priests" (Hirschheim & Klein, 1989; Hirschheim & Newman, 1991) to influence decisions of non-IT employees in UMaint. However, it consciously tried to limit the extent of influence IS professionals would have on the decision process:

...we didn't want to influence anyone's decision. ... this isn't supposed to be about us ... this is about their achievement, this is their product in the end, we're just charged with implementing it. They're the ones who gonna have to use it every day and live with it. We wanted it to be about them, and that's why we've been so big about having their involvement....

Nevertheless, being in such an influential position, the IS manager noticed the benefits of being able to influence people's decision, as he was trying to choose the most adequate system for the organization:

...in many ways, we put our hands on the wheel. Because we basically had to come to a decision that works. Everybody, they're just gonna do whatever we say. So we better make sure that what we're saying is really what is the greatest good for everyone ... and that's one thing that has been very confident for us that we have always been striving for the absolute biggest bang for the buck.... Whatever we could get. The most we could do.

Finally, a decision was made by the Executive Committee based on a number of criteria, which included the weighted scores, reference calls conducted by evaluation team members, and finally, consideration of the budget. Before the final decision was made, an informal vote (that would not influence the final decision) was held. Interestingly, the selection made was not consistent with the results of the vote. The vendor that ranked first in the informal vote was not considered for selection, since its product did not meet the budget criteria. The vendor that ranked second had the newest technology and offered a highly customizable product; however, it was regarded as being too risky, as universities were considered an entirely new market for the vendor and was thus dropped from the selection. The vendor that was finally selected ranked third in the last informal vote.

Even though the vendor selected scored very high in terms of meeting the requirements, functions, and features, the decision found only partial support by many organizational members, including the IS Department members. The IS group was not at all convinced that the product had necessary technological and functional capabilities. In a vendor review demonstration, an IS Department member mentioned that the vendor's *"technology is severely outdated and does not offer any customization for the user."* Other departments also were not satisfied with the selection, with a member of the accounting team stating, *"...just the one that got chosen... we wished that it hadn't."* The storeroom, a very powerful unit of UMaint's Administrative Services Department, shared the same thoughts: *"we were looking at [Vendor A] and [Vendor B]. These were what we thought were the two best, but other factors came into play in the decision making...."*

CURRENT CHALLENGES
FACING THE ORGANIZATION

When the final decision was announced to the vendors, it was also not received too well by the vendors whose products had not been selected for implementation. One of the unsuccessful bidders decided to lodge an appeal with the state's Department of Information Services. The members of the Department of Information Services, who had not been at all involved during the selection process, took only a very short time to review the decision and mandate the termination of the project based on procedural errors. As contended by the unsuccessful bidder, the change in scoring criteria for the first stage could not be considered a harmless error, as vendors might have otherwise submitted different proposals. Furthermore, asking the vendors to re-submit cost proposals could be considered a "best-and-final offer" procedure, which was explicitly ruled out in the initial RFP. On the other hand, not asking the vendors for a renewed cost proposal would have led to the elimination of all vendors but one during the first evaluation stage, rendering the RFP process "just a paper chase" (Ciotti, 1988, p. 48). Making a new selection, as recommended by the state's IS Department, would not only mean reviewing the original decision, but also reverting to the beginning stages of the project. In order to select a different vendor, the entire selection process would have to be started anew, including sending out a request for proposals, inviting candidates for product demonstrations, scoring the different products, and making reference calls. As this process would most likely continue well into the following year, funding for the project was not automatically valid any more. UMaint would have to reapply for funding, and in light of the state's tight budget situation, getting funds for such a project a second time seemed highly unlikely.

As of 2003, no computerized maintenance management system was implemented at UMaint. Frank (the IS manager), reflecting on the sequence of events in the project, still remains perplexed about why the initiative turned out to be a disaster, despite all his and his colleagues' efforts to consciously manage stakeholder input and thus avoid failure, so he approached an IS academic to find out the reasons for the failure of the current project, how UMaint's current project could be "salvaged," and how similar problems could be avoided in future projects.

REFERENCES

Ciotti, V. (1988). The request for proposal: Is it just a paper chase? *Healthcare Financial Management, 42*(6), 48-50.

Gustin, C. M., Daugherty, P. J., & Ellinger, A. E. (1997). Supplier selection decisions in systems/ software purchases. *International Journal of Purchasing and Materials, 33*(4), 41-46.

Hirschheim, R., & Klein, H. K. (1989). Four paradigms of information systems development. *Communications of the ACM, 32*(10), 1199-1216.

Hirschheim, R., & Newman, M. (1991). Symbolism and information systems development: Myth, metaphor, and magic. *Information Systems Research, 2*(1), 29-62.

Howcroft, D., & Light, B. (2002, December 15-18). A study of user involvement in packaged software selection. In L. Applegate, R. Galliers, & J. I. DeGross (Eds.), *Proceedings of the 23rd International Conference on Information Systems,* Barcelona, Spain (pp. 69-77).

Hwang, M. I., & Thorn, R. G. (1999). The effect of user engagement on system success: A meta-analytical integration of research findings. *Information & Management, 35*(4), 229-239.

Land, F. F. (1982). Tutorial. *The Computer Journal, 25*(2), 283-285.

Lucas, H. C., Jr., & Spitler, V. (2000). Implementation in a world of workstations and networks. *Information & Management, 38*(2), 119-128.

manageStar. (n.d.). manageStar — Products. Retrieved May 3, 2004, from http://www.managestar.com/facility_mgmt.html

Piturro, M. (1999). How midsize companies are buying ERP. *Journal of Accountancy, 188*(3), 41-48.

Ragowsky, A., & Stern, M. (1995). How to select application software. *Journal of Systems Management, 46*(5), 50-54.

Raouf, A., Ali, Z., & Duffuaa, S. O. (1993). Evaluating a computerized maintenance management system. *International Journal of Operations & Production Management, 13*(3), 38-48.

Schwab, S. F., & Kallman, E. A. (1991). The software selection process can't always go by the book. *Journal of Systems Management, 42*(5), 9-17.

Weber, C. A., Current, J. R., & Benton, W. C. (1991). Vendor selection criteria and methods. *European Journal of Operational Research, 50,* 2-18.

West, R., & Shields, M. (1998, August). Strategic software selection. *Management Accounting,* 3-7.

ENDNOTES

1   Names of the university and its divisions have been replaced by pseudonyms. Further, the identities of the employees of the university and other stakeholders of the system have been disguised to ensure confidentiality.

2   For the sake of authenticity, the quotations have not been edited.

3   That is, technical or infrastructure related requirements such as scalability or technical potential for the future

APPENDIX A

Organizational Chart of UMaint

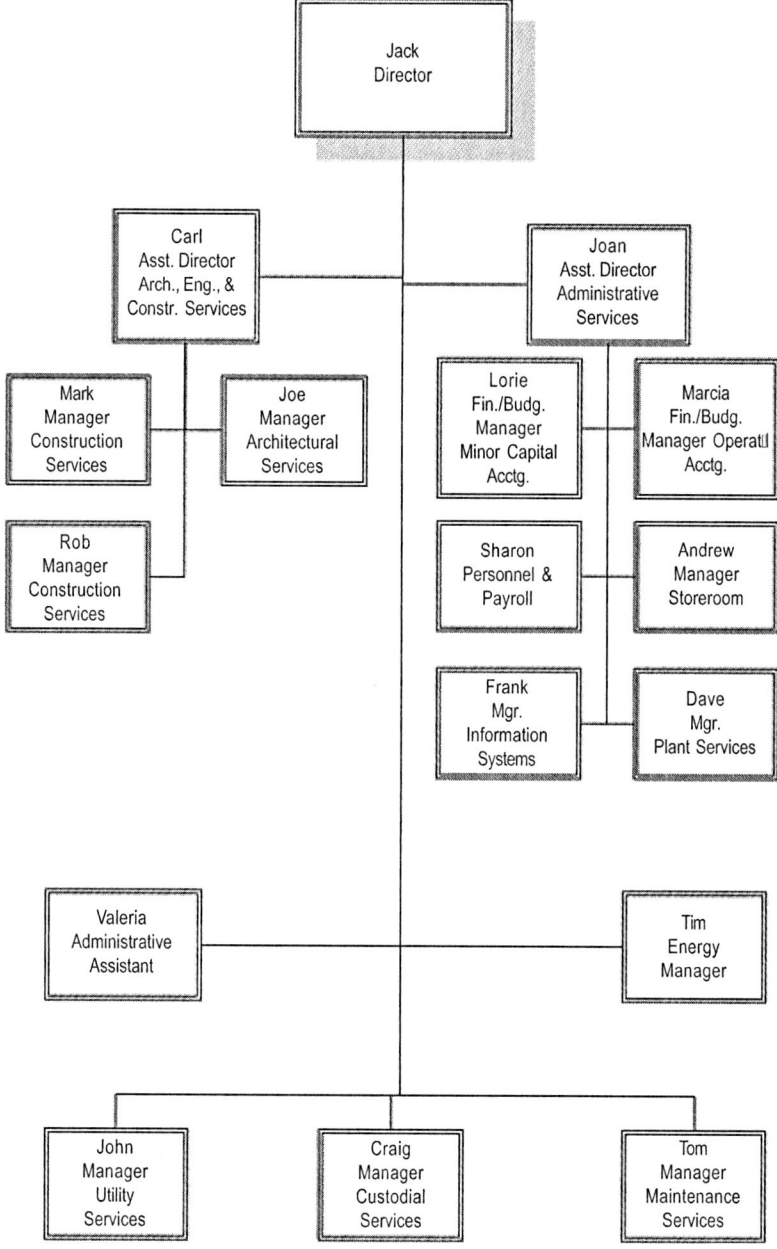

APPENDIX B

Typical Features of a CMMS

Work Order Management
- Receive and route web-based work requests.
- Obtain approvals as part of the workflow if necessary.
- Receive alerts on critical issues in your workflow.
- View a comprehensive list of work in process — work plans, schedules, costs, labor, materials, assets and attached documents.
- View overdue work, or sort work orders on a place, space, asset or engineer basis.
- Set-up predetermined workflow processes or create them on the fly, assigning work orders to available personnel.
- Link related work orders.
- Send clarifications that are tracked in the message history.
- Attach documents, including drawings, specs, and more.

Asset Management
- Click on an asset to launch a work order.
- Maintain all critical asset information.
- Do preventive maintenance.
- Track and get alerts on asset contracts or leases.
- Track assets, costs, histories and failures.
- Link assets to a work order, place, space, project or contract.
- Drill down in your organization to locate assets.
- Get reminders and alerts on any element of the asset.
- Attach documents, such as diagrams, to the asset.

Inventory Management
- Receive minimum/maximum alerts on inventory levels.
- Allow employees to request or order products.
- Add any products to a service delivery, e.g., a new computer for a new employee set-up.
- Kick-off automatic purchase orders to pre-approved vendors.
- Track Bill of Materials, SKU, price, stock, description, vendor and transaction information.
- Connect inventory with specific budgets to track exact costs.

Project Management
- Kick-off new projects with a few clicks.
- Monitor project schedules and milestones on real-time interactive Gantt charts.
- Track budgeted vs. actual spending on a project-by-project basis.
- Manage resources by viewing project analyses.
- Issue service requests within projects and build into Gantt charts.
- Draw relationships between projects, people, places, things, contracts, POs, inventory, vendors and more.
- Attach and share documents.

Procurement Management
- Route POs and invoices automatically.
- Receive approvals and responses automatically.
- Set amounts that require approval and workflow automatically obtains it.
- Integrate with financials.
- Automate all procurement of services, independent contractors, vendors, etc.
- Create and broadcast requests for proposals.
- Track and compare all out-bound and in-bound proposal and bids.
- Negotiate online with vendors.
- Attach proposals to people, places or things as well as projects and contracts.
- Send proposals out in workflow for review and approval.

Adapted from manageStar (n.d.)

APPENDIX C

Timeline of the CMMS Project

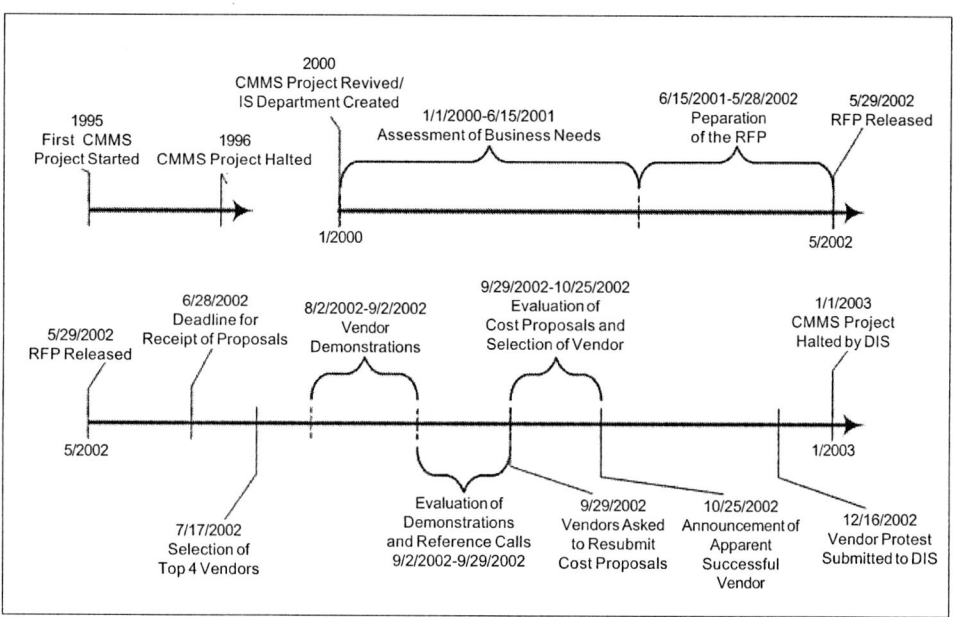

APPENDIX D

Organizational Chart of BigU

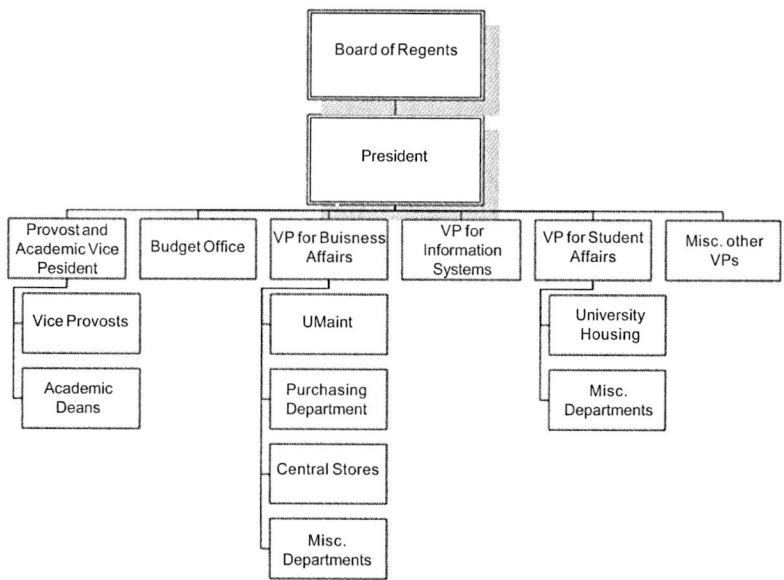

APPENDIX E

Stakeholders of the Proposed System

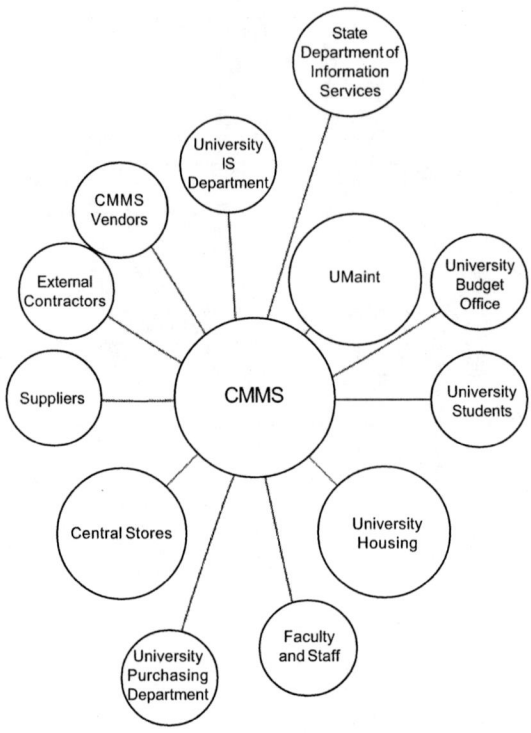

APPENDIX F

Vendor Selection Process

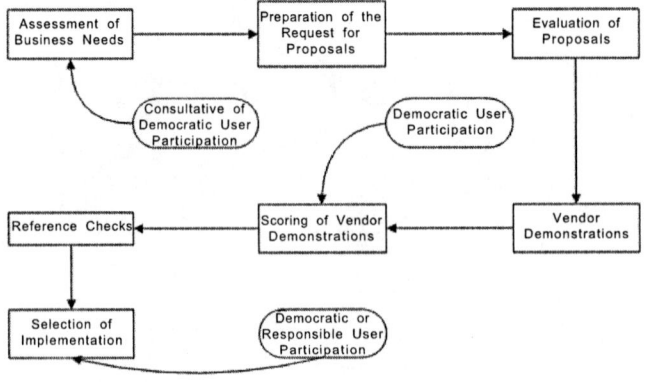

Adapted from West and Shields (1998) and Land (1982)

APPENDIX G

Contents of the Proposal

---

1. Proposal Contents

The sections of the vendor proposal should be as follows:

  Section 1 Transmittal Letter *(signed paper copy must also be included)*
  Section 2 Administrative Requirements *(see 3)*
  Section 3 CMMS Requirements, Functions and Features *(responses in Attachment x)*
  Section 4 Technical Requirements and Capabilities *(responses in Attachment xx)*
  Section 5 Vendor Qualifications *(responses in Attachment xxx)*
  Section 6 Cost Proposals *(responses in Attachment xxxx)*

All vendors must use the RFP templates and format provided on the CMMS website.

2. Section 1 - Transmittal Letter

The transmittal letter must be on the vendor's letterhead and signed by a person authorized to make obligations committing the vendor to the proposal. Contact information for the primary contact for this proposal must also be included.

3. Section 2 - Administrative Requirements

This section of the proposal must include the following information:

a) A brief (no more than three pages) executive summary of the vendor's proposal including:
   1. A high-level overview of your product and the distinguishing characteristics of your proposal.
   2. Indicate the number of universities using the system as proposed. "System" is defined as the vendor's current version of the software with all the functionality proposed in the response to this RFP.
   3. Describe how closely the proposed system matches UMaint's needs.
   4. Discuss what attributes of your proposal offer BigU a distinguishing long-term vendor relationship.
b) A specific statement of commitment to provide local installation for the system.
c) A specific statement warranting that for a period of five years after acceptance of the first application software, that all application software will continue to be compatible with the selected hardware and system software, and will be supported by the vendor. This does not include the database software and hardware to be selected by BigU.
d) For proposal certification, the vendor must certify in writing:
   1. That all vendor proposal terms, including prices, will remain in effect for a minimum of 180 days after the Proposal Due Date.
   2. That the proposed software is currently marketed and sold under the vendor's most current release only (or be added within six months)
   3. That all proposed capabilities can be demonstrated by the vendor in their most current release of the system.
   4. That all proposed operational software has been in a production environment at three non-vendor owned customer sites for a period of 180 days prior to the Proposal Due Date (except in cases where custom or new functionality is designed for BigU)
   5. Acceptance of the State's Department of Information Services (DIS) Terms and Conditions.

(continued on next page)

4. Section 3 - CMMS Requirements, Functions, and Features

Section 3 responses of this RFP contain mandatory and desired features for specific CMMS application modules. This requires coded responses only.

5. Section 4 - Technical Requirements and Capabilities

Section 4 responses of this RFP ask the vendor to identify the computer system software environment for the CMMS, and to provide additional technical information regarding system interfaces, hardware requirements, and application flexibility/enhancements. Answers to these questions should be provided with both narrative and coded responses.

6. Section 5 - Vendor Qualifications

Section 5 responses of the RFP requires the vendor to provide information about the vendor organization and customer base; and to propose support capabilities in terms of design, installation, data conversion, training, and maintenance of the proposed system. This information is to be provided in narrative format.

7. Section 6 - Cost Proposal Instructions

Section 6 of the RFP contains the formats and instructions for completing the cost proposal. Section 6.2 (Instructions) requests five-year summary costing for:

a) Application software (including upgrades and customization).
b) Software maintenance.
c) Services (installation, training, interfaces, project management, conversions, etc.).

Christoph Schneider is a PhD student in MIS and graduate assistant/instructor at Washington State University. Prior to starting the PhD program at WSU, he studied at the Department of Business Informatics at the Martin-Luther University in Halle, Germany.

Suprateek Sarker is an associate professor of information systems at Washington State University. His research interests include virtual teams, BPR, IS implementation, and qualitative research methods.

This case was previously published in the *Journal of Cases on Information Technology*, 7(2), pp. 50-66, © 2005.

Chapter IV

The Information Plan for Celerity Enterprises, Inc.:
A Teaching Case

Laurie Schatzberg, University of New Mexico, USA

EXECUTIVE SUMMARY

Celerity Enterprises competes in the semiconductor manufacturing industry. At the start of the case, business conditions are favorable for them to launch a new production facility to manufacture flash memory. The new facility must achieve exceptionally ambitious productivity and cost goals. A facility-level strategic planning process reveals opportunities to substitute information for other more-expensive resources. By the end of the case, just a few months later, worldwide economic conditions change radically and the future of the new facility is in jeopardy. The case describes the participants, the planning process and findings. It provides a rich setting to discuss aligning information and business planning, realities of the volatile industry, outsourcing for IS planning leadership, and using a combination of top-down and bottom-up planning.

BACKGROUND

The Case

This case seeks to address the dynamics of information planning in a highly volatile manufacturing industry. To portray the dynamics with realism, the case also highlights: (1) the need for a firm to align its information strategy with its overall business strategy, and (2) the need to engage in a planning process such that employees participate and embrace it, and ultimately own the resulting transformation.

Information planning is, itself, a part of an information system development cycle. That development cycle is a major component of the organization's own life cycle. Finally, the organization itself is a player within its industry. Industry dynamics and worldwide economics play a major role in influencing the direction of an organization, and its IS planning in turn. As readers analyze this case, it will be useful to keep in mind this interdependence among the IS planning process, the culture of the organization, and the dynamics of that organization within the industry.

Celerity Enterprises

Celerity Enterprises competes in the semiconductor manufacturing industry and is based in the USA. Like many such companies, it was formed in the 1960s by a dynamic team of scientists and engineers. Celerity now is a Fortune 100 company with six manufacturing sites and four assembly sites in the USA, Asia, Europe, and Latin America. Celerity boasts 70,000 employees across these facilities, with nearly 80% of them in manufacturing or assembly positions. While all research and development, transnational e-mail, business transactions and internal conferences are conducted in English; daily operations are conducted in the local languages.

Celerity enjoys state of the art information services, thanks to two separate but related internal organizations: Automation and Information Support. Automation is responsible for manufacturing-related information services while Information Support is responsible for business-related information services.

Automation's purview includes production planning and scheduling, process control and real-time data collection on the factory floor. Automation is on Celerity's critical path at all times, since information flowing within, to and from the factories forms the lifeblood of the company. Automation must approve any and all proposals for changes that impact manufacturing or communications with manufacturing.

The Automation hierarchy spans from the Director of Automation (reporting to the Vice President of Manufacturing) to Automation specialists dedicated to specific machine tools in each factory. Since the manufacturing processes are closely guarded corporate secrets, no significant Automation work is outsourced.

Supporting Celerity's business information requirements, the Information Support functions include worldwide administrative and technical communications, sales and customer service, a suite of intranet-based office and research utilities, and office-automation in at least a dozen different languages. Celerity has a Vice President for Worldwide Information Support, and often outsources the Information Support functions.

All members of the combined information services organizations are trained in the basic manufacturing processes, and are expected to keep their own technical training current. Top management reasons that the best information services will be delivered when the staff members understand their (internal) customers. Similarly, the manufacturing staff members must keep their computing and information processing skills current. Celerity provides the time, resources, and mentors needed to enable each employee to progress along his or her personal development path.

Consistent with Celerity's commitment to cross-training, and for the benefit of the readers who do not already possess a background in basic semiconductor manufacturing, the remainder of this section provides an overview of the manufacturing context in which this case unfolds.

Celerity designs, manufactures and markets a wide variety of microcomputer components and related products. Their principal components consist of silicon-based semiconductors. These principal components are described below:

1. Microprocessors, also called central processing units (CPUs) or logic chips, control the central processing of data in personal computers (PCs), servers, and workstations. Celerity manufactures related chipsets, and sells CPUs, chipsets, and/or motherboards to original equipment manufacturers. Celerity's Sentinel brand of products are logic chips.
2. Random Access Memory (RAM) is used as primary memory in computing devices. This memory is volatile and holds information only as long as the electrical current to the chip is maintained. Their contents are lost when the device is turned off or during an unintended electrical outage. Celerity manufactures both SRAM (faster) and DRAM (less expensive), but does not have a unique brand identity for these products.
3. Flash memory provides programmable nonvolatile memory for computers, cell phones, cameras and many other industrial and consumer products. Unlike SRAM and DRAM embedded in CPUs, flash memory is more permanent and its contents are preserved even when power is off. Celerity created the Cel brand to market their flash memory. Cel is the focus of this case.

To nonspecialists, the manufacturing processes for these three categories of chips seem similar. Each process requires hundreds of manufacturing and testing steps. The engineering and design differences that define each category differ markedly, however. In turn, these design features define the markets, profit margins, and cost philosophies for the products. These resulting strategic differences then drive the operational production and quality management processes. For example, logic chips costing about $22 to manufacture sell for about 10 times their cost, while flash chips that cost about $1.50 to manufacture are sold for nearly four times cost.

Throughout the industry, production facilities are generally dedicated to only one of these product lines. The major customers for Celerity's flash products include original equipment manufacturers of computer systems and peripherals, and makers of telecommunications, automotive and railway equipment.

The company is organized into several divisions: Manufacturing, Research and Development, Sales and Marketing, Finance, Information Services, and Corporate and Community Relations. The Division Managers, based at the corporate headquarters, comprise the executive management organization. Corporate headquarters is in Florida, USA.

The Major Product Lines: Sentinel and Cel

Celerity Enterprises began as a RAM manufacturer and has very effectively applied organizational learnings from RAM to both their highly profitable Sentinel (logic chip) line and to their newly burgeoning Cel (flash chip) line. The Sentinel product line has put Celerity Enterprises on the map and is projected to keep Celerity in the Fortune 100 for years to come.

In recent years, Celerity has retrofitted some aging RAM producing factories to produce the more complex and potentially profitable, flash memory. They now have two facilities dedicated to the production of the Cel product line.

The new Peak plant will manufacture the Cel — a commodity product used in hundreds of consumer electronic devices. Demand for the Cel has been steadily increasing for the past 15 years. Demand is expected to increase over 50% worldwide within about four years, as new consumer products requiring Cel's functionality are introduced. Cel had captured nearly 75% of the world market in early years and has been holding steady at about 60% for the past five years. As the demand for flash memory explodes worldwide, so too does the pressure to contain costs and prices. Profit margins, slim as they are already, continue to diminish.

Worldwide, RAM manufacturers are racing to fill the increasing demand for flash memory. They all hope to deploy their expertise with RAM production in the flash market. In this quest for greater share of flash market and greater profitability, firms first maximize productivity from their existing facilities. To achieve greater output, they then make the choice between building a new facility from the ground up or re-engineering existing facilities to meet the aggressive new requirements. This was the situation facing Celerity. Management decided to convert an aging flash factory (whose older technologies severely limited productivity) into a state-of-the-art flash factory that would be designed to achieve far greater productivity levels.

SETTING THE STAGE

The Industry

While the semiconductor market is well-known for its volatility, at the time this case begins, flash memory manufacturers were experiencing a boom in demand. OEMs using flash memory were concerned that their supply of flash would fall far short of their own production plans for cell phones, laptop and palm-top computers, automotive products, and the emerging markets for digital cameras. The manufacturing process for these types of flash memory products is quite different from the dynamic random access memory (DRAM), although there is about an 80% overlap in the basic manufacturing process. While the manufacturing technology to produce flash memory is related to that of DRAM, very little manufacturing capacity is actually dedicated to these more highly complex products.

The OEMs consuming flash products are communications, computing, consumer, industrial, and other. In the three years from 1995 to 1998, factory revenues nearly tripled from $1.8 to $5.9 billion, with expectations that in 2000, the revenues will have doubled again. Meanwhile, with improvements in technology, the actual OEM price per bit of computing will have fallen from $.0026 to a mere $.0006 (Brinton, 1997).

Celerity

About three months ago, Celerity's strategic management team decided to shift the production capacity at the Peak facility to the newer Cel flash memory. Not only would Peak produce new products here, but Peak would also launch new production and decision-making processes. This approach will save, literally, over a billion dollars (AMD, 1997; Intel, 1998) in new construction and infrastructure development. Longer term, this approach also minimizes the risks of owning under-utilized real estate.

Shifting production capacity is not without significant costs, however. This initiative involves major new capital equipment, as well as development of new business practices. Of major significance to the industry are the productivity and cost targets established for the new Peak team. Peak's new management team is charged with achieving 300% improvements productivity, while reducing labor costs by nearly 50%, and cutting cycle times in half. These targets are ambitious and can only be achieved with radical redesign of all business processes. Such redesign must challenge standard (and proven) operating procedures, and must create a highly effective organization that can anticipate and react nearly instantly to changing demands. The ambitious targets, never before achieved in the industry, are nevertheless consistent with industry expectations for the next generation flash facility.

From a global perspective, Celerity is striving to achieve new technological goals. From a local, human perspective Celerity is also forging a new organization. For the technology goals to be achieved, the individuals comprising the new organization must be committed to the goals and engaged in the process of the organization's design. Further, there must be mechanisms in place to allay fears and to manage natural resistance to change.

For Celerity, a company long admired for its proactive employee participation programs, this organizational challenge seems to be understood. In years past, management has successfully guided the firm through many product life cycles, facility life cycles, and through workforce downsizing and outsourcing.

The Human Dimension

Celerity employees expect that their careers will be dynamic, and they understand they must be proactive to continually improve their skills and abilities. Even hourly employees refer to the company as a meritocracy. While Celerity has many programs in place to engage employees in decision making and to smooth the seemingly continuous organizational evolution, two are particularly salient to their creation of the new Peak organization: (1) constructive confrontation, and (2) employee development. The remainder of this section summarizes each of these programs and their relevance to the creation of a new Peak organization.

Constructive Confrontation of Issues. Organizational transformation ultimately rests on the collective abilities and motivations of the participants. Celerity has a long and strong history of employee involvement that supports and enables them to achieve corporate goals. They are also among the leaders in using information technology to support this principle. For example, each quarter top management hosts an hour-long status report meeting during which they present updates to the employees, recognize individuals and teams for notable achievements, and field spontaneous questions from the floor. These meetings are held in each Celerity facility worldwide, within the same week, and on each shift.

When issues of strategic importance are presented, the presentations are made by executives and transmitted by satellite to the facilities, so that everyone receives exactly the same information. Employees are encouraged to participate and are paid for their time. Whether broadcast or run locally, these quarterly meetings are closed to contractors and other outsiders and have the reputation for being highly interactive and brutally frank.

From the top, Celerity sets the tone for direct and ongoing constructive confrontation of employee issues. This business practice is invaluable for easing employees through the many transitions they experience during their Celerity careers.

Employee Development. Each Celerity employee maintains a career path plan and is evaluated annually on progress toward achieving that plan. With input and guidance from their supervisor and any mentor they choose, employees track their progress-to-plan on an ongoing basis, and they refine their plans as part of the annual review process.

Each plan includes sections for continuing training and formal education, technical proficiency, communication skills, team-development and leadership skills, safety training, and personal enrichment. In the annual evaluation process, the employees and their supervisors assesses progress, areas of needed improvement and any additional resources needed to achieve those goals. All efforts are made to ensure that each employee's plan is challenging and achievable for that individual, and meaningful for their success at Celerity.

An interesting information systems project called SkillFull enables Celerity to chart their supply and demand of critical skills by linking certain Human Resources (HR) databases and processes to those in Manufacturing. To chart the supply baseline for the coming three years, the application pulls data on skills and employee development plans from the HR database. Then, these supply projections are compared to the skills demand projected by the Manufacturing in their production goals and plans.

In many cases, these projections indicate a likelihood of shortages or overages of particular critical skills in the future. These projections are communicated company-wide. Although formally an operationalization of Celerity's employee development principle, this communication clearly reinforces Celerity's commitment to confront issues constructively, and to engage all employees in the solutions to serious business issues. The frank communication helps stimulate existing Celerity employees to focus on developing "in demand" skills. The communication strengthens employees' trust that they are informed of all relevant business issues — both positive and negative. (The projections also provide a basis upon which Celerity seeds the pipeline for schools and universities to train Celerity's next generation of employees).

Selecting Peak Employees. How did Celerity identify and select the employees for the new Peak organization? The executive management team selected Peak's top manufacturing managers after internal job postings and candidate interviews. Once selected, the Peak management team then created internal job postings for their own authorized direct reports. In turn, those direct reports began to fill Peak's remaining manufacturing, engineering and supervisory positions.

The internal job posting information system provides both "push" and "pull" job notices. Some employees set up their profiles such that they'll be informed whenever there is a new internal posting that matches their skills. Others simply browse the current postings when they are interested. Employees are assured confidentiality of their interest for both parts of the postings system. When mutual interest is established, the hiring manager and the releasing manager develop the employee's transition calendar. No employee is released to their new organization until after a full transition of their duties and a handoff of their ongoing assignments.

With these examples, readers can begin to appreciate how Celerity earns and creates a trusted environment within which tremendous changes occur. Celerity implements programs that tie directly to their basic values and principles. The decision to launch Peak and the process of staffing the new organization illustrates how Celerity operationalizes their commitments to constructive confrontation and employee development.

The Technical Dimension

Celerity's current production cycle for flash memory — from the time the silicon substrate reaches Celerity until the memory chips are produced, tested, packaged and shipped to OEMs — is about two-and-a-half months. When they reach steady-state operation, the newly developed facility will cut this cycle time to five weeks. Such a change requires major changes to infrastructure, major equipment, and production processes.

Infrastructure re-engineering involves the specification of a transition plan — to move the Peak facility from its current production methods and products to its new methods and product mix. Concomitant with this volatile organizational environment, the information services for Peak must be analyzed, specified, and implemented. Consider that daunting challenge: information services must be in place to support the business and manufacturing requirements — and yet these requirements appear to be evolving.

Since flash memory is a commodity with strong customer demand, Celerity will tolerate no production downtime during the transition. Thus, both Automation and Information Support services must accommodate existing and emerging needs. Celerity management has some experience rolling out new facilities using a staged approach, and they expect to apply the previous lessons learned in the Peak transition.

Their schedule mandates that they are in production mode within nine months, and that they achieve steady state within four months of reaching production stage. During the nine month "facility build," Peak management must achieve intimidating transition and performance goals.

For the ensuing nine months, the entire Peak team is focused on meeting their productivity goals. While business cycles are well-known within the industry, all efforts are directed at the Official Peak Plan. There is no second-guessing or hedging on the plan. They assemble their teams, develop the specifications for manufacturing and material handling systems, determine which of the existing equipment can be used or retrofitted, acquire new equipment, define their operating principles and business practices, and develop the "esprit de corps" (enthusiastic commitment to the goals of the new Peak organization). Eventually all this and more will enable them to produce world-class product in record time.

CELERITY, INC.

The Setting

It almost never happens, but the vision for Celerity Enterprise's new manufacturing organization (code-named Peak) actually calls for ongoing strategic information planning. Peak management wants the initial strategic information plan completed within the next couple of months, coinciding with the Peak business planning.

The newly assigned management team for Peak is experienced enough to know that they can only meet Peak's production goals by fully exploiting reusable information resources. Building on Celerity's world-class resources and expertise in information systems and manufacturing, Peak management is expected to achieve breakthroughs in several performance measures, including manufacturing productivity and labor content.

While Peak management doesn't yet have a clear idea what an information plan is they have a gut feel that they need a resource assessment and a road map to their future. They must achieve unheard-of productivity levels in record time within a budget. This challenge also enables Celerity to test the leverage and the limits of its renowned information resources, and to establish some new benchmarks.

The new management faces major organizational challenges in creating the new Peak culture, as well. Most of the management is moving from the logic-chip manufacturing divisions of Celerity. The strategy driving that division, the flagship for Celerity, is one of innovation. They repeatedly lead the industry in the introduction of faster, more powerful logic chips. In pursuit of that strategy, capital resources are relatively plentiful. Logic chip manufacturing always receives the highest priority in Celerity's board room, since well over 90% of the firm's net income results from the sales of the Sentinel line (see Figure 1).

Thus, the new Peak management team has to create and adapt to a completely different divisional strategy: low-cost production. The shift from pursuing goals within

Figure 1. Sources of Celerity Net income, 1995-1998

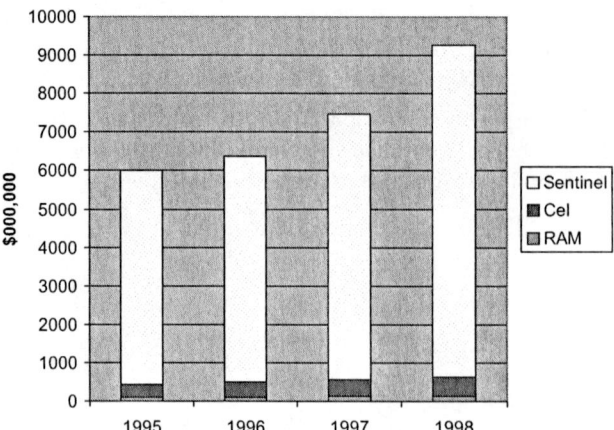

a resource-rich innovation strategy to pursuing goals driven by cost slashing is, perhaps the most difficult conceptual leap for the team. They are charged with selecting new organizational members (from the existing Celerity talent pool), most of whom have also been nurtured in the innovation-oriented culture. They must then transform their guiding principles and operating procedures to support the mandate for a low-cost production environment.

Thus, organizational transformation must occur at both the strategic level and the operational level. The strategic level transformation establishes the goals and parameters within which daily decisions are made. At every level below that, problems are framed and solved from a new highly integrative perspective — one that's informed by an urgent need to leverage each resource, especially the employees.

Strategically, Celerity's Peak team must achieve a sustainable competitive advantage in the about-to-explode flash market, by honing their quick-response capabilities and developing new skill in achieving world-class quality, with a meager overhead structure, and lightning-fast organizational learning.

Peak also requires an operational level of organization transformation — a deliberate effort to improve the organization's strategic performance by changing the behaviors of the key individuals responsible for meeting the organizational goals (King, 1997). They must learn to solve problems and apply learnings broadly across the Peak organization. They must learn to recognize and integrate key information into daily operating decisions, while creating new processing.

One major focus of the strategy is to achieve 300% productivity improvements. One way that productivity is measured is by the labor required to produce each 1,000 wafers of memory chips. (Each good wafer may contain a few hundred chips, depending on the diameter of the wafer). This productivity measurement, an industry standard for comparison, takes into account the "touch labor" of technicians, and engineering, management and other support staff involvement. Cost factors include not only the time actually dedicated to producing product, but also the time required for training, safety programs, maintenance, vacation, breaks, formal continuing education, and team-building activities. Where traditionally, Celerity may have involved 150-200 people per 1000 wafers, they are now expected to achieve 50-70 people per 1,000 wafers.

In an environment of innovation, such achievements would be remarkable, but would be clearly aided by a tendency to invest heavily in a new and potent infrastructure that would embed emerging technologies needed to replace intelligent people. In the quest for low-cost production standards, however, investments will attempt to capitalize on proven technologies rather than to lead the industry with new technologies. Some cushion does exist for the new Peak team, as they will benefit from most corporate-wide productivity improvement programs that may be undertaken during their start up.

At the same time, Celerity is facing intense price competition for the commodity product line (the Cel) that Peak will produce. Celerity clearly expects Peak to demonstrate how information can replace constrained, consumable resources — in the pursuit of continuous, high-quality production of a product mix that brings only slim profit margins.

Strategic Considerations

Celerity has enjoyed great technological and financial success using a strategy of innovation and market leadership. They are the dominant players in this market. Their strategy has fueled their steady growth over several decades. To achieve market dominance with their highly profitable Sentinel line of electronics, Celerity has also committed to research, development and manufacture of electronics, software or technology that stimulates demand for Sentinel. In some instances this commitment has resulted in strategic partnerships or acquisitions of other firms; in other instances, Celerity relies on in-house resources.

It was this commitment to nurture demand for Sentinel that drove Celerity to produce Cel. Now, however, there is demand for Cel that goes well beyond support for Sentinel. And that independent demand is projected to surge in the coming years.

Since Cel is a commodity item, OEM customers are not loyal to Cel per se. However, these customers do value Celerity's reputation for on-time delivery and unparalleled quality.

In contrast to Celerity's overall strategy of innovation and market dominance, the business unit responsible for Cel practices a variation on "low-cost producer" strategy. It is in this environment that Celerity launches the Peak manufacturing organization. And it is the low-cost spirit that motivates their interest in exploiting information resources.

Major Players and Organization Structure of Peak

Mark Pitcher is the Senior Manufacturing Manager for Peak, and is officially the top manager for the factory. He reports to the Worldwide Manufacturing Manager and his counterparts are the Manufacturing Managers of the other Celerity factories; most directly, his cohorts are the Flash Manufacturing Managers. Mark is a Mechanical Engineer by training and education. He earned both a Bachelor's and a Master's degree before beginning his career with Celerity 15 years ago. During his tenure with Celerity, he worked initially in the Sentinel production, but found more opportunities for greater management growth in the RAM and Cel lines. He has been with various Cel factories for the past ten years. He has won several achievement awards attesting to his creativity, engineering abilities, general problem-solving skills, and his terrific people skills.

Mark has two shift managers to assist with the more tactical level decision making. Chris Martinsen covers the first half of the week: Sunday through Wednesday, and overlaps on Wednesday with Oliver Wilshire who covers the second half: Wednesday through Saturday. Their schedules are demanding during this conception and start-up phase, since the organizations are comprised of both day and night shifts. Chris is an industrial engineer with about 10 years in the semiconductor manufacturing industry. Oliver, relatively new to Celerity, has degrees in both mechanical and industrial engineering.

The manufacturing organizations that report to Chris and Oliver are managed in matrix fashion with Chris and Oliver providing the lead management direction, while manufacturing functional area managers provide the lead technical direction. Shift managers and functional area managers routinely have formal technical training, and only informal managerial and team-developing training. Figure 2 illustrates the reporting structure to shift and functional area managers. Each manufacturing employee and team belongs to one of the 24 cells in the table. Employees in each cell report to the manager of the functional area shown in his/her corresponding row heading. Employees are also responsible to the shift manager in the corresponding column heading. Thus, at all times an employee has access to functional area expertise and management decision-making authority.

Figure 2. Matrix manufacturing organization at Celerity

Shift Management Structure				
	First ½ of week		Second ½ of week	
	Days	Nights	Days	Night
Litho				
Thin Films				
Etch				
Implant				
Testing				
Packaging				

This complex structure enables functional area, engineering, and management decision making to occur around-the-clock. The 24×7 calendar is motivated both by internal costs (manufacturing equipment is enormously expensive) and by the opportunity to sell at a profit every good chip that is manufactured.

Professor Leah Schiffman from South East University (SEU) also participated in the start-up phase of Peak. She was brought in as a faculty fellow because of her manufacturing-related research, consulting and current activities in management information systems. Leah would bring fresh ideas and approaches to the team. Leah was brought in as a faculty fellow, working full time with Mark for three months in the summer. Leah not only had research experience in manufacturing-related areas, she had also spent her previous sabbatical year working with another Celerity flash factory. In that assignment, she had worked the interface between the manufacturing and information systems organizations to roll out a new production scheduling approach in the factory. Thus, she had not only the perspective of an experienced outside consultant, but also had already established contacts and an understanding of Celerity as an organization. These characteristics proved invaluable for the current assignment.

Adam Parisi, manager of automation systems for the Florida factory, was keenly interested in this new project. His primary focus was on the information systems that directly support the production floor, such as process control, production scheduling, and specialized procurement. His staff directly supports the "interface" between information systems and manufacturing. They do lead the evolution of information services to the factory, but do not have the resources to conduct strategic planning for Peak. Based on his awareness of her earlier projects, however, Adam felt confident that Leah would not only understand his organization's concerns, but support his efforts to educate the manufacturing organization, as well. Adam's manager is the Worldwide Manufacturing Manager (just as Mark Pitcher does) — an executive level position.

Jason Mountain is the manager of information support worldwide, which happens to be based in the Florida facility. Interestingly, Jason had been an MBA student of Leah's and was now responsible for developing standards for Celerity's worldwide information infrastructure. In contrast to Adam's factory-floor focus, Jason's primary focus was on information systems outside of the factory. Jason was responsible for the computing and communication systems that tie the engineering and managerial groups within and among the facilities spanning the globe. Jason's manager is the CIO.

INFORMATION PLANNING

The Initial Stages: Weeks 1 and 2

Mark Pitcher conceived of the information planning project shortly after he was appointed to the new Peak organization. Mark dwelled on the challenge facing him: His new team was expected to achieve productivity and cost goals that had never been achieved in the industry. As he mulled over many of the good practices that had helped his teams achieve ambitious goals in the past, he understood that standard (even Celerity standard) problem solving would not succeed in this situation. Peak needed fundamental and radical changes from the mission and goals to the operating practices, job descriptions, reward structures, and communication and decision-making rhythms.

The corporate culture at Celerity rewarded calculated risk-taking in their environment of persistent uncertainty. Mark saw clearly that his new challenge was yet another example of this corporate phenomenon. No one at Celerity knew how to achieve the goals for Peak, they only knew that it was critical to achieve those goals. Celerity's success in flash manufacturing depended on Peak's ability to demonstrate how. Mark believed that a major part of the solution involved better use of information, but didn't know anyone from within Celerity with the background and perspective to either correct him or validate that intuition and help him plan the organization accordingly.

It was fortuitous that Celerity had established good working relationships with the engineering and management faculty at SEU. It was also fortuitous that summer was coming. It didn't take long for Mark to find and then contact Leah for initial discussions of the challenge he faced. Together, they defined a scope of work for Leah: to develop an information plan for Peak. Based on experience and belief that business and information planning should be concomitant and ongoing activities, Leah agreed that by taking a strong information-centered approach, Peak had the best chance to achieve its productivity and cost goals.

There was little time to develop the plan, but Leah had a few weeks of "thinking time" finishing the semester before she would even start the project. Driven both by the academic calendar and the ambitious start-up schedule, Leah would need to complete the plan within eight weeks of starting. The opportunity to guide an organization to develop an information plan at start up was delightful, however. Leah was more accustomed to working with firms who either (1) had no information planning process at all and were constantly battling technology issues, or (2) were wrestling, after the fact, with stresses resulting from a mismatch in the business plan and the information plan. Lack of integration between information and business planning creates problems (Bartholomew, 1998; Dempsey & Koff, 1995). Nonetheless, it remains unclear just how to effect the appropriate linkages between them (Reich & Benbasat, 1996). Now, even with a short timeframe and the realization that her involvement would be this "one shot" (leaving the evolution of the plan to Celerity upon her return to SEU), the assignment was a welcome challenge.

Once the assignment officially started, Leah spent the first week defining a reasonable process to get the involvement and buy-in from the right people. To be successful, she needed both manufacturing and information systems management to commit to a strategic information planning process. If the process were sound, she reasoned, the results would be the best they could all produce. If the process were sound, the results would also be rather dynamic — evolving over time as the Peak organization evolves. Her goals then, were to develop and implement a process that would (1) generate immediate results for the fledgling Peak organization, and (2) have the potential to become institutionalized throughout Celerity. That is, she sought to define a new business process to continually evolve the information strategy and infrastructure in concert with Celerity's business strategy process.

The process she chose was based on experience as well: find and involve credible champions, enlist their guidance to identify and invite lower-level participants. Once the team was formed, she would host a meeting of the whole team to present their "charge," field questions, and sketch out expected outcomes. The final task in that meeting would be to schedule individual follow-up workshops to review each participant's ideas. Leah wanted to understand the individual ideas and concerns — to see how they fit with the operating principles and how they complement one another.

Mark cautioned Leah that the bottom-up strategy would likely result in her hearing about issues that, while important to the individual, were tangential to the information plan. They agreed to face that risk and to avoid the mire off-topic issues. With Mark's input and perspective to help, Leah would decide which issues were relevant and which would be tabled.

Ultimately, Leah decided to use a combination of top-down and bottom-up investigation. By tapping into the issues pressing top management, she would better understand their major concerns and she would use them develop the boundaries and internal "structure" of the information plan. She could gain a lot of insight from strategy documents already published internally, and she used those documents as her starting point.

Leah's first 10 days onsite resulted in an initial proposal. During that time she also re-established contact with both Adam Parisi and Jason Mountain from the Florida and Worldwide Information Systems groups to inform them of the initiative and, to a lesser extent at this point, gain their perspectives on the overall direction of Celerity's IS organizations. Leah intended to keep the two IS groups fully informed during the entire planning process, since they would greatly influence any eventual implementation. To facilitate earning their trust and cooperation, Leah would encourage their full participation throughout the project and include them in important updates.

Leah first presented her proposal to Mark in a formal meeting with him. Mark listened intently and felt that Leah's approach would work within their organization. He was initially concerned about the broad base of involvement, an involvement that might cause delays in converging upon a feasible plan — if not the "best possible" plan. In response, Leah agreed to weight progress ahead of full-participation as she planned the work.

Mark further expected to see a full budget associated with any plan. That expectation was harder to promise and Leah told him up front that detailed budget development would be outside the scope of their initial process. She clarified for Mark that the first cut information plan would provide direction and suggest possible alternatives for

Figure 3. Information resources strategic plan outline

1.   Acknowledgements to Champions and Participants
2.   IRSP Overview—Background and Motivation
3.   Executive Summary
4.   Information Architecture
a. Current Architecture
b. Recommended Architecture
5.   Specific Recommendations for Functionality
a. Overview/Themes
b. Work in Process Monitors
c. Capital Equipment Monitors
d. Maintenance Monitors
e. Performance to Goals
f. Supplement Materials Monitors
g. Human Resources & Payroll Linkages
6.   Specific Recommendations for Procedural Changes
a. Overview/Themes
7.   Appendices
a. PEAK Operating Principles
b. Raw Data Collected

moving in those directions. While the alternatives could be stated in terms of resources requirements, there was not time to fully develop budgets. That work would need to follow in the subsequent design work.

Overall, Mark concurred with the proposal and, by the end of their discussion, he was eager to hear the input from his shift managers and IS managers. Using Celerity's calendar management resources, Leah convened a subsequent meeting with Mark, Jason, Adam, Chris and Oliver. She found and reserved a half-hour period that Wednesday during which none of them had previous commitments. In that meeting, she would present her proposal, consider modifications driven by the manufacturing or IS issues, and then revise and ratify the proposal as the roadmap to develop a strategic information plan for Peak.

Leah made several revisions based on her meeting with Mark. The Information Planning Proposal now included (1) a goals statement, (2) an outline of the planning process (top-down and bottom-up) that would ensue in the coming weeks, (3) suggested participants (by role), (4) time line, and (5) an outline of the plan itself. The proposal did not include full budget estimates, nor did it contain design details. See Figure 3 for the detailed Information Planning Proposal Outline.

Everyone Leah invited participated in the Wednesday meeting. There was some good discussion around the whole idea of information planning and some healthy skepticism that a meaningful product would result from such a process. The IS representatives were concerned that any plan be "feasible" from their perspective, since ultimately their organizations would be called upon to deliver desired functionality.

Adam was also concerned that scope of the proposed planning process went well beyond the decision making and resources available to Mark's or his organization. He suggested that a narrower scope — focusing directly on the manufacturing floor and its support — was more appropriate. Since his organization would carry most of the design and implementation responsibilities, the scope needed to fit within his purview. Adam's urgings led to a very productive activity: the participants enumerated the business processes (and related organizations) that the plan should encompass and explicitly named those that should be excluded. Using a spontaneous brainstorming method, the existing organizations were identified and then classified into one of two groups "within" or "outside" the scope of the planning process. The activity afforded much greater clarification and generated consensus. The scope would now include all manufacturing planning and control, and communications with manufacturing technician groups, among technicians, managers and engineering. The scope would not directly include communications with other groups (such as payroll, external suppliers, and other factories). The scope would also not include process control, which is executed at a machine level nor the tools and methods available to engineering or management groups. This scope refinement also resulted in some changes to the participant list.

When the meeting ended and everyone dispersed, Mark caught up with Leah and congratulated her on having run a very effective meeting. He was particularly pleased by the scoping work and Jason's support of the work and his eagerness to remain involved. To Mark that indicated this effort would be watched closely by IS Worldwide, and that pleased him. When IS Worldwide is in on the same initiative as the Production IS group, terrific synergies result.

Weeks 2 through 6

In the ensuing weeks, Leah worked on several subtasks of the planning process. The work involved meeting with staff from all levels in the Peak organization as well as industry research and analysis.

Meetings and Debriefings. Leah needed to collect input from a large number of individuals, and wanted to ensure that the results would be straightforward to organize at the end, and would fit well with Peak's business planning results. She acquired copies of Peak's mission, goals and objectives with which to organize the information plan and constructed an "If only ..." information gathering instrument. On the back of the form was a complete numbered list of all Peak's operating principles. The form was available in hard copy and on Peak's intranet. Figure 4 is a mock-up of the form. For simplicity, this mock-up contains only the top-level operating principles; the actual form contained detailed subheadings for each of the operating principles shown. The whimsical design is intended to trigger and encourage the free flow of creative ideas.

She also chose to use parallel interviewing schedules with the technician-level participants and the exempt (engineering, IS, and management) participants. In doing so, she intended to make each group comfortable expressing areas of needed improvement. Leah used a combination of e-mail, electronic calendaring, and phone calls to make the initial contacts and schedule meetings.

To reach the engineering, IS and manufacturing managers, Leah convened a "kickoff" meeting to outline the process and the input she needed from them. She reviewed and distributed a sample copy of the "If only ..." form for capturing their input. She informed them of the top management support (some of whom were in attendance) for the work and answered their questions. Then she scheduled individual or small-group

Figure 4. "If only ..." form

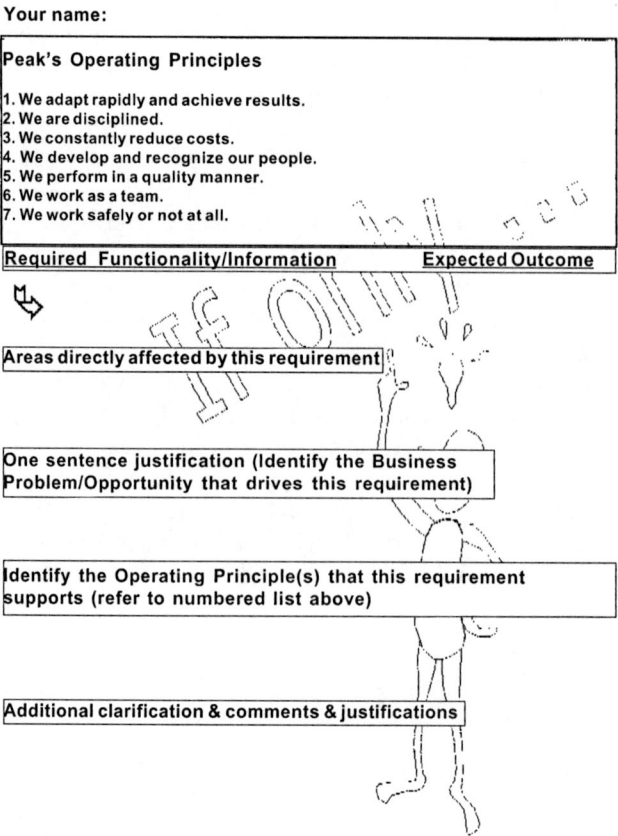

Your name:

Peak's Operating Principles

1. We adapt rapidly and achieve results.
2. We are disciplined.
3. We constantly reduce costs.
4. We develop and recognize our people.
5. We perform in a quality manner.
6. We work as a team.
7. We work safely or not at all.

Required Functionality/Information   **Expected Outcome**

Areas directly affected by this requirement

One sentence justification (Identify the Business Problem/Opportunity that drives this requirement)

Identify the Operating Principle(s) that this requirement supports (refer to numbered list above)

Additional clarification & comments & justifications

meetings with them to review the input they would construct over the ensuing two weeks. Their task was to complete as many "If only..." forms as they had suggestions for improvement. As noted on the form itself, for each recommendation they made, they needed to identify the specific goals and objectives it supported. In this manner, Leah would be able to tie the resulting information systems plan to the Peak business plan (Horner & Benbasat, 1996).

To involve the technicians, she planned 10 individual meetings with the 10 representatives from all shifts and all functional areas. She had relied on the Shift Managers and Functional Area Managers to identify the representatives. Her goal was to talk with individuals who are opinionated, committed to the organization, and would be willing to brainstorm and share their ideas with her. She invited each representative to a meeting at their convenience, which meant that the meetings occurred on their shift, in their meeting area, and at the time that was best for them. Such an approach clearly signalled their importance to the process.

At least a week prior to each scheduled meeting, Leah provided the technician with the background of her project and asked them to think about their work and what could make it better. They were to analyze the tasks assigned to them and/or their teams and identify things that currently block better individual, team, or area performance. If they had previous experiences in the industry or in other industries, she asked them to identify the things that really made their work easier or of a consistently higher quality.

Then, as with the exempt staff, she asked them to complete as many "If only ..." forms as they had ideas to contribute. When they met, she would ask them to explain their ideas so she was certain to understand their suggestions.

In meetings with both the exempt staff and the technicians, Leah approached the task in the "language" and jargon of her customers. She took it upon herself ask for input on business processes and then to do the translations into information resources. Such an approach freed up the respondents to speak to her in the nearly the same language they used to brainstorm in the cafeteria.

To ensure she really did understand the issues raised in the discussions and enable the participants to see the "information impact" of their concerns, Leah would suggest business process or information systems ideas in response to a gnawing problem raised. These types of dialogs eased the participants into addressing their business problems by harnessing information in creative new ways.

As an example, one "night shift" technician noted that process adjustments were often delayed even after they felt the root cause had been identified. Why? Engineering staff were much less available to them at night and their approval was needed for any process modifications. Leah then described the concept of real-time exception notification alerts, where based on knowledge of who had authority and responsibility for exception codes, that person's pager would be signalled when they were needed by the technicians. Coupling such a notification system with mobile two-way conferencing and probe equipment, the decision makers could choose between gowning up to go into the clean room factory, or viewing the problem while discussing the details remotely with the cognizant technician. In either case, problem analysis and an ultimate decision could be made quickly.

To a person, the technicians were enthusiastic about their participation and optimistic about the future of their new organization. The managers had done a good job selecting the technician sample and were, themselves, good contributors to Leah's data-collection efforts.

Background Research

Leah sought two types of information to build her understanding of the current state of information resources at Celerity: industry standards and an internal inventory of information resources (i.e., the major functionality of existing applications). These sources provide the "as publicly planned" and "as is" perspectives, respectively.

For industry standards, Leah combed the archives of SEMATECH for any published works on information management in the industry. SEMATECH is a consortium of U.S. firms involved in semiconductor manufacturing who pool resources and exchange some expertise on the generic issues facing them. SEMATECH provides several research and development venues that are available only to its member firms. Other published works are available to the general public. Most of the work is in the basic sciences or engineering — the foundations of semiconductor manufacturing. Leah found very little on management and no works directly on the subject of strategic information planning.

For an understanding of the current Celerity — and more specifically, Peak — resources, Leah met individually with Jason and with Adam. In both meetings she asked for an inventory of information resources currently available. Reviewing such an inventory would give her the baseline from which to build. Clearly, the baseline was insufficient to achieve the productivity and costs, but she didn't really know what the baseline was.

To her dismay, no such inventory exists — even in paper form. Production IS is responsible for managing hundreds of applications and yet there is no systematic method to discover what those applications are, the data they contain, how and when they are used, or by whom. From a hardware perspective, it is possible to summarize the resources — but such a summary would be derived from data in a Property Management organization, not from within Manufacturing.

Worldwide IS faces a similar situation. Leah was interested to view a summary of the management, engineering, and support applications — to examine the types of information services that these staff depend on. While Jason could list a number of applications that are widely used, neither an inventory nor a summary could be produced with any reasonable effort.

The meaning of this finding is that, as an enterprise, Celerity has not maintained searchable information resource metadata (Hsu et al., 1994; Hsu & Rattner, 1994). Searchable metadata would not only allow IS to manage a definitive catalog of their applications and better predict the impact of changes to underlying databases, but also it would enable users to browse for existing applications that might meet their information needs. With such browsing capabilities, Celerity will minimize the number of redundant and obsolete "small applications," of which IS estimated there to be thousands. Peak will need to get a handle on its metadata in order to achieve the organizational speed it is pursuing. A fully functioning metadatabase that stores and manages details of Peak's information resources would provide part of the infrastructure to begin to substitute information for more costly human resources.

In the absence of a metadatabase, Peak will still rely on individuals to discover the location and access methods to organization data they require. When the search is too onerous, Peak will still rely on individuals to recreate it. Thus, any new state-of-the-art information delivery system would still be crippled by a constricted flow of data to them.

Technically, this finding was significant. Some years earlier, Leah had seen metadatabase research and implementation (Hsu & Rattner, 1994; Hsu et al., 1993) tested in other manufacturing industries. The concepts certainly fit the Celerity's current needs as well.

Weeks 4-10: Completing the Information Resources Plan

Throughout the data collection period, and during her subsequent analysis and plan development, Leah scheduled weekly status meetings with Mark. During these meetings, she would discuss specific "If only ..." contributions that seemed to have significant bearing on the direction that Peak information resources strategy should take.

In one such meeting, Leah explained the need to manage metadata to Mark. She illustrated the concept by tracing the problem-solving method currently used to address mis-processed wafers in the Etch area. When the steps of the process are shown, the process is reasonable. When the time for each step are made explicit, however, it is instantly clear that "searching for the data" to generate an initial solution takes hours. The evaluation needed to identify the root cause, however, requires days or weeks or months of searching for data. That lengthy and costly search period results from having no systematic metadata management.

Another major finding that emerged over time was that Peak's lack of standardized communication products limited its organizational speed. Different e-mail systems, directories, calendaring packages, pager systems, and electronic notification systems severely hampered any effort to distribute time-sensitive information quickly and systematically.

Leah participated in several shift-change meetings on the factory floor. These meetings are the forum for one shift to communicate to the next shift the status of products and machines. Status information includes upcoming scheduled maintenance, "bubbles" of product moving through the production line, and special expedited orders or problems that had not yet been resolved.

During these meetings, each Functional Area representative made a brief report. Some shift-change meetings were well run, and all participants focused on the issues raised by each representative. Other shift-change meetings were poorly run, with several simultaneous small-group discussions going on. At the same time as Functional Area Managers were conducting these meetings, technicians might be discussing issues specific to their tool-set with the incoming technician. For such an exchange to occur, however, required that either the "ending" technician stay late or the "beginning" technician arrive early enough to share such valuable information. There were no direct rewards in place for participating in such exchanges, so only the most highly motivated technicians would routinely engage in them. In this manner, critical information for troubleshooting and pattern recognition would be lost or else conveyed in an inconsistent manner — again hampering attempts to recover quickly and solidify organizational learning.

The Environment Shifts. In her meeting with Mark during week 7, Mark introduced some major uncertainty to the project. Worldwide demand for flash memory had softened. Due in large measure to economic problems in the far east, Celerity's executive management had reopened the analysis that lead the launching Peak. No one was certain at this point if the plans for Peak would be scrapped, modified, or maintained as originally planned.

A final decision-date had not yet been made, but the expectation was that an announcement would be made internally within two weeks. Some time was needed to evaluate the now, most likely economic and demand scenarios, and their implication for Celerity— including, but by no means dominated by the plans for Peak.

This news explained a lot of nuance that Leah had picked up in the past week or so from her previously upbeat and enthusiastic plan participants. While they had not at liberty to discuss the internal status of Celerity with Leah (a contractor), several meetings that week had been postponed for vague reasons.

Mark informed Leah that she was not at liberty to discuss this uncertainty with non-Celerity individuals. As a contractor, she was not fully informed of the details, and Mark requested that she not actively seek out details. He promised to continue to keep her informed on a "need to know" basis.

The no/no-go decision for Peak would be included in the formal announcements that were expected in the next few weeks. Until that time, everyone at Celerity was expected to continue operating according to the "active plan of record." In Leah's case, that simply meant to continue developing the information resources plan according to the proposal that had been ratified weeks earlier. Regardless of the eventual outcome for Peak, the emerging findings and recommendations would be valuable to Celerity Enterprises.

Understandably, this new information gradually led to the realization that Celerity was unlikely to launch a new facility in the face of declining market demand. This realization, in turn, took away some measure of enthusiasm and focus on "If only ..." from Leah's participants who began to focus on their more-personal "What if ..."

In her meeting with Mark during Week 8, Leah suggested that she cancel the presentation they had scheduled (for week 10) for her to review the findings and recommendations. Instead she would complete the study and provide copies to each participant. While the results might well be meaningful for Celerity, it would be awkward and inappropriate to spend an afternoon discussing the vision for a new facility — that may never be. If appropriate in the future — even if it occurred after her formal assignment with Celerity had ended — Leah would return to more publicly present the results.

Closure. Except for the final presentations, Leah completed the study as planned. The information resources plan included a metadata management foundation, intranet-based communications systems, and a variety of integrated services. The rationale for each piece of the plan included a direct link to the Peak organization's goals and objectives. The plan also included an additional section, not initially conceived of when the project started: recommended changes to business processes and procedures. Several of the "If only..." contributions had raised potentially important issues that were outside the scope of the information plan. The new section at least ensured that the issues had been documented for later evaluation.

REFERENCES

AMD. (1997). *AMD signs loan agreement for Dresden megafab* (Press Release). Retrieved from http://www.amd.com/news/corppr/9728.html

Bartholomew, D. (1998, March 2). Getting on track. *Industry Week, 247,* 22+.

Brinton, J.B. (1997, February). Flash faces possible shortfall. *Electronic Business Today,* 57-60.

Clark, C., Cavanaugh, N., Brown, C., & Sambamurthy, V. (n.d.). Building change-readiness capabilities in the IS organization: Insights from the Bell Atlantic experience. *MIS Quarterly, 21*(4), 425-455.

Dempsey, J., & Koff, W. (1995). Increasing IS productivity: A holistic approach. *Information Strategy, 11,* 5-12.

Hsu, C., Babin, G., Bouziane, M., Cheung, W., Rattner, L., Rubenstein, A., & Yee, L. (1994). A metadatabase approach to managing and integrating manufacturing information systems. *Journal of Intelligent Manufacturing, 5*, 333-349.

Hsu, C., Pant, S., & Rattner, L. (1994). Manufacturing information integration using a reference model. *International Journal of Operations and Production Management, 14*(11), 52-72.

Intel. (1998). *Intel opens 1st 0.25 micron microprocessor production factory in Europe* (Press Release). Retrieved from http://www.intel.com/pressroom/archive/releases/AW051198.HTM

King, W. (1997). Organizational transformation. *Information Systems Management, 14*, 63-65.

Lederer, A., & Sethi, V. (1998). Seven guidelines for strategic information systems planning. *Information Strategy, 15*(1), 23-28.

Reich, B. H., & Izak, B. (1996, March). Measuring the linkage between business and information technology objectives. *MIS Quarterly, 20*, 55-81.

Segars, A., & Grover, V. (1998). Strategic information systems planning success: An investigation of the construct and its measurement. *MIS Quarterly, 22*(2), 139-163.

Laurie Schatzberg, PhD, is an associate professor of management information systems at the University of New Mexico's Anderson Schools of Management. Prior to joining the faculty in 1991, Dr. Schatzberg earned a PhD from Rensselaer where she had also been a project manager at the Center for Manufacturing Productivity and Technology Transfer. Dr. Schatzberg has published numerous articles concerning manufacturing and information systems. These articles have appeared in DataBase, Production and Operations Management, IEEE Transactions on Systems, Man & Cyernetics, Journal of Intelligent Manufacturing, *and* Informing Sciences Journal, *and elsewhere. Such corporations as ALCOA, Digital, GM, Intel, Johnson and Johnson, and UTC have sponsored her work. Dr. Schatzberg has been the secretary/treasurer of ACM SIGMIS and department editor for* DataBase.

This case was previously published in the *Annals of Cases on Information Technology Applications and Management in Organizations*, Volume 2/2000, pp. 187-213, © 2000.

Chapter V

Project MI-Net:
An Inter-Organizational E-Business Adoption Study

Pankaj Bagri
Indian Institute of Management Bangalore (IIMB), India

L. S. Murty
Indian Institute of Management Bangalore (IIMB), India

T. R. Madanmohan
Indian Institute of Management Bangalore (IIMB), India

Rajendra K. Bandi
Indian Institute of Management Bangalore (IIMB), India

EXECUTIVE SUMMARY

This case gives a detailed description of the adoption of an e-business initiative by Miracle Industries Limited (MIL), a fast-moving consumer goods (FMCG) organization in India. The initiative involved linking up with key distributors so as to get important sales-related data on a real-time basis. The case describes how the decision to adopt the project was taken after a comprehensive analysis involving a detailed cost-benefits study, and an examination of the roles of various stakeholders — the distributors and the Territory Sales Officers. It also illustrates how the organization proactively managed the changes introduced by the adoption by communicating extensively about the benefits of the project to the stakeholders, and by providing training and incentives to them. The role of the existing IT infrastructure and unambiguous support from the top management in enabling a smooth rollout is also discussed. Finally, the dissatisfaction of some distributors in the post-implementation stage has been captured.

THEORETICAL BASIS FOR THE STUDY

Electronic business, or e-business, is a new business paradigm increasingly being adopted by organizations around the world to capitalize on the potential of new technologies such as the Internet and the World Wide Web (WWW) to "rethink business models, processes, and relationships along the whole length of the supply chain, in pursuit of unprecedented levels of productivity, improved customer propositions, and new streams of business" (Feeny, 2001). The sheer scale of changes brought about by the Internet and related technologies have thrown up a host of possibilities for organizations, in terms of how they conduct their businesses, that were simply not practical, or even possible, before (Straub, Hoffman, Weber, & Steinfield, 2002). As Evans and Wurster (1997) put it, the traditional tradeoff between "richness" and "reach" is now being disputed by the Internet. The use of the *public Internet protocol standard* (TCP/IP) allows rapid growth in *internal and external links* at much lower costs than was possible earlier, say with Electronic Data Interchange (EDI), which is typically proprietary and more expensive. The *many-to-many communication* model made possible by the Internet is also a radical change from the traditional one-to-many broadcasting paradigm (Straub et al., 2002).

According to Porter (2001), while by itself the Internet might not provide a sustainable competitive advantage, what would differentiate the winners from the losers would be the ability to use the Internet as a complement to traditional ways of competing. The key to success for conventional firms, as Gulati and Garino (2000) put it, is in carrying out the integration between virtual and physical operations. Ross and Feeny (2000) posit that relatively speaking, the technology aspect of an e-business strategy is now easy, and that "vision and holistic thinking rather than technology rollout are the key requirements." The literature on IT adoption also suggests that, increasingly, implementation failure — and not technology failure — is responsible for many organizations' inability to achieve the intended benefits of the IT they adopt (Klein & Sorra, 1996).

Zaltman et al. (1973) propose a two-stage adoption process for innovations in organizations — a firm-level decision to adopt the innovation (primary adoption), followed by the actual implementation, which would include the individual adoption by users (secondary adoption). In the context of adoption of inter-organizational systems, the secondary adoption phase can be considered as involving not just the users within the organization, but also external entities in the value chain. Though there are a number of studies on adoption of inter-organizational systems, the most commonly studied interorganizational system, EDI, is fundamentally different from e-business in a number of ways that impact the attitude of both the initiating organization and its partners. EDI has evolved around industry standards, whereas e-business relies on more open standards. On the other hand, the Internet is a ubiquitous public network that provides many advantages over value-added networks (VANs), including low costs, worldwide connectivity, platform-independence, and ease of use infrastructure. This makes the adoption of e-business practices typically a less expensive proposition for the partner organizations. Also the fear of getting locked into a particular system can be handled more effectively in the e-business scenario. To illustrate, a buyer has to invest in a single personal computer (PC) and Internet connection in order to access any number of supplier sites (or extranets) to check its order status or statement of accounts. In fact the varying diffusion rate for EDI and e-business practices is itself an indicator of the difference between the two technologies and its implication for adoption by organizations.

Premkumar and Ramamurthy (1995) posit that in case of implementation of inter-organizational systems, the secondary adoption can be achieved in two ways: (a) through coercion by exploiting the partner's dependence on the adopter, or (b) through inducement by providing appropriate incentives and training. They suggest that the latter might be a better option because these partners should be considered as "extended users" and the support provided for internal users should be extended to these firms, at least initially, for successful implementation. Among other things, this would call for a proactive approach towards change management, as the success of the implementation process will depend on managing multiple stakeholders spanning organizational boundaries. Boddy et al. (1998) propose, with reference to supply chain partnering, that such an initiative "implies changes to the social system of at least two separate organizations, so the scope for resistance is considerable." Underestimating the scale of change involved in the adoption may create avoidable problems during the implementation process.

This case study chronicles the adoption and implementation of an inter-organizational, e-business system by an FMCG organization in India. The focus of the case is on understanding the implementation process and the extent of proactive change management practices employed by the organization. The case brings out issues like the importance of building value propositions to attract the stakeholders, and providing the requisite training and incentives to commit them to the system. It also demonstrates how effective communication can bring down resistance/reluctance to change, not only within the organization, but also in partner organizations. Finally, it highlights the importance of post-implementation support to the stakeholders to maintain the tempo built during the implementation.

ORGANIZATION BACKGROUND

Miracle Industries Limited (MIL) is a leading Indian fast-moving consumer goods (FMCG) organization, with a turnover of Rs. 7 billion (US$142 million) in the financial year ended March 2002. Appendix 1 provides a brief account of the Indian FMCG industry. MIL operates in two broad segments — Nature Care (coconut oil, value-added hair oils, anti-lice treatment, fabric care) and Health Care (refined edible oils, low sodium salt, processed foods). It has consciously concentrated on "unique/ethnic Indian product categories" where large multi-national companies (MNCs) do not have a strong presence. Consequently, MIL has grown over the years from a largely two-brand company in 1990 to a nine-brand company operating in select product categories in 2002. It has a wholly owned subsidiary in Bangladesh, and exports its branded products to several South Asian and Middle East nations. Three of the brands are market leaders in their respective categories, while three others are number two. Appendix 2 provides a snapshot of MIL's financial performance over the past six years. MIL has six manufacturing units and also sources products from 15 contract manufacturers.

MIL's head office is located in Mumbai. There is a regional office in each of the four sales regions covering northern, eastern, southern, and western India. The organization structure of MIL is shown in Appendix 3. The total employee strength is about 1,000. The organization is divided into three profit centers — Nature Care, Health Care, and International Business — and three service centers — Finance (including IS), Human Resources, and Sales. MIL has a nationwide distribution network comprising 32 carrying

& forwarding agents (CFAs), 970 distributors, and 2,500 stockists. The typical distribution network for FMCG companies in India is as follows:

- Manufacturer → Distributor → Retailer (for urban areas)
- Manufacturer → Super Distributor → Stockist → Retailer (for rural areas)

Appendix 4 provides an indicative summary of MIL's distribution network in India. Every month, more than 35 million consumer packs (approximately 125 Stock Keeping Units (SKUs)) from MIL reach approximately 18 million households through 1.6 million retail outlets serviced by the distribution network. At present, the network covers every Indian town with a population greater than 20,000. The rural sales and distribution network of MIL ranks among the top three in the industry and contributes 24% to the company's topline.[1]

MIL is a relatively young organization, with an average employee age of 32 years. The organization is professionally managed, with the Chairman & Managing Director (CMD) being the only member of the promoter group in an executive position. People are encouraged to take initiatives, and there is emphasis on teamwork. The compensation system has a high variable component, linked to the performance of the individual. Additionally, the compensation is also linked to the performance of the organization as a whole. Thus, 50% of the annual profit achieved over and above the budgeted figure (known as Super Profit) is distributed among the employees. However, with a steady growth over the years, need is being felt to put systems in places. According to a senior personnel member who has been associated with the organization for more than a decade in several capacities, the last being as project manager for the Enterprise Resource Planning (ERP) system implementation (PM-ERP), "…we have a very informal culture. But there is a very thin line between informality and casualness. We were not much systems driven earlier, but now with the ERP in place, we are getting the systems in place."

MIL is a very sales-focused organization, as is evident from the existence of Sales as a separate division responsible for "leveraging the economies of scale" and "allow[ing] for more distribution opportunities." Thus the Sales division provides service to all three profit centers, and carries out various initiatives to improve the productivity of the selling process.

The IS team consists of 19 people at the headquarters, headed by the General Manager-Information Systems (GM-IS), who reports to the CFO. Each manufacturing unit has one IS person looking after the systems. Additionally, each regional office has two IS persons — one to look after ERP-related issues and the other to look after distributor software issues. The IS orientation at MIL underwent a sea change in 1998, from being a back room cost center to a strategic partner in achieving organizational objectives. Talking about the role IS people should play, the GM-IS remarked, "The only job of (IS) head is to look at the linkage between business goals and IT capabilities." On another occasion, he said, "Time should come when you cannot make out what is IT Department. Diffuse it."

Corporate planning at MIL is carried out on a yearly basis. The three profit centers as well as the service centers draw their own strategic plans. Then, through a process of interaction and discussion, the individual plans are revised to accommodate the objectives of the business divisions and the capabilities of the service centers, and an overall organization strategy document is prepared. The entire process takes between two and three months.

SETTING THE STAGE

In typical FMCG parlance, a sale is deemed a *primary sale* when the manufacturer sells the goods to its direct customer, the distributor. The sale becomes *secondary sale* when the distributor in turn sells it to a retailer. When the retailer sells it to the end consumer, the sale is called *offtake*. In an ideal scenario, in order to accurately forecast demand and hence plan various manufacturing and supply-chain-related activities, an FMCG organization should be able to obtain consumer offtake data in a timely and reliable manner. However, given the scale of operations (e.g., MIL's products are sold through some 1.6 million retail outlets), it is virtually impossible to capture the offtake data for supply chain decisions, which have a much shorter timeframe (higher frequency). Hence most FMCG organizations make their forecasting based on primary or secondary sales data.

MI-Net (Miracle Industries Network) was conceptualized as an Internet-enabled application that would "establish a network between MIL and its distributors through a Web interface." The prime motivation behind the project was to obtain the secondary sales data at the distributors' end in a timely and accurate fashion, so as to render it actionable. The idea originated in September 2000 when MIL appointed an IT consultancy firm to chart out an IT strategy for the organization. One of the "laundry list" items suggested by the consultants was SellNet, a project that visualized connecting the organization to its distributors using Web technologies. However, it was felt that MIL as an organization was not ready to embark on such a project, as it did not have the requisite IT infrastructure, and the idea was kept in abeyance. Instead, MIL focused on two other IT projects that were to play a crucial role in MI-Net — a standard software package to capture secondary sales, MIDAS, and an ERP solution for the organization.

Project MIDAS

Traditionally, MIL had relied on Town Monitoring Reports (TMRs) for collecting secondary sales data. The data was gathered by the territory sales officers (TSOs) of MIL, who would periodically visit the distributors in their territory and write down data on opening and closing stock levels for various SKUs, performance of sales schemes, etc. Depending on the category of market (large/small), this data would be collected over a seven-day or a 10-day period. Each TSO would then collate the data for all the distributors in his territory and send it to his area sales manager (ASM). Each ASM would then collate the data from the six TSOs reporting to him, and send it to the regional sales manager (RSM). Finally, each RSM would collate the data for five ASMs in his region and send it to the corporate headquarters. Even though the report was passed on to the next higher level by fax/e-mail, typically the consolidated data for a particular cycle would reach the forecasting personnel only by the middle of the next cycle. As a result, the whole exercise was not much better than a "post-mortem." According to the CEO (Sales), "Most of the time (the data) is collated and forgotten."[2] Moreover, since the data was collected manually by the TSOs, its accuracy was also questionable. Hence, MIL started exploring options on increasing both the timeliness as well as the accuracy of secondary sales data.

The base conceptualization of MIDAS (Miracle Industries Distribution Automation Software) took place sometime in late 1999. The project was initiated by the Sales division, and was provided support by the IS personnel. One particular sales development manager (SDM) championed the concept and rallied support for the project. The idea was to provide the distributors with software that would capture information related to various bills and invoices, dispatches, receipts, etc. Instead of making entries in a ledger book/billing book, the entries would be made directly into the computer. Two options were considered for providing the software to the distributors. One, have the software on MIL's central server, and a Web-based system where a distributor would log on to the server, and update his account on a periodic basis. However, this would require the distributor to stay connected to the server for a long duration (largely due to bandwidth problems), which was clearly not feasible. Hence the second option was chosen — provide the distributors with the software, which would reside on their PCs and work in a standalone fashion. This would not only reduce paperwork at the distributor's end and make his account keeping easier, but it would also ensure easier data capture for MIL's TSOs.

The next step was to identify a software package that could meet the requirements of MIL. The project team identified a particular vendor and started working on customization of the software. At this stage, the project was rather low key, largely a Sales-IS initiative, and top management was kept informed about the progress, but was not actively involved. About six months down the line, the MIL team realized that the vendor was not in a position to deliver the goods, and changed over to a larger vendor whose software was closer to the needs of MIL. The new package, primarily an off-the-shelf billing and stock management software, was customized and then demonstrated to top management. However, top management pointed out a number of features in the product that would be unacceptable to the distributors. As the GM-IS put it, "We made the mistake of thinking of the distributors as MIL." The typical distributor was a small entrepreneur who might not want all transaction details to be reflected in the system. Also, the package had certain built-in controls (e.g., if a distributor credit limit was breached, no further transaction would be allowed) that did not reflect the ground realities. Consequently, the MIL team (along with the vendor) went back to the design board and came up with a toned-down version of the package, christened MIDAS. According to the SDM championing the project, "It was like changing from version 2002 to version 1999."

Top management approved MIDAS, and the formal rollout began in November 2000. MIL decided to use a pull approach, rather than a push strategy, to obtain the cooperation of distributors, as it believed the former to be more effective in the long run. The distributor community within a region is fairly close-knit, and typically there are a few large distributors (known as "greenball" distributors) whose actions are closely followed by the smaller distributors. Hence, MIL decided to first target these large distributors in each region.

Since the attempt was pull based, a lot of emphasis was put on communicating the benefits that MIDAS could offer to the distributors, and on how both MIL as well as the distributor community could be better off. Among the projected benefits that would interest a distributor were faster accounts reconciliation and less paperwork. MIL felt that a TSO who interacted on a daily basis with the distributors would understand their language better than any person from the central office. Hence six TSOs from across the country were roped into work full time on communicating with the distributors. Also, the MD himself sent out letters to all distributors explaining the concept and the potential benefits. The tone of the letter was that of a partner talking to another and managed to convey top management's involvement in the project. While a distributor needed to invest in the hardware, mostly a desktop PC, the software was given free to the distributors.

The initial response of the greenball distributors was mixed. Most of them realized the need for automation as their transaction volumes grew, and many already had some sort of distribution software in place. However, they were also a bit skeptical about the ability of MIL to carry out a nationwide project, since no other FMCG company in India had managed to have a single software package for all its distributors. More importantly, the distributors that were already using some package were reluctant to change over to MIDAS. There were also concerns about the ability of the package to handle the transactions related to other FMCG companies. Clearly the distributors did not want to use separate software for each of their clients.

Taking into account these reactions, the MIDAS team again made changes in the software, converting it into a multi-user, multi-company product so that distributors could manage products from multiple companies. It also built in features allowing data transfer from MIDAS to some popular accounting packages (e.g., Tally) used by many distributors. For distributors not using any software, MIL also provided basic training for a period of one week, during which a TSO along with a technical person would regularly visit the distributor and handhold its staff through the entire process of operating the package.

These steps generated a lot of confidence among the distributors, and the rollout was reasonably smooth thereafter. However, a few distributors were still unwilling to convert to MIDAS, and MIL terminated its relationship with them. By March 2001, MIL had covered all its class 'A' distributors (turnover greater than US$100,000 per year) in the urban areas. The rural rollout started in November 2001, and was over by February 2002. In all, 330 class 'A' distributors, accounting for almost 75% of MIL's primary sales, were provided MIDAS. The rest were not covered, as MIL's low business volume with them made it financially unviable.

ERP Implementation

At around the same time, MIL launched an ERP implementation project in association with leading global vendor, SAP AG. The major motivation behind the initiative was to root out the inefficiencies in the distribution process that resulted time and again into lost sales on one hand and inventory build up on the other. There was no distribution planning system in place — previous attempts to use some systems had failed. Systems integration was missing, with there being "islands of automation" all operating on different platforms. There were problems in back room and front room coordination, and the MIS reports from various departments would often be inconsistent.

MIL decided to connect all its offices, manufacturing units, and warehouses across the country using a SAP R/3 package. In addition, it also decided to install the SAP BW (data warehouse) and SAP APO (demand planning and forecasting), in the process becoming the first Indian company to do so. A big-bang approach towards implementation was taken, using SAP's ASAP (Accelerated SAP) methodology, and connecting the head office, five manufacturing units, five contract manufacturers, four regional offices, and 30 depots. The R/3 implementation started on June 1, 2000, and the system went live on April 2, 2001. The APO implementation started on August 15, 2000, and was operational on May 2, 2001.

The ERP implementation was characterized by an "over-commitment" of resources, as the PM-ERP put it. MIL's track record in managing IT initiatives prior to ERP implementation was rather poor. The sales force was especially notorious for not accepting various IT applications. However, change management was especially emphasized during the ERP implementation. In addition to a core team consisting of 24 people from MIL and 20 from the implementation partners, various process owners and users were identified and involved in their respective phases. A lot of emphasis was placed on training the depot operators, as they were relatively less familiar with IT systems. The project also had high credibility as the top management in internal communications highlighted its importance from time to time. As a part of the ASAP methodology, two surveys were conducted among the targeted users to assess their perceptions regarding the credibility of the initiative, the amount of risk it posed for the organization, and the amount of risk it posed for the individual. The results of the second survey, carried out in April 2001, were more favorable in all three categories, as compared to the first survey (August 2000), especially in the matter of credibility. The change in attitude was greatest in the case of sales people.

CASE DESCRIPTION

Even as the urban distributors started using MIDAS, and TSOs obtained the secondary sales figures directly from the distributor's PC, MIL realized that only a part of the initial problem was being addressed. The data available now was more accurate, but it was still not available in time. The earlier system of manual copying had been replaced by copying on a floppy drive, but the process of distributor-level data reaching the sales headquarters through multiple collations still took a long time. In order to make the optimum use of the MIDAS effort, it needed to be complemented with another effort that would now make the secondary sales figures available on a "real-time" basis. This would render a host of benefits, including capture of stock availability with every distributor, distributor sales team performance, information about outlets not being served, and new product launch monitoring, among others.

This was what the SellNet concept was all about. While the idea had earlier been shelved as one "whose time had not come" (as put by Sales IT manager), a number of things had changed since then. For one, the MIDAS software had been developed and already implemented successfully in urban areas. Equally importantly, MIL had recently finished its ERP implementation, and had gone live across all regional offices, sales depots, and the corporate headquarter on April 1, 2001. As the CEO (Sales) put it, "Very obviously, the next logical step was to integrate these into one system to give online access to one and all on both the primary level and secondary level information."[3]

The morale within the organization was also very high. The MIDAS initiative was the first of its kind in the FMCG industry (implementing one standard software package across all distributors) in India, and its successful rollout generated a lot of goodwill for the Sales division within the company and among the distributors. On the other hand, the ERP installation was the first IT project of this scale (in the organization) to have been implemented successfully within schedule and without any cost over runs. As the GM-IS explained, "In a previous project, when we were attempting to connect four of our manufacturing units with four of our depots, we took two and a half years to install eight VSATs. The ERP installation required 30 VSATs to be installed in one and a half months! So obviously there was a lot of skepticism regarding our being able to meet the schedule. But when we achieved that, we won the confidence of one and all."

Figure 1. Basic MI-Net diagram[4]

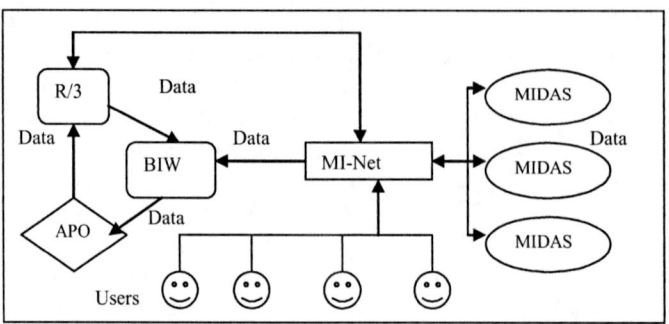

However, what was still required was conceptualization of the entire functionality for the initiative, and building up of a business case that would objectively and quantifiably evaluate the costs and benefits of the project. The GM-IS said, "We are a conservative organization. FMCG is not a capital-intensive industry. A new factory can be set up for as little as US$2 million. Hence we need to justify any IT investment and build a detailed business case for it. As far as achieving benefits are concerned, we believe that you can't control what you can't measure."

In May 2001, MIL appointed another consultancy firm to explore the feasibility of SellNet and to chalk out a roadmap for connecting the organization with its distributors. After a joint four-week intensive strategy workshop, which included key people from the Sales division as well as the IS Department, the consultancy firm delivered a detailed business case with quantifiable costs and benefits, a scalable and extensible technical architecture, and a structured implementation plan. The network was to be called "MI-Net," short for Miracle Industries Network. The MI-Net conceptual architecture is as shown in Figure 1.

The objective of MI-Net was to provide "real-time, comprehensive, and easy-to-use information resulting in quantitative business benefits for MIL."[5] To that end, MIL envisaged an Internet-based system where the distributors could log in and supply the necessary data. With the availability of MIDAS, it was possible to build an application that would reside on the distributors' PCs and would automatically transfer the data from MIDAS to MIL's central server every time the distributor logged in.

However, as with MIDAS, here too MIL wanted to "pull" the distributors to the new system rather than impose it on them. So the project team started thinking of ways in which the project could deliver benefits to the distributors too. With the ERP implementation now complete, it was possible to make relevant data available to each distributor from a central server. Among the benefits identified for distributors were information on order status, statement of accounts, stock of goods at MIL's depots, and details of various promotion schemes available to a distributor. All the above information could be provided if the MI-Net system was linked to the ERP at MIL. Hence it was decided to build a Web site that would be linked backwards to the ERP server. An application program would automatically download the data from the ERP system into the MIDAS package in the distributor's PC, and also upload the data from the package to the ERP. This would require some modifications in the existing MIDAS package.

The secondary sales data obtained from MI-Net was expected to result in better performance through reduction in lost sales due to stockouts, lower inventory, and better distributor credit management, among other benefits. Moreover, due to backward linkage to the ERP system, it was now possible to provide information to the TSOs also about stock levels at various depots, as well as order status of distributors, distributor performance, etc. The information from MI-Net would not only provide a TSO with more current information at lesser effort, but would also free up a lot of time that was earlier spent in data collection and collation for preparing reports. This extra time could be utilized to improve the overall sales productivity. Providing both TSOs and distributors with information on stock position at various depots would also improve their mutual relationship, as there would often be an element of distrust on the part of a distributor that a TSO was denying him stock replenishment while supplying them to some other distributor. So there were clear benefits in getting TSOs to use the system

Once again, the project team started brainstorming on creating a value proposition for TSOs to "pull" them to MI-Net. What emerged was the concept of a "virtual office" for TSOs. MIL had 180 TSOs who were constantly "on the move," visiting various distributors and retail outlets. They would typically spend more time out in the field than in the area office. The extent of their interaction with regional offices or the corporate headquarters was minimal. This not only created a problem of "bonding" with the organization, but also rendered certain basic procedures, like getting a loan/leave application sanctioned, a cumbersome and time-consuming task. A TSO would typically visit his area office once in four to five days, make a loan application to the area manager (who would pass it on to the head office), and would be able to find the status of the application only on his next visit. Also, there were times when a TSO would learn about a new TV advertisement campaign for a product he was selling from a distributor or a retailer rather than from the company itself. Issues like these could be addressed by creating a "virtual office" for the TSOs on the MI-Net site. A TSO would log in and would be able to access data related to his territory. In addition, he would be able to make applications on the site, check out new commercials for various products, check out the corporate bulletin board, etc. In order to build a community feeling, the site would also have features like chat, daily announcements of birthdays and anniversaries, contests, and so on.

Thus in July 2001, a comprehensive business plan was presented to the top management at MIL, with a projected payback period of two and a half years. Top management approved the project and the CEO (Sales) was made the principal sponsor for the project. The SDM — who had earlier championed the MIDAS project and had been involved in planning MI-Net — was appointed the overall project in-charge, with the project manager reporting to him. Next followed the selection of an IT company for building the Web site and the application that would fetch data from the ERP system as well as the MIDAS package. Initially around 25 vendors were screened and five were short-listed for a detailed study. The MI-Net team spent one day with each vendor assessing the capability of the vendor and how well it had understood the project requirements. Finally, it chose a reputed Indian IT major to do the project. The work on the software started in September 2001, and the project was ready for rollout in February 2002.

To create awareness about MI-Net both within the organization as well as among the distributors, and to communicate about the progress, the project team came out with an in-house magazine — *MI-Ne(t)ws*. The inaugural issue was launched in October 2001, and contained addresses from both the CMD as well as the CEO (Sales). The CMD endorsed the project as being "the most vital link in integrated information network." He also emphasized the importance of planning in no uncertain terms: "This mega venture will require an arduous process of planning, and effective usage. And above all change management at all levels."[6] The address from the CEO (Sales), on the other hand, clearly explained the concept behind MI-Net, why the initiative was needed, and what were the potential benefits (and responsibilities) for both TSOs and the distributors. Thus, the TSOs were told, for example, how the availability of data at the click of a mouse would improve their productivity, as they would be able to identify under-performing distributors (and their agents) much faster than before. A TSO would also be free from the tedious job of collecting data "just because your boss wants it, because it is available, at the same point in time, to your boss also."[7] The facilities of the virtual office were also highlighted—status of (loan) sanction request, leave balance status, online approval of leave, claiming of medical benefits, and so on. All a TSO needed to do, the address said, was to ensure "that your MIDAS distributors enter all transactions in MIDAS and just once a day, yes, only once in a day, log into MI-Net and click on one button (for upload-download of information) and the rest will be taken care by the system."

The MIDAS implementation had addressed issues like reluctance of distributors to use computers/change to a new package, and had also generated a lot of confidence and trust among distributors regarding MIL's ability to implement an IT project. So this time, the real challenge was in convincing them about the benefits of MI-Net. As the GM-IS put it, "They are businessmen. If they see value, they adopt faster (than TSOs)." Hence the column by the CEO (Sales) in the inaugural issue of *MI-Ne(t)ws* addressed the issue clearly under the heading "What is in MI-Net for a Distributor?"[8] MI-Net promised resolution of "all the pain areas" for a distributor, like information regarding credit note status, stock position at depot, order status, account statement for tax purposes, claims settlement, and so on. All a distributor had to do, the column said, was to "enter all transactions in MIDAS and just once in a day, yes only once in a day, log into MI-Net and click on one button (for upload-download of information) and the rest will be taken care by the system." Another address by the SDM emphasized the "mission" of MI-Net as "PIONEER the process of optimizing the performance of our key resources, 'Our People' by using information technology," with "our people" including "our distributor sales representatives (DSRs), our distributors, our field force," in that order.[9]

Thus, right from the start, there were very clear communications not only about the benefits from, and ease-of-use of MI-Net, but also that the initiative had a very strong backing from the top. Hence, the credibility of the initiative was never in doubt. The distributors and the TSOs were also given a "feel" of the system in terms of its functionalities, even before the application was fully developed. While the various RSDMs were given the responsibility of taking the TSOs through the detailed functionality (during December 2001 and January 2002), for the distributors, this task was handled by the ASMs, who carried the presentation in their laptops whenever they visited a distributor.

But it was not enough to ensure that the initiative was taken seriously — the users also had to be provided the necessary skills to use MI-Net effectively. To this end, both distributors as well as TSOs were given usage training by the project team members. The TSOs were given training in batches of 10 for two days. Prior to that, they were taught a brief computer/Internet appreciation course by a leading IT training company. The usage training was held throughout February across different sales regions. The distributors were trained on the application for one day, along with their computer operators. The less time spent with distributors was presumably because of lesser functionalities for them (compared to virtual office for TSOs). The fact that MI-Net was easy to run — the distributor would just have to log on to the site and then the upload/download of data would take place automatically — made the process smoother.

In addition to providing training to the TSOs, efforts were made to encourage self-learning. To that end, MIL provided browsing allowances enabling one hour of browsing every day. Additionally, it also tied up with a bank to provide soft loans to those TSOs who were willing to buy a PC and surf from their home. *MI-Ne(t)ws* regularly carried interactive features like crossword competition about MIL, quiz on MI-Net, and so on, which gave visibility to active participants. Even some distributors would at times participate in these events.

In the meanwhile, two sites were selected for pilot run — one in a metro city (Delhi), and the other in a smaller city (Raipur). The choice of sites was dictated both by the desire to test the systems in both large and small places, as well as due to enthusiastic response from both distributors and TSOs in these territories. Both the application and the Web site were ready by the first week of February. The pilots commenced on February 15, 2002. While it was heartening to see that TSOs had picked up the training "very fast," the real pleasant surprise was the involvement of distributors. The RSDM conducting the pilot at Delhi reported, "All the distributors were much more evolved in computer usage than I had expected. The distributors in general had better exposure to computer usage than anticipated."[10] The RSDM handling Raipur pilot also reported similarly, "The distributors' involvement was above expectations." Additionally, gestures on the part of MIL, like giving gifts to the distributors and their operators, as well as TSOs, also generated a feel-good factor. As the Raipur RSDM noted, "The gifts were appreciated by all, especially distributors and operators (who did not expect to be gifted anything)."[11]

The pilot stage results were encouraging enough for the project team to decide on a "big-bang" approach towards rollout. Usage training for the TSOs as well as the distributors was completed by the third week of March. All the class 'A' urban distributors were connected to MI-Net by mid-April. The rural distributors were then brought into the fold. MI-Net was officially launched on July 1, 2002, with the system connecting 330 class 'A' distributors, together accounting for nearly 75% of its primary sales, with the central server. More than 70% of the TSOs availed of the loan facility and bought PCs to be used from home.

The entire implementation approach was characterized by proactive change management in terms of high levels of communication about the benefits of MI-Net and provision of requisite training. Top management kept a close eye on the project progress. As the project manager put it, "Every second day, (CMD) used to call me." When asked what prompted such a proactive approach, the project manager said, "We wanted to succeed. We tried to imagine all possible sources of failure, and then acted on it." In a somewhat similar vein, the SDM in charge of the project said, "We were proactive with change management because we planned very carefully." The PM-ERP attributed the attitude towards change management to the organization culture. He said, "We have always appreciated the need for change management because of our democratic value systems. We would rather take people along, not force things on them. But you need a healthy balance between democracy and firm decision from the top."

CURRENT CHALLENGES/PROBLEMS FACING THE ORGANIZATION

MI-Net system usage is being monitored in terms of frequency of usage by each distributor (daily, once in two days, etc.). MIL believes it is too early to start measuring benefits, but expects immediate benefits in terms of reduction in distributor transaction processing costs and costs on communication related to inventory and order status. From the data available for the month of July 2002 (first month of usage), around 86% of the distributors are using the system at least once daily, with the range being between 79% and 92%. MIL has set a target of getting at least 95% of the distributors using MI-Net to connect to the system daily. Explicit measurement of benefits through cost savings and increased sales force productivity are planned, once the system usage stabilizes.

However, not all distributors were as enthusiastic about MI-Net. While they did admit grudgingly that the new system was advantageous to them, especially for tracking order status and accounts statements, they were not too happy about having to shell out the extra telephone charges that they incurred due to connecting to the Internet. There were also some reported glitches in providing support to the distributors in case of any system trouble. One of the key distributors located in Mumbai, which is also the corporate head office of MIL, complained about lack of support from MIL's IT Department in case of problems with the system. It complained about being kept on hold on the phone for a long time, and the IT staff not coming promptly to address the problem. It also claimed that other distributors were also facing the same problems, and were actually now approaching it for solutions before approaching the IT Department, as this distributor was relatively more IT savvy and an early adopter.

Thus, in spite of a detailed implementation plan that was executed smoothly, there seems to be a lack of preparedness in handling post-implementation issues. Did the organization go wrong somewhere in the planning stage? Is technology the easier aspect of adopting Internet-based applications? Should the organization have included the distributors in the conceptualization stage of MI-Net? If the distributors do not perceive any long-term benefits from the system relative to the costs incurred, how can MIL ensure high usage by them? It is in a position to use coercive tactics, given its position as a dominant partner, but would that be an effective solution in the long term?

REFERENCES

Boddy, D., Caitlin, C., Charles, M., Fraser-Kraus, H., & Macbeth, D. (1998). Success and failure in implementing supply chain partnering: An empirical study. *European Journal of Purchasing and Supply Management, 4,* 143-151.

Evans, P. B., & Wurster, T. S. (1997). Strategy and the new economics of information. *Harvard Business Review, 75*(5), 71-83.

Feeny, D. (2001). Making business sense of the e-opportunity. *Sloan Management Review, 42*(2), 41-51.

Gulati, R., & Garino, J. (2000). Get the right mix of bricks & clicks. *Harvard Business Review, 78*(3), 107-114.

Klein, K. J., & Sorra, J. S. (1996). The challenge of innovation implementation. *Academy of Management Review, 21*(4), 1055-1080.

Porter, M. (2001). Strategy and the Internet. *Harvard Business Review, 79*(3), 63-79.

Premkumar, G., & Ramamurthy, K. (1995). The role of interorganizational and organizational factors on the decision mode for adoption of interorganizational systems. *Decision Sciences, 26*(3), 303-336.

Ross, J. W., & Feeny, D. F. (2000). The evolving role of the CIO. In R. W. Zmud, & M. F. Price (Eds.), *Framing the domains of IT management: Projecting the future through the past* (pp. 385-402). Cincinnati, OH: Pinnaflex Education Resources.

Straub, D. W., Hoffman, D. L., Weber, B. W., & Steinfield, C. (2002). Measuring e-commerce in Net-enabled organizations: An introduction to the special issue. *Information Systems Research, 13*(2), 115-124.

Zaltman, G., Duncan, R., & Holbeck, J. (1973). *Innovations and organizations.* New York: John Wiley & Sons.

ENDNOTES

[1] Source: MIL Annual Report 2001-02
[2] *MI-Ne(t)ws*, October 2001, p. 2
[3] *MI-Ne(t)ws*, October 2001, p. 2
[4] *MI-Ne(t)ws*, October 2001, p. 3
[5] *MI-Ne(t)ws*, October 2001, p. 2
[6] *MI-Ne(t)ws*, October 2001, p. 1
[7] *MI-Ne(t)ws*, October 2001, p. 3
[8] *MI-Ne(t)ws*, October 2001, p. 3
[9] *MI-Ne(t)ws*, December 2001, p. 2
[10] *MI-Ne(t)ws*, March 2002, p. 3
[11] *MI-Ne(t)ws*, March 2002, p. 3

APPENDIX 1

The FMCG Industry

The Indian FMCG industry, estimated at US$8 billion, comprises segments like personal care, soaps and detergents, skin care, oral care, health and hygiene products, agro products, and branded food. Following the liberalization of Indian industry in 1991, the FMCG sector grew rapidly, registering high double-digit growth over successive years throughout most of the nineties. This was attributable to a number of reasons like buoyancy in rural income growth, distribution expansion to smaller towns and rural areas, and fall in prices of FMCG goods due to reduction in import duties and indirect taxes.

However, the growth has slowed down considerably, to single digit levels, over the past three to four years, in both urban as well as rural areas. Consequently, the competition among players has become even more fierce, with price cutting and attractive discount schemes being the order of the day. At the same time, most of the companies have taken measures to optimize their internal processes and minimize costs. Now the effort has shifted towards identifying areas outside the company where efficiencies can be improved, most notably, the distribution channel and logistics. For example, recently 10 FMCG majors, with an aggregate turnover of US$4 billion, have come together to launch an Efficient Consumer Response (ECR) initiative, aimed at "fulfilling consumer demand better, faster, and at a lower cost by eliminating unnecessary expenses and inefficiencies from the supply chain." ECR is based on the principle that there are a lot of non-competing areas where collaboration can lead to lower costs and better supply chain management. ECR started in the USA in the early nineties as a result of work done by large retail chains like Wal-Mart, and has since spread across most of Europe, Australia, and Asia. In India, ECR India has been set up as a voluntary body, and there are currently four workgroups looking at the following aspects:

- Stockouts — to measure the actual level of stockouts in the industry and evolve common solutions,
- Logistics — to draw up logistics standards and drive possible cost savings through collaboration,
- Dataflow — to standardize data definition and flows between various constituents of the supply chain, and
- FMCG policy for organized retail — to draw up a policy guideline for FMCG transactions with organized retail.

MIL has chosen not to be a part of the ECR India initiative.

(Source: Businessworld, February 18, 2002)

APPENDIX 2

Key Financial Numbers (in Rs. Million)

Particulars	FY97	FY98	FY99	FY00	FY01	FY02
Sales & Services	4100	4900	5510	6500	6710	6960
Profit before Tax	280	370	440	430	500	580
Net Profit (PAT)	220	300	380	360	460	500
Net Cash Profit (PAT + Depreciation)	250	340	430	440	550	640
Earning per share-Annualised Rs.	14	21	26	25	32	34
Book value per share (Rs.)	55	68	83	98	118	136
Net Worth	800	980	1210	1420	1710	1970
ROCE %	35	41	42	33	33	32

Key Financial Numbers (in US$ Million)

	FY01	FY02
Sales Value	137.86	142.90
Profit Before Tax	10.30	11.88
PAT Before Deferred Tax	9.40	10.88
PAT After Deferred Tax	9.40	10.28
Cash PAT	11.23	13.21
Net Worth	35.19	40.50

APPENDIX 3

Organization Chart

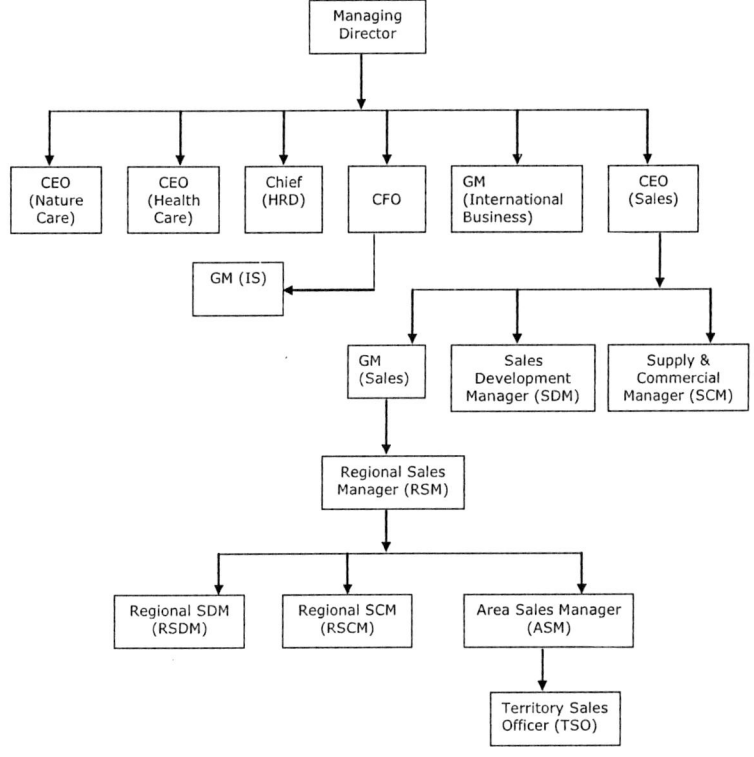

APPENDIX 4

MIL's Distribution Network in India

	Urban	Rural	Total
Sales Territories	150	30	180
Towns Covered (000s)	3.1	12.3	15.4
Distributors	850	0	850
Super Distributors	0	120	120
Stockists	0	2500	2,500
Retail Outlets—Direct Reach (000s)	310	220	530
Retail Outlets—Indirect Reach (000s)	400	620	1,020
Total	710	840	1,550

Pankaj Bagri is a doctoral candidate at Indian Institute of Management Bangalore (IIMB), India. He also holds a Post Graduate Diploma in management (equivalent to an MBA) from IIMB, and has worked in the financial services sector for two years. His research interests include the adoption of IT-based innovations, organizational development and change management.

L. S. Murty has a bachelor's degree in mechanical engineering from JNT University, India, and doctoral degree in management, from the Indian Institute of Management Ahmedabad, India. Currently he is a professor of operations management at the Indian Institute of Management Bangalore, India. He worked in the aerospace industry before moving into academia. During his 15 years of teaching, he taught in graduate and doctoral programs in management and executive education. His research interest is in the field of planning systems (MRP, MRP II, ERP), and other interests include teaching methodology and research methodology.

T. R. Madanmohan is an associate professor of technology and operations at the Indian Institute of Management Bangalore, India. He also holds an adjunct research position at Carleton University, Canada. His areas of research interest include evolution of e-com business models, competitive agents, online auctions rules, standard setting and technology communes, and strategic behavior in the pharmaceutical industry. He has published more than 20 articles in journals such as the International Journal of Technology Management, International Journal of Technology Transfer and Commercialization, Technology Analysis and Strategic Management, Productivity, *and a book published by Kluwer Press.*

Rajendra K. Bandi is an associate professor of information systems and chairperson of the Center for Software Management at the Indian Institute of Management Bangalore,

India. An engineer by training, he earned his PhD in business administration (computer information systems) from the Robinson College of Business, Georgia State University, Atlanta, USA. His research interests are in social impacts of computing, knowledge management, IT in government, software engineering, object-oriented systems, and software development process models. He has published/presented several research papers in various international journals and at various conferences such as IEEE Transactions on Software Engineering; International Conference on Object-Oriented Programming, Systems, Languages & Applications (OOPSLA); International Academy for Information Management; and International Conference on the Impact of ICT Applications on Relocation of Work, Working Conditions, and Workers.

This case was previously published in the *Annals of Cases on Information Technology*, Volume 6/ 2004, pp. 387-405, © 2004.

Chapter VI

An Innovative Adaptation of General Purpose Accounting Software for a Municipal Social Services Agency

Andrew Schiff, University of Baltimore, USA

Tigineh Mersha, University of Baltimore, USA

EXECUTIVE SUMMARY

Organizations with unique characteristics and transaction processing requirements, such as government agencies, often satisfy these requirements by (a) acquiring software from vendors who have developed applications for that particular type of organization, or (b) developing software internally from scratch. When either of these approaches is taken, the development costs are spread over a relatively small number of organizations, and the resulting system can be very expensive. Also, due to the uniqueness of the application and the relatively small number of users, it may take a long time to identify and correct any processing errors. An alternative is to acquire general-purpose software that has been developed for a wide range of organizations, and to adapt it for the agency in which it will be installed. However, this alternative approach is frequently not undertaken because it is often believed that general-purpose software is unable to provide all of the information required by the organization.

When the required information can be provided, though, general-purpose software can be less expensive and less time-consuming to implement. This case presents a successful use of the latter approach. It describes how a general-purpose accounting software package was successfully adapted to meet the information processing needs of the foster care program in a large municipal government agency located on the Eastern seaboard of the United States. The primary contribution of this case is to explain the agency's information processing needs and how the application software was modified to meet them, since it is often believed that general-purpose software cannot be customized to meet the needs of organizations which are not typical merchandising, manufacturing or service businesses. Therefore, this case should be useful as a reference for others who are involved in, or who are considering, similar projects. In addition, as will be discussed briefly below, this approach yielded the additional benefits of significantly reducing the time and the cost required for system development.

BACKGROUND

One of the important roles of government is to provide a safety net to certain sectors of society that need assistance. While a variety of services are provided to needy families and children by social service agencies in the United States, ensuring the safety and well-being of at-risk children has always been a focus of attention among government officials, community leaders and charitable organizations. Among the many programs designed to serve the needs of children in most jurisdictions, foster care programs are the most popular.

The primary purpose of the foster care program is to provide alternative care to children who cannot remain at home due to maltreatment, abandonment or neglect. In the city on which this case study is based, the state develops policies that govern the foster care program and also maintains administrative oversight while local social service agencies are charged with the task of administering the program in accordance with the law. This includes identifying care providers and arranging temporary and permanent out-of-home placement for children believed to be in imminent risk.

Upon removing children from their homes in response to reported incidents of maltreatment or neglect, social workers first explore the possibility of placing the children with relatives. Typically, relative care providers are not paid for their services although they may receive food and medical assistance for the children placed in their care. If relatives who are willing and able to care for the at-risk children cannot be identified, the agency places the children in foster homes.

Two types of foster home resources are available: individual foster homes or institutional care facilities. Among the 4,000 children in the city's foster care system, approximately 80% are placed in individual foster homes. Both individual foster care providers and group care facilities receive payment for caring for the children placed with them. The payment rates vary depending on the age and medical condition of the child. Caregivers who take care of children with special needs receive higher payment for their services. Annual payments by the city to foster homes and institutional facilities total about $75 million. Much of this funding comes from the state, although federal funding is also available to cover some of the costs of caring for the children placed in the foster care program.

The System Used to Pay Foster Care Providers

Upon placing a child with an individual or institutional care provider, the child's social worker completed a placement authorization and payment (PAP) agreement form, indicating (a) when the child had entered the home and (b) the authorized daily or monthly payment rate. This form was also used to indicate the transfer of the child from a particular service provider's home or facility to another. Hence, each placement or removal of a child required completing the PAP form.

A copy of the PAP form was then sent to the city's Foster Care Accounting Unit (FCAU) where a file was opened for the child and the information was entered into the FCAU's data system. The FCAU was responsible for paying the foster care providers, and for maintaining historical records of disbursements. The FCAU also maintained information about the children receiving foster care, the type of care given to each child as well as the care provider in order to be able to properly pay the caregivers.

On the basis of information contained in the PAP form, a payment authorization form would be completed by FCAU and sent to a state agency in another location where the checks were prepared. The checks were then sent back to FCAU where they were reviewed and mailed to the caregivers. Care providers expected to receive payment within 10 working days of submitting the PAP form.

Institutional care providers were required to submit an invoice each month to get their reimbursements. However, a different procedure was used to generate payments for individual care providers. If a child remained in the care of a particular individual care provider for more than 60 days, the caregiver(s) would be placed on what was called a "payroll system." Placing the caregiver in the payroll system enabled automatic generation of payment checks to the individual foster parent every month. As a result, individual care providers were not required to submit monthly invoices. Such payments would continue until the state office that was printing the checks was advised that the child was no longer in the care of the foster parent.

Problems with the Existing System

As indicated above, payment to caregivers was triggered or discontinued by completing the PAP form for each foster care case by the social worker in the field. At times, however, social workers did not complete these forms or, if completed, the forms were not received in a timely manner by FCAU. Meanwhile, a service provider could continue to receive checks even though the child was no longer under his/her care which resulted in overpayments. Getting refunds of the overpayments from individual caregivers was often very difficult and involved incurring additional collection costs.

In addition to the regular "payroll" checks, manually prepared checks were issued to caregivers to cover emergencies. To ensure that such payments were included in the payment history to the provider, the amount for which a check was cut manually would be entered into the computerized system with the understanding that the computer generated duplicate checks would later be voided. On occasion, however, these duplicate checks were erroneously sent to caregivers and were cashed. Recovering the overpayments from institutional caregivers was relatively easy since the overpaid amount could be deducted from future invoices. However, it was not always possible to recover overpayments from individual foster care providers. If the amount was successfully recovered, it would often be after a long time and at a significant additional cost.

Another problem of the current payment system concerned the lack of adequate automation to facilitate the management of the enormous amount of information in the FCAU. For example, upon completing the data entry and check preparation process, the state agency would send to FCAU lengthy hard-copy printouts detailing historical information on each child as well as on each care provider who received payment. There was no automated system for verifying the accuracy of the information provided to the state agency for check processing, and at times payment authorizations submitted by the FCAU were rejected by the automated check printing system at the state agency. Moreover, there was no automated procedure for verifying that checks that were issued at the state agency tallied with the amounts authorized for payment by FCAU.

When authorized payments were rejected by the state system that was printing checks or when caregivers called demanding payment, FCAU personnel were compelled to spend substantial amounts of time manually searching through hard-copy printouts sent to them monthly by the state agency. This was done in order to (a) identify the cause(s) of the error(s); (b) answer questions from foster care providers who had not been paid at all or who believed that their checks were short; (c) identify, correct and resubmit transactions that have been rejected by the system; and (d) search for potential duplicate payments. This process was cumbersome, time-consuming and ineffective.

SETTING THE STAGE

Many of the shortcomings outlined above were noted in an audit of the FCAU conducted in the late 1990s. More specifically, the audit report disclosed the following shortcomings:

1. Delays in processing provider payments,
2. Inadequate documentation to confirm that payments had been made to foster care providers,
3. Insufficient procedures to verify that children listed on invoices were actually in the care of the providers submitting the invoices, and
4. Duplicate payments to providers.

The administration of the social service agency was diligently seeking to improve the effectiveness and efficiency of the service delivery system. It sought, and acquired, outside funding sufficient to develop an automated system for managing the foster care payment process and to purchase the required hardware. The FCAU and the manager of finance also wished to introduce new technologies to manage this process. Consequently, the authors were requested to assist the head of FCAU to create a system to be installed within the FCAU which would:

1. Provide online access to data about children who had been placed in the foster care system and the type of care that each child received,
2. Provide online access to data about foster care providers,
3. Automatically calculate the amount to be paid to each provider for each child every month,

4.   Maintain online historical information about payments both by child and by foster care provider,
5.   Automatically detect possible duplicate payments prior to disbursement, and
6.   Print disbursement checks to foster care providers with supporting documentation indicating the children and the time periods for which payment was being made.

The individual payment amounts were based upon the period of time for which each child received foster care, and the type of care received. The payment amounts would then be sorted and accumulated by foster care provider, and the check made payable to the provider. This made the FCAU's request specially challenging because, in contrast to a typical payroll system, there would not be a single paycheck made payable to each "employee" (the child), but instead a grand total check would be made payable to a provider for all of the "employees" (children) under the provider's care. Moreover, in contrast to a typical accounts payable system, payment history would have to be accumulated and made available not just for each "vendor" (provider), but for each unique line item (each child) included in one or more payments.

CASE DESCRIPTION

Three approaches to developing a foster care payment-processing system were considered. The first approach was to create a system from scratch using an RDBMS, such as Microsoft Access. The second approach was to acquire a system from a vendor who had developed software for this particular type of organization. The third approach was to select a general-purpose accounting software package and modify it to meet the specific needs of the FCAU.

The last of these three approaches was chosen for several reasons. First, it was believed that developing a custom system from scratch would take substantially more time and resources than acquiring and modifying general-purpose software, which was already available. Moreover, the time that would be required to develop a custom system could not be estimated with any reasonable degree of precision. The cost of software developed for government social service agencies was significantly (as much as 10 times) greater than that of general-purpose accounting software. This was well beyond the budget of the FCAU. Finally, the transaction processing and reporting requirements of the FCAU system closely resembled those which could be met by general-purpose accounting software.

The Selection Process

To implement the last approach, the authors obtained information about various general-purpose middle-market accounting software packages through professional publications, vendor literature, and other sources. Three packages were identified which appeared able to meet the needs of the FCAU after appropriate modifications were made. These packages were Great Plains Dynamics, Solomon for Windows, and MAS 90.

The authors then contacted the local value-added resellers (VARs) of these three accounting software packages. After demonstrations and discussions with the VARs, it was determined that one package, Great Plains Dynamics, could be modified in a way which would enable it to fulfill the transaction processing and reporting requirements of the FCAU. This package was then demonstrated to the FCAU administrators and staff, who also approved its use.

System Design

As mentioned above, individual payment amounts were based upon the length of time and the type of care provided to each child, but the checks were made payable to the providers. In order to accomplish this, the Great Plains payroll module and the accounts payable module were used. This was done because no single module was able to perform these functions, and to provide online historical information about payments both by child and by foster care provider, entirely by itself.

First, the accounts payable module was modified to capture critical information about each foster care provider, including the provider's identification number, address and phone numbers. Figure 1 shows the main screen for entering this information into the accounts payable module (called Provider Maintenance), after it had been modified from the standard accounts payable format. Additional addresses were entered in a screen that was accessed by pressing the Address button on the main screen. (Further information that might be needed for a typical accounts payable application, and which would be accessed by pressing the other buttons, was not required by the FCAU).

Next, the payroll module was modified to capture critical information about each child, including:

a. The child's case number,
b. The type of foster care being provided (emergency, institutional, etc.),
c. The date the child was placed in foster care,
d. The rates to be paid for basic care and any special services, such as education or medical care,

Figure 1. Provider maintenance screen

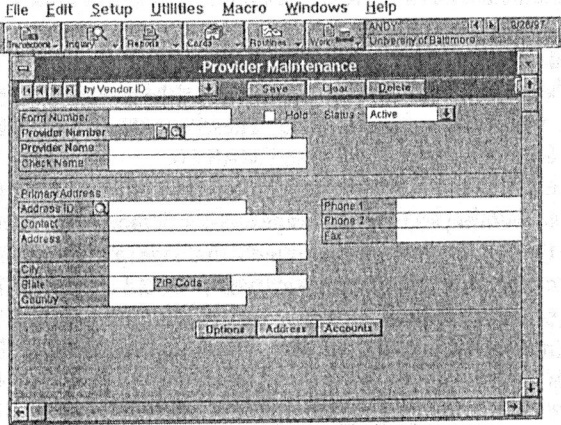

Figure 2. Child maintenance screen

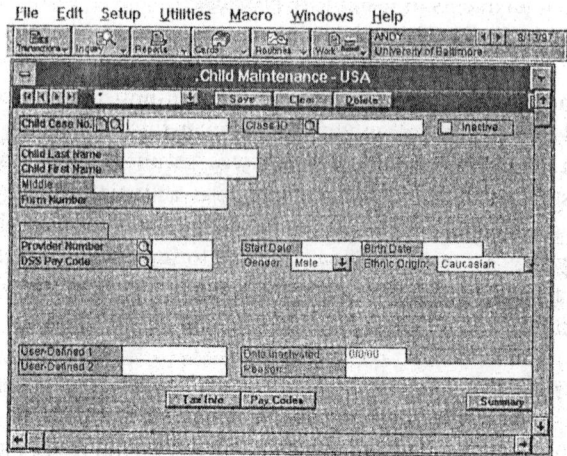

e.   The child's foster care provider, and

f.   The federal or state program under which the child was eligible for funding.

Figure 2 shows the main screen for entering this information into the payroll module (called Child Maintenance), after it had been modified from the standard payroll format. The rates to be paid for basic care and any special services were entered in a screen that was accessed by pressing the Pay Codes button. The federal or state program under which the child was eligible for funding was entered in a screen that was accessed by pressing the Tax Info button. (Additional information that would be needed for a typical payroll application, such as marital status and exemptions, was not required by the FCAU).

After the above information was entered for each provider and each child, the payroll module was used to calculate the amount to be paid to the child's provider for the length and type of foster care given each month. This was accomplished through a modification of the data entry procedure and screen normally used to record information about hours worked and payroll dollars earned. The modified data entry screen, which is shown in Figure 3, contained three rows of information for each child within a monthly pay period. This included, on the left side of the screen, the child's case number and name, the provider's identification number, the pay period start and end dates, the type of foster care being provided (DSS Code), and the state or federal program under which the child was eligible for funding (Pay Stat). On the right side of the screen, spaces for recording a code and the rate to be paid for basic care or for a special service, the number of days for which this care was provided, and the calculated payment amount, were presented three times. These spaces were presented three times because basic care and up to two special services might be recorded for a child during the monthly computation of payments to foster care providers.

Information for all of the children under the care of one provider was entered into the above screen, which could be scrolled down to display additional lines for entering

Figure 3. Payment entry screen

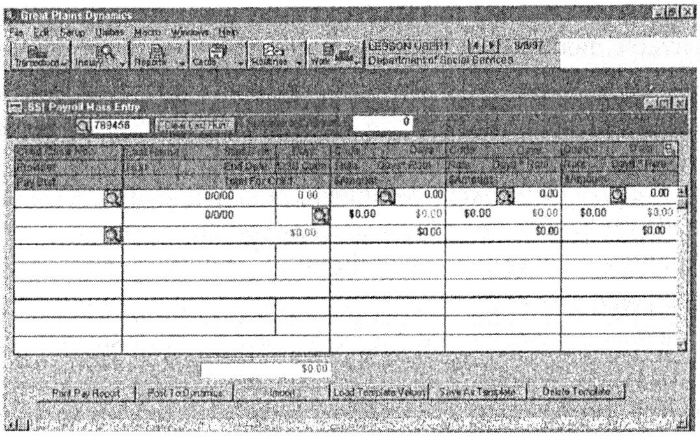

more data. After this information was recorded, it was saved into the payroll module and similar information for the next provider was entered. An additional benefit was that all of the information for a specific foster caregiver could be memorized as a template, thereby greatly reducing the amount of time required to enter this information in subsequent months (only the changes, if any, would have to be recorded). The payroll module also accumulated a historical record of payments by child, which was accessible online.

Transferring the Payment Information to Accounts Payable

Next, a short program (macro) was written and included in the software package to transfer payment information for each child from the payroll module to the accounts payable module. A feature of the Great Plains software called Open Integration Modules was used to reformat the data produced by the payroll module so that it could be properly read by the accounts payable module. The items of information that were transferred to the accounts payable module were:

a.   The child's case number,
b.   The provider's identification number,
c.   The code for basic care or for a special service,
d.   The number of days for which the basic care or special service was provided, and
e.   The calculated payment amount for the basic care or special service.

These five items were transferred for each child from one to three times since, as noted above, basic care and up to two special services might be recorded for a child during the monthly computation of payments to foster care providers. The accounts payable module then reorganized this information by foster care provider. This made it possible to print one check to a provider for all of the children under the provider's care, to list the details of each individual payment on the check advice, and to maintain a historical record of payments by provider which could be accessed online. As mentioned previously, both the payroll module and the accounts payable module were used because no single module was able to perform all of the above functions, and to provide online historical information about payments both by child and by foster care provider, completely by itself. A flowchart of the major processing activities in the FCAU system is shown in Figure 4.

Figure 4. System flowchart

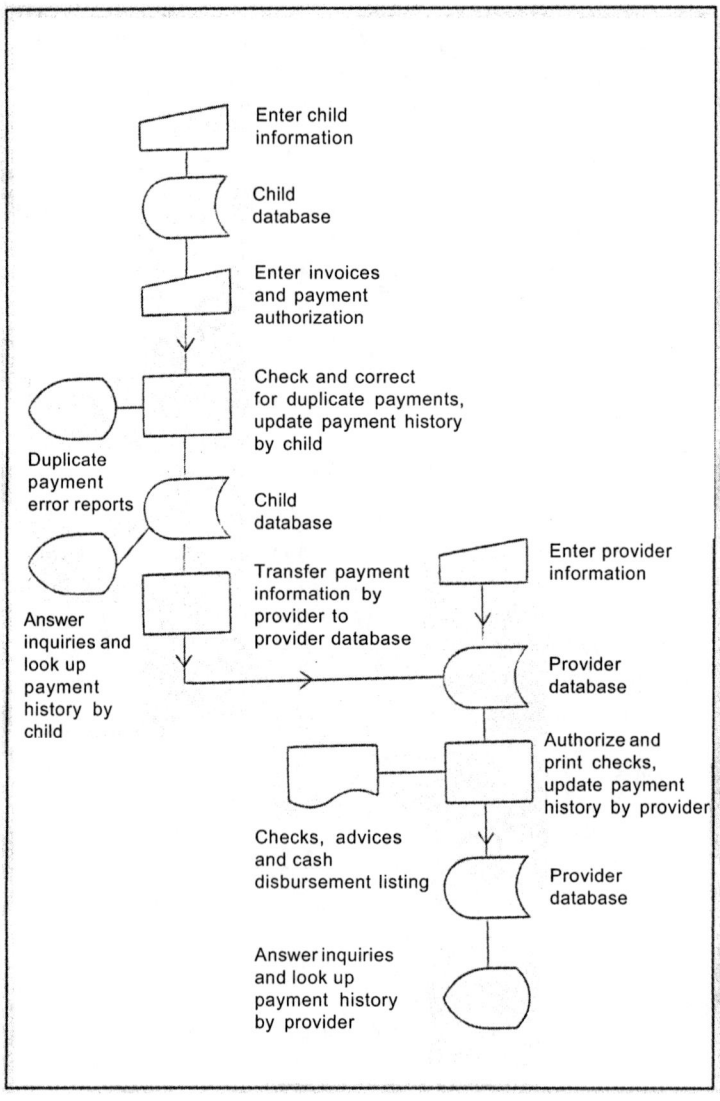

System Implementation and Operation

Implementation of the new system required several steps. First, it was necessary to enter information about each child in the foster care system into the payroll module, and information about each foster care provider into the accounts payable module. Historical information about payments already made to providers for the children under their care would be entered through an automated process, after converting the information from the format of the existing system to the format of the new system.

After this was accomplished, data about the length and type of care given to each child would be entered into the payroll module at the end of each month from invoices or other documentation supplied by foster care providers and/or social workers (such as a placement authorization and payment form). The payroll module would then calculate the amount to be paid for each child. Next, the key information about each payment discussed previously, i.e., the child's case number, the child's foster care provider, the amounts to be paid for each category of care, and the total payment for the child, would be transferred from the payroll module to the accounts payable module using the macro noted above. The accounts payable module would then reorganize this information by foster care provider and print the checks to the providers. These steps are illustrated in Figure 4. The FCAU installed and operated the new system on a local area network, which had been set up within the FCAU.

CURRENT CHALLENGES FACING THE ORGANIZATION

The above system eliminated the need for FCAU personnel to search through hard-copy printouts to correct and resubmit transactions that had been rejected, since possible errors (such as a child case number that had not been entered into the payroll module) could be detected and corrected immediately online. Because historical information was also available online, the above system also made it possible for FCAU personnel to quickly respond to questions from foster care providers about payments received and not received. The ability of the payroll module to record payment history by child made it possible to detect:

a. Whether the same provider was going to receive more than one payment for the same child for the same period of time, or
b. Whether two or more providers were going to receive payment for the same child for the same period of time

by sorting on provider within child within pay period. Searches for duplicate payments could be performed before the checks were printed, either through visual inspection or through a macro written for this purpose. Thus, duplicate payments could be virtually eliminated.

These features, along with its other transaction processing capabilities, enabled the new system to achieve all of the objectives listed above that had been established by the FCAU. There were also benefits in the form of time and cost savings. In contrast to similar systems that had been installed in other municipalities, this project was scheduled to be completed over a period of months rather than years, and for less than 25% of the cost of using software which had been custom developed for government social service agencies. Moreover, the payroll module was able to accumulate payments based on the federal and state programs under which funding was to be received. The accumulated totals made it possible for the FCAU to invoice the related federal and state agencies with a higher level of accuracy than had previously been possible.

The organization found successful implementation of the new system somewhat more challenging than initially expected. This was primarily due to employee turnover at FCAU. The supervisor of FCAU was the most computer literate person in the unit. It was this individual who was first trained in using the system, with the understanding that he would provide in-house training to other members of the unit. This was designed to cut costs and facilitate an "on-the-job" approach to staff training. However, the supervisor resigned unexpectedly, thereby causing a temporary setback in getting the system up and running as scheduled. It was necessary to wait until a new supervisor was hired, at which point it was possible to resume the process of staff training and system operation. This process is currently ongoing.

FURTHER READING

Boockholdt, J. (1999). *Accounting information systems: Transaction processing and controls* (5th ed.). New York: Irwin/McGraw-Hill.

Collins, J. (1999, August). How to select the right accounting software. *Journal of Accountancy*, 61-69.

Cushing, B., & Romney, M. (1994). *Accounting information systems* (6th ed.). Reading, MA: Addison-Wesley.

Great Plains Dynamics. (1996). *Payroll-USA*. Fargo, ND: Great Plains Software.

Great Plains Dynamics. (1997). *Payables management*. Fargo, ND: Great Plains Software.

Mersha, T., Bartle, S., & Wilson, J. (1996). Case flow efficiency improvement in a social service agency. *Proceedings of the International Association of Management Annual Conference* (Vol. 14, addendum).

Milakovich, M. E. (1995). *Improving service quality: Achieving high performance in the public and private sectors*. Delray Beach, FL: St. Lucie Press.

The Taylor Group (1997). *Great plains dynamics open integration modules*. Bedford, NH: The Taylor Group.

Andrew Schiff, PhD, CPA, is an associate professor of accounting at the Robert G. Merrick School of Business, University of Baltimore. He earned a PhD from Rutgers University. Prior to joining the University of Baltimore, Dr. Schiff held a variety of positions including internal audit manager, assistant controller and controller for organizations ranging in size from $6 million to more than $600 million per year. Dr. Schiff has published in a number of academic and professional journals including Behavioral Research in Accounting, Journal of Management Systems, International Journal of Case Studies and Research, CPA Journal, *and the* Massachusetts CPA Review. *He is also the co-author and/or editor of two textbooks.*

Tigineh Mersha, PhD, is a professor of management at the Robert G. Merrick School of Business, University of Baltimore. He earned an MBA and a PhD from the University of Cincinnati. Prior to starting his academic career in 1982, Dr. Mersha held managerial positions in Ethiopia and at the Walnut Hills-Evanston Medical Center in Cincinnati. He has been a full-time faculty member at the University of Baltimore since 1984. He has taught graduate and undergraduate courses in production and operations management, service operations management, and quality and productivity improvement. His research interests include service sector management, quality and productivity improvement, and managing operations in developing countries. Dr. Mersha has published in many academic and professional journals.

This case was previously published in the *Annals of Cases on Information Technology Applications and Management in Organizations*, Volume 2/2000, pp. 249-262, © 2000.

Chapter VII

The Rise and Fall of CyberGold.com

John E. Peltier, Georgia State University, USA

Michael J. Gallivan, Georgia State University, USA

EXECUTIVE SUMMARY

This case study describes the life cycle of CyberGold, a start-up "dot-com" firm that rose to prominence in the world of online currency and micro-payments. The case describes the inception of the firm, the talent base of its senior executives, and its innovative and patented business model, known as "Attention Brokerage." The case focuses on a specific decision problem faced by CyberGold's team of senior managers early in its life cycle: how to modify the company's business model and communication with its members in order to encourage repeat visits to its site and to provide a clearer understanding of where CyberGold credits may be spent by members.

BACKGROUND[1]

CyberGold, founded in 1995 (Bank, 1998), was an Internet marketing firm created to harness the power of the World Wide Web to profile user demographic information and provide targeted marketing services to advertisers. CyberGold aimed to add value to the Web advertisements of its clients by allowing the client to offer small cash payments for viewing their ads. Along with collecting these payments in individual members' CyberGold accounts, the company built profiles of Web surfers that it sold to its advertisers in order for the advertiser to more directly target its audience. It also enabled CyberGold to more precisely target client advertisements shown to its members.

It was the fall of 1997. Nat Goldhaber stood up with a look of concern when his co-founders, Regis McKenna and Jay Chiat, presented recent news articles from the press during the morning's board meeting about the firm's performance. Goldhaber, as CEO and co-founder of pioneering Internet marketing company, CyberGold, took a personal interest in public perception of his company at a time when perception influenced stock market valuation as much as or more than actual performance. The company had gained attention, but not all the press the company received was good. Jay read aloud from an article by Kenneth Hein (1997): "There were only two problems. The advertisers weren't biting and consumers were unclear as to what the points could actually be used for." Responding to the assertion about advertisers, Goldhaber observed "I want 2,000. And I want at least as many small merchants as big ones. One of the reasons I started this business was to help the little guys" (Hein, 1997).

As the discussion continued, the group began to explore other ways to make the company more competitive and to increase revenue as Internet usage continued to grow exponentially. Chiat, recruited to the company primarily for his marketing expertise, suggested that because stock prices were being affected more by user count than by profit, the company should consider offering Web surfers more money to view ads — by doubling or tripling the amount the client company contracted to pay per click. Since most clients agreed to pay each user for just one viewing of an advertisement, the increased expense would be directly tied to new user harvesting. Goldhaber told the group that his vision of the company involved Web surfers (customers) returning to CyberGold for more frequent visits, and involved the company branching out into the electronic currency market before other players captured the market share.

CyberGold's revenues were drawn from its advertising partners only. The advertiser offered CyberGold a set amount per ad viewed (or "click through"), and CyberGold kept a portion and offered the rest as an incentive for the Web surfer to view the advertisement. Surfers who registered for CyberGold accounts were able to accumulate CyberGold points and spend the value at a participating merchant or convert the points to frequent flier miles (Glasner, 2001). As consumers learned, the companies using CyberGold for advertising services were not necessarily the same as the companies accepting CyberGold as a form of payment. The value that CyberGold attempted to add to the advertiser's business was to lower customer acquisition cost — which was especially high for start-up Internet firms (Hoffman & Novak, 2000). CyberGold proposed that it could lower any business' cost of acquiring new customers by about 25% (Garber, 1999). The consumers themselves paid no fee to be a member of CyberGold, but were required to allow their Web surfing habits to be tracked and, in some cases, they were required to complete surveys to demonstrate their attention before being paid for viewing an ad.

Advertising rates peaked in the late 1990s, before the realization that it would take more than just being "first mover" to establish a successful business overtook the Internet euphoria on Wall Street. Even during this time, a strictly Internet-based business founded upon providing incentives for online advertising was not winning over the Wall Street analysts. Drew Ianni of Jupiter Communications observed "We haven't seen any positive case studies in [providing incentives for] online viewers to look at ads" (Borland, 1998).

After its 1995 founding, CyberGold had a short life as a publicly traded company. Its initial public offering (IPO) was completed in September 1999 at $9 per share. The stock peaked at $24 per share in 1999, but by April 2000 the stock was trading at $5.64, and the company was acquired by MyPoints.com (Fowler, 2002). See Appendix B for a chart of CyberGold's stock performance.

Technology veteran and politician Nat Goldhaber founded CyberGold, with assistance from marketers Regis McKenna and Jay Chiat (Semilof, 1996). Goldhaber had previously served as secretary of energy for Pennsylvania in the late 1970s, and in the mid-1980s, he developed the "Transcendental Operating System" (TOS) to allow DOS and Macintosh computers to be able to transfer files easily (Garber, 1999). He also served as Vice President of Sun Microsystems after Sun purchased his TOS product, and subsequently served as CEO of Kaleida Labs, which was an IBM/Apple joint venture. After MyPoints' purchase of CyberGold in April 2000, Goldhaber was named Vice Chair and Director of MyPoints, and ran for U.S. Vice President on the Natural Law Party ticket with Presidential candidate John Hagelin (Gunzburger, 2000).

The company's cofounders had extensive experience in start-up consulting and in advertising. Regis McKenna's background includes his work for his own firm, The McKenna Group, specializing in strategy consulting. He consulted on several highly successful industry-defining technology startups before the Internet age, including Apple, 3Com, Lotus and Microsoft. Jay Chiat held the team's main advertising industry experience. His firm Chiat/Day was a leading advertising agency, which had been named Agency of the Year twice in the 1980s by *Advertising Age* magazine, and also won that publication's Agency of the Decade award for the 1980s (Cuneo, 2002).

SETTING THE STAGE

Along with providing marketing data to advertisers, CyberGold was positioning itself to be a force in two related industries: online currency and micro-payments. Online currency systems are envisioned as an alternative to credit cards, the established leader in online electronic purchase payments, which would allow individuals to make purchases with an electronic form of payment not associated with a credit card account. Consumers have long expressed concern with providing personal information — especially financial — over the seemingly wild cyber frontier.

The Economist (2000) lists five reasons that electronic forms of payment could play a part in the growth of electronic commerce: (1) Consumer anonymity. A credit card identifies a consumer, while an online payment form may not be as easily traced to an individual. (2) Lack of credit cards, especially outside the United States. In many countries such as China, credit cards are not as commonly used as in the United States (Farhoomand, Ng, & Lovelock, 2000; Martinsons, 2002), which presents a challenge to firms attempting to engage those potential consumers in electronic commerce transactions. (3) Users too young to obtain credit cards. Younger Internet surfers may not have a credit card, but could potentially hold a CyberGold account or some other form of online payment. (4) The need for a system to enable payments between individuals for auction-type commerce. This need has already subsequently been identified and filled by firms such as PayPal and BillPoint.

The fifth reason cited by *The Economist* concerns micro-payments. Micro-payments are payments of very small amounts — ranging from 1/10 of a cent to $10 — that are envisioned as the way to charge users for delivery of small-scale product, such as listening to a single song (*Economist*, 2000). Credit cards, which generate fees primarily by a direct 2-4% fee charged against the transaction amount, price themselves out of the micro-payments market by also charging transaction and verification fees on a per transaction basis — in some cases, these could be greater than the cost of the purchase transaction (Guglielmo, 1999). CyberGold's primary business was actually executing the micro-payment principle in reverse by handling micro-payments to the consumer from the advertiser. One firm that has achieved success in the micro-payment arena focusing on consumer-to-consumer transactions is PayPal, which established itself as the handler of transactions on the immensely popular eBay, and was subsequently acquired by eBay (Glasner, 2002).

CASE DESCRIPTION

CyberGold's business model, in the framework of atomic business models presented by Weill and Vitale (2001), is primarily that of an Intermediary (see Appendix C for a schematic diagram of CyberGold's business model). CyberGold reverses the role of the intermediary as described by Weill and Vitale by paying the customer and retrieving information, whereas in the Weill and Vitale model the intermediary receives money and provides information. CyberGold primarily offers aggregated consumer information to its paying customers, the advertisers. This type of information product is not directly offered to consumers, but is directly offered to the businesses that choose CyberGold as their advertising partner. However, CyberGold only adds value for that partner when it creates a meeting between an advertiser and an interested consumer, and in this function CyberGold is acting as an intermediary.

Soon after it was founded, CyberGold conducted a test hosted by an established portal which was designed to spotlight the rationale behind its innovative business model. CyberGold was based on the premise that an advertisement that offered the viewer a reward (money) to click on it would be viewed more times than an advertisement which did not offer a reward. The company ran a test of banner ads on the PathFinder portal that demonstrated a banner ad was clicked 13 times more often when $5 was offered than when no money was offered, and four times more often when $1 was offered (Hein, 1997). This appeared to support the company's hypothesis.

CyberGold provided its members a categorized listing of partners that offered CyberGold for various promotions (see Appendix A). Outside of banner ads on other sites that offered CyberGold, the member was encouraged to return to the site regularly to take advantage of offers as soon as they were added. Each advertisement by a particular vendor could be "redeemed" for CyberGold one time (Metcalfe, 1997); after that initial connection between the consumer and the advertiser, CyberGold was no longer part of that relationship. This meant that the consumer could return to the merchant's site many times in the future, but CyberGold would neither record nor profit from such follow-up visits. In order to verify that value was being provided to the advertiser, the viewer was often required to complete a short questionnaire that verified his comprehension of the advertising being presented in order to earn CyberGold (Metcalfe, 1997). This was to prevent consumers from simply clicking on the merchant's banner ad to collect the reward, but not paying attention to the message.

Growth

As the Internet grew in quantity of users and prominence, CyberGold's membership numbers grew but the company was plagued by the perception that its message was unclear. This perception led the company to establish two major strategic alliances in order to better its market position.

First, CyberGold paired with CyberCash Inc. in a scheme to award CyberCoins. CyberCash had already gained attention as a possible "future currency of the Internet" with its virtual wallet technology, and had already established marketing relationships by which its CyberCoins could be redeemed at more than 50 Web sites (Hein, 1997). This partnership was designed both to move CyberGold into the business of online currency and to clarify for consumers the confusion over where the CyberGold they earned could actually be used. The intent was to provide customers with more options for spending CyberGold credits. However, whether adding yet another level of abstraction (in the form of an additional cyber partner) truly minimized customer confusion is open to question.

CyberGold formed a second alliance with Inference, called Cash2Register, which paid customers small amounts of CyberGold to complete product registration forms online instead of by mailing them in (Lucas, 1999). CyberGold's primary business model was an online advertising medium that aimed to persuade Internet users to be more receptive to advertisements in order to assist in profiling existing users. This partnership with Inference had the potential to bring CyberGold to a different market: people who may not already be active online, but who learned about CyberGold based on the offers printed on the conventional physical product registration forms, and who might be somewhat more motivated to complete such registration forms than they would be ordinarily, by the opportunity to receive a cash reward for doing so. The degree to which people not actively online were brought to CyberGold by printed product registration forms is the degree to which this partnership created new customers, but this was not directly discernable from any reports publicly released by the firm.

As the company grew its user base and initiated alliances, it also grew its staff. In its early growth period in mid-1997, the staff of CyberGold numbered 24 (Poole, 1997). During its prime time period of late 1999, the company reached a staffing level of 124 individuals, 77 of whom were in sales (CyberGold, 2000). By the summer of 2001, after the merger with MyPoints, the company was said to be negotiating to keep some of its 10 employees (*East Bay Business Times*, 2001).

Financial Results

In the years 1997 through 1999, CyberGold saw its revenues rise, but also saw its net loss grow faster. Appendix D provides selected financial figures for the years ending in December 31 of 1997, 1998 and 1999 from the company's 10-K filing. Notable among these figures is the growth of the business in the year 1999, which was the year the company went public. In 1999, total revenues increased from the $1.0 million realized in 1998 to $3.3 million. Unfortunately, at the same time sales and marketing expenses rose from $2.7 million to $8.3 million — more than a tripling of advertising expenditures. Sales and marketing was the largest component in the company's overall financial statement, which saw the company lose more money each year. In 1997, the net loss attributable to common stockholders was $3.8 million; in 1998, the loss grew to $5.4 million, and in 1999 the loss more than doubled to $11.6 million. Due to these excessive advertising outlays, CyberGold lost increasingly more money each year, with net losses growing from $3.8 M to $5.4 M to $11.6 M between 1997 and 1999. Appendix E provides an additional view of the company's performance, focusing on the first quarter of 1999 and the first quarter of 2000, from the subsequent 10-Q filing.

CURRENT CHALLENGES/PROBLEMS FACING THE ORGANIZATION

The Competition

CyberGold was one of three major firms providing this type of marketing research and Web surfer reward programs. NetCentives, one of the well-known players in the market, teamed up with AOL in January of 2000, in a partnership utilizing NetCentives as the reward provider for AOL AAdvantage, an AOL/American Airlines partnership, as well as for ICQ (Hu, 2000). The market penetration and strategic advantage this provided NetCentives may have been responsible for a drop in the market value of CyberGold in the spring of 2000.

MyPoints.com was another leading reward marketing company which had similar membership numbers and a similar business model, but which was reporting higher revenue numbers. According to figures published in *Direct* magazine in early 2000, MyPoints reported 8 million members and 1999 revenues of $24.1 million. In contrast, CyberGold claimed 7 million members and reported 1999 revenues of $5.3 million.

CyberGold had demonstrated membership growth through its short history, growing sharply from the 30,000 registered members the company claimed in September 1997 (Hein, 1997). Subsequently the company had reported reaching 1.5 million in March 1999 (Guglielmo, 1999), and then reached 3 million members in October 1999 (Lucas, 1999).

In addition to CyberGold, NetCentives, and MyPoints, there were a few smaller competitors. One of these, Webstakes, a third competitor, is a company that performs similar marketing functions by arranging sweepstakes for Web surfers to enter, and, by focusing on a slightly different market niche with a different value proposition to the consumer, it has maintained its activity to this day.

Intellectual Property

Goldhaber observed that the technology and business model his company was based on were not only imitable, but already being imitated, by its competitors. The company's leaders reasoned that by obtaining legal protection for their method of doing business, they would be able to charge royalties from other companies wishing to implement a similar process.

As the board meeting began in the fall of 1997, CyberGold's board members discussed the challenges faced by the firm. One suggestion they raised was the possibility of seeking a patent for CyberGold's business model. The board eventually concluded that patent protection would be a good idea to pursue. Goldhaber was uneasy with this plan, since he knew that because patents were not regularly awarded for Internet business models, the odds were not in the company's favor. However, word was out that other "dot-com" firms were applying for similar patents on their business models, and Goldhaber thought it was worth the effort to try for a "business method" patent, as well. He next suggested that the company ought to have a strategy for what course of action to take if the patent were to be granted.

Prior to the late 1990s, patents were generally granted for designs for physical devices. A decision in 1998 in the widely publicized case of *State Street Bank and Trust Co. vs. Signature Financial Group* served to widen the scope of the Patent and Trademark Office to include methods of doing business and abstractions such as computer software. Only 170 applications were filed for this type of patent in 1995, but by 1999 the number had increased to 2,700 — and 583 of those patent applications had been granted (Allison & Tiller, 2002).

CyberGold applied for such a patent and, in 1998, it obtained a patent on its business model of providing incentives to customers for viewing ads and completing surveys — which it termed "attention brokerage" (Bank, 1998). Essentially the firm was pursuing a type of "blocking strategy" (Afuah & Tucci, 2002) by preventing other start-up firms from emulating its business model. Goldhaber stated that CyberGold "made the patent application as broad as we possibly could," (Bank, 1998) presumably in order to position the company to exact royalties from companies performing similar marketing functions. In effect, CyberGold was seeking the exclusive right to award Web surfers any form of incentives — whether cash, points, frequent flyer miles, or any other form of compensation — in exchange for viewing online advertising. NetCentives, one of CyberGold's competitors, received a similar patent on a closely related model of rewards based mainly upon online purchases. Goldhaber commented that, "our objective in obtaining this patent was not to stifle the market. We want to foster the practice of providing incentives online. Within limits, we are willing to license it out" (Borland, 1998).

CyberGold adopted a passive strategy for enforcing its business method patent at first, in order to maintain growth of the market and presumably to project a positive public relations image. Goldhaber explained, "We are interested in licensing CyberGold technology to any company offering rewards or incentives for consumer actions on the Web. Right now we are studying the infringement issues, but our attitude is 'Come on in.' We want to encourage the continued growth of the incentives market by making attention brokerage affordable for companies currently offering online rewards that may be infringing on our patent" (*InternetNews.com*, 1998). The company pursued licensing deals based on its patented business method but did not become involved in any lawsuits regarding the technology. No significant stream of revenue appears to have been derived from the patent, and it does not appear to have been either beneficial or detrimental to the company's performance.

Meanwhile, additional investment capital was acquired in 1998 (Gimein, 1998) from two companies whose primary interest lay in CyberGold's ability to enter the micro-payment market. The investors indicated that they were not investing in the "old" business model of incentives for Web browsing, but rather in the alternative model of micro-payments.

As Goldhaber, Chiat and McKenna met in the board room, they contemplated the steps they should take to ensure that consumers and advertisers alike understood the value of the services CyberGold offered to them. They further discussed modifying their business model to build ongoing revenue from the customer relationships they helped initiate.

REFERENCES

Afuah, A., & Tucci, C. (2002). *Internet business models and strategies: Text and cases* (2nd ed.). New York: McGraw-Hill.

Allison, J., & Tiller, E. (2002). Internet business method patents. *Texas Business Review*. Retrieved January 11, 2003, from http://www.utexas.edu/depts/bbr/tbr/Oct_02.pdf

Bank, D. (1998). CyberGold claims the patent rights to surveys on Web-users attention. *The Wall Street Journal, 232*, p. B6.

Barr, S. (2000). Waiting for the dough. *CFO Magazine, 16*, 62-8.

Borland, J. (1998). Web company patents online ads with payoff. *TechWeb News*. Retrieved January 11, 2003, from http://www.techweb.com/wire/story/ TWB19980824S0009

Cuneo, A. (2002). Jay Chiat dies: Legendary adman helped revolutionize business. *AdAge.com*. Retrieved from http://www.adage.com/news.cms?newsId=34540

CyberGold awarded patent for attention brokerage. (1998, August 25). *InternetNews.com*. Retrieved January 11, 2003, from http://www.internetnews.com/IAR/article.php/ 10131

CyberGold Corp. (2000, March 30). *Form 10-K*. Securities and Exchange Commission. Retrieved April 6, 2003, from http://www.sec.gov/Archives/edgar/data/1086937/ 0000950149-00-000693-index.html

CyberGold Corp. (2000, May 15). *Form 10-Q*. Securities and Exchange Commission. Retrieved April 6, 2003, from http://www.sec.gov/Archives/edgar/data/1086937/ 0000950149-00-001174-index.html

CyberGold site to close this month. (2001, August 8). *East Bay Business Times*. Retrieved April 7, 2003, from http://eastbay.bizjournals.com/eastbay/stories/2001/08/06/ daily33.html

E-Cash 2.0. (2000, February 18). *The Economist, 354*, 67-9.

Farhoomand, A. F., Ng, P., & Lovelock, P. (2000, January). *Dell: Selling directly, globally* (Case HKU 069). University of Hong Kong, Center for Asian Business Cases.

Fowler, G. (2002). Internet IPO darlings: Where are they now? *Startup Journal*. Retrieved January 11, 2003, from http://www.startupjournal. com /financing/public/20020611- fowler.html

Garber, J. R. (1999). Show me the money. *Forbes, 163*, 142.

Gimein, M. (1998). CyberGold finds attention doesn't pay. *Industry Standard*. Retrieved January 11, 2003, from http://www.thestandard.com/article/0,1902,1790,00.html

Glasner, J. (2001). Online cash ain't worth squat. *Wired News*. Retrieved January 11, 2003, from http://www.wired.com/news/business/0,1367,46180,00.html

Guglielmo, C. (1999). Cybergold: Big market for small-ticket sales. *Inter@ctive Week*.

Gunzburger, J. (2000). Presidency 2000. *Politics1.com*. Retrieved May 10, 2003, from http://www.politics1.com/nlp2k.htm

Hein, K. (1997). CyberGold's back from the dead. *Incentive, 171*, 28-29.

Hoffman, D. L., & Novak, T. P. (n.d., May/June). How to acquire customers on the Web. *Harvard Business Review*, 3-8.

Hu, J. (1999). Are 'registered user' figures worth anything? *CNET News*. Retrieved January 11, 2003, from http://news.com.com/2100-1023-229863.html?tag=rn

Hu, J. (2000). AOL inks deals with Netcentives, American Airlines. *CNET News*. Retrieved January 11, 2003, from http://news.com.com/2100-1023-236272.html?legacy=cnet

Hu, J. (2002). Yahoo tacks fees onto e-mail, storage. *CNET News*. Retrieved January 11, 2003, from http://news.com.com/2100-1023-865570.html

Lucas, S. (1999). CyberGold and Inference launch Cash2Register. *Adweek Eastern Edition*, *40*, 124.

Martinsons, M. (2002, August). *E-commerce in China: Theory and cases*. Presentation to the Academy of Management Organizational Communication and Information Systems (OCIS) Division.

Metcalfe, B. (1997, October 13). CyberGold spurns spam by paying people to read ads on the Internet. *Infoworld*, *19*, 41.

MyPoints buys rival for $142 million. (2000, May 31). *Direct*. Retrieved January 15, 2003, from http://www.directmag.com/ar/marketing_mypoints_buys_rival/

Parker, E. (2001). *Websites/CyberGold*. Retrieved October 29, 2002, from http://www.elisabethparker.com/web_sites/cybergold.htm

Parker, P. (2000). MyPoints.com acquires CyberGold for $157 Million. *InternetNews.com*. Retrieved January 11, 2003, from http://www.internetnews.com/bus-news/article.php/3_342041

Poole, G. A. (1997). Panning for Web gold. *Wired.com*. Retrieved April 7, 2003, from http://www.wired.com/wired/archive/5.05/rebels.html?pg=6 &topic=

Richtmeyer, R. (2002). Opinions split on Yahoo! Turnaround. *CNN/Money*. Retrieved January 11, 2003, from http://money.cnn.com/2002/10/08/technology/yahoo/index.htm

Semilof, M. (1996). A new way of mining for customers. *Communications Week, 617*, 23.

United States Internet Council and International Technology & Trade Associates, Inc. (2000). *State of the Internet 2000* (Section 1). Retrieved October 29, 2002, from http://usic.wslogic.com/section1.pdf

Weill, P., & Vitale, M. (2001). *Place to space: Migrating to e-business models*. Boston: Harvard Business School Press.

ENDNOTE

[1]   The board meeting depicted in this case study is a dramatization illustrating the challenges facing the organization. The authors prepared this case based on press releases and publicly available sources of information in order to provide material for class discussion. The authors do not intend to illustrate effective or ineffective handling of managerial situations with the case.

APPENDIX A

The CyberGold Homepage

Source: http://www.elisabethparker.com/web_sites/cybergold.htm

The CyberGold Shopping Page

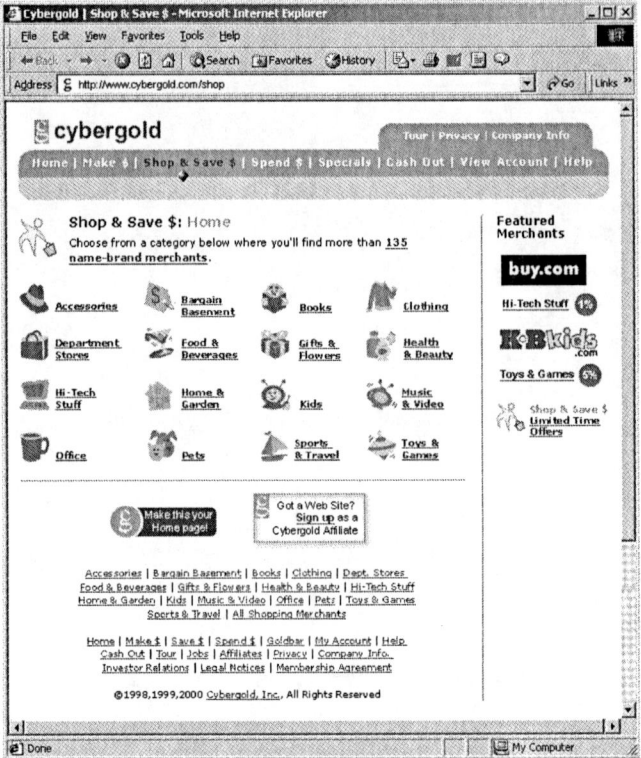

Source: http://www.elisabethparker.com/web_sites/cybergold.htm

CyberGold Gifts and Flowers Shopping Page (Drilldown)

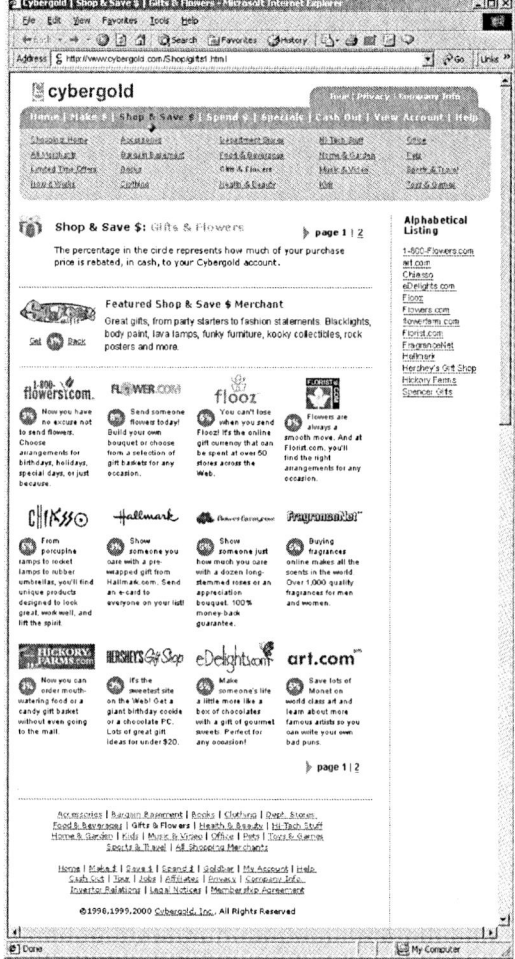

Source: http://www.elisabethparker.com/web_sites/cybergold.htm

APPENDIX B

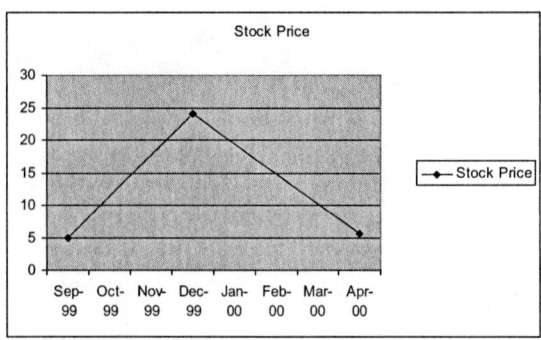

APPENDIX C

CyberGold Business Diagram

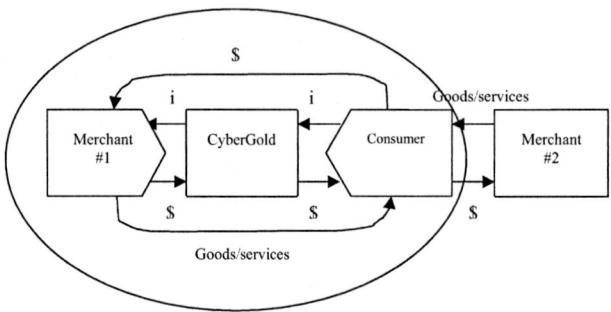

Derived from Weill and Vitale (2001)

- In this model, the typical interaction involving consumer and merchant, orchestrated by CyberGold, is depicted in the oval. The merchant (#1) offers CyberGold a fixed amount per new *paying* customer introduced or per advertisement viewed. CyberGold offers a reward of a specified amount of CyberGold (roughly equivalent to cash) to the consumer for viewing the advertisement or taking action. In exchange, the consumer provides demographic information to CyberGold, which allows CyberGold to market that information to merchants in their role as advertisers. The services of the merchant are provided to the consumer directly, in exchange for any appropriate form of payment.
- The direct lines from merchant #1 to consumer and the reverse depict the relationship that develops after the introduction engineered by CyberGold. After the relationship is established between merchant and customer, the customer purchases additional services from the merchant directly, as no additional CyberGold

is offered to the consumer once he has already earned CyberGold from the associated merchant.

- In the area on the right, the consumer's redemption of accumulated CyberGold is depicted. The consumer forfeits an amount of CyberGold in exchange for products or services to a participating merchant, who may or may not be one of the merchants contracting with CyberGold for marketing information.

APPENDIX D

	YEAR ENDED DECEMBER 31,		
	1997	1998	1999
REVENUES:			
Transaction.................................$	457,074	$ 628,350	$ 3,315,871
Custom marketing services and other.........	74,342	376,583	1,987,090
Total revenues........................	531,416	1,004,933	5,302,961
COST OF REVENUES:			
Transaction.................................	256,123	292,865	1,593,811
Custom marketing services and other...........	37,048	173,253	278,628
Total cost of revenues................	293,171	466,118	1,872,439
Gross margin.........................	238,245	538,815	3,430,522
OPERATING EXPENSES:			
Product development...........................1,190,047		1,700,421	2,670,737
Sales and marketing...........................2,162,413		2,694,601	8,312,130
General and administrative.....................	739,816	791,837	2,515,787
Amortization of deferred compensation.........	--	198,288	734,614
Total operating expenses..............4,092,276		5,385,147	14,233,268
Loss from operations.................(3,854,031)		(4,846,332)	(10,802,746)
Interest Income (Expense), net...................	(15,292)	78,381	739,930
Net loss...........................(3,869,323)		(4,767,951)	(10,062,816)
Dividend Attributable to Preferred Stockholders.	--	(660,430)	(1,570,307)
Net Loss Attributable to Common Stockholders...$(3,869,323)		$(5,428,381)	$(11,633,123)
NET LOSS PER COMMON SHARE:			
Basic and diluted(1)........................$	(0.97) $	(1.35) $	(1.40)
WEIGHTED AVERAGE COMMON SHARES OUTSTANDING:			
Basic and diluted(1).........................	3,979,489	4,020,393	8,308,482

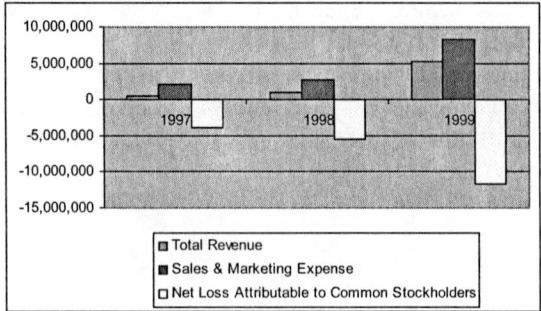

Source: CyberGold Corp, 10K, March 30, 2000

APPENDIX E

	THREE MONTHS ENDED MARCH 31,	
	1999	2000
Revenues:		
Transaction	$ 333	$ 2,269
Custom marketing services & other	170	2,021
Total revenues	503	4,290
Cost of revenues:		
Transaction	159	1,094
Custom marketing services & other	85	78
Total cost of revenues	244	1,172
Gross margin	259	3,118
Operating expenses:		
Product development	513	1,630
Sales and marketing	1,024	5,410
General and administrative	307	2,302
itarget.com acquisition costs	0	2,142
Amortization of deferred compensation	327	113
Total operating expenses	2,171	11,597

(continued on following page)

Loss from operations	(1,912)	(8,479)
Interest income, net	11	563
	--------	--------
Net loss	$(1,901)	$(7,916)
Other comprehensive loss	0	(117)
	--------	--------
Comprehensive loss	$(1,901)	$(8,033)
	========	========
Net loss per common share,		
basic and diluted	$ (0.37)	$ (0.38)
	========	========
Weighted average common shares,		
outstanding, basic and diluted	5,175	20,642
	========	========

Source: CyberGold Corp. 10-Q, May 15, 2000

John E. Peltier recently completed the Master of Science program in computer information systems at the Georgia State University Robinson College of Business in Atlanta. He also earned a Bachelor of Business Administration in management from the same university. Prior to enrolling in the MS program, Mr. Peltier worked in retail management with AMC Theatres. He currently works in the software development field as a quality assurance analyst for PracticeWorks, Inc., a healthcare information systems provider in Atlanta.

Michael J. Gallivan is an assistant professor in the CIS Department at Georgia State University Robinson College of Business in Atlanta. He conducts research on human resource practices for managing IT professionals, as well as strategies for managing effective IT implementation, IT outsourcing, and inter-organizational alliances. Dr. Gallivan received his PhD from MIT Sloan School of Management and served on the faculty at New York University prior to his present appointment at Georgia State University. He has published in Database *for* Advances in IS, Information Systems Journal, Information Technology & People, Information & Management, *and* IEEE Transaction on Professional Communications. *Before joining academia, Dr. Gallivan worked as an IT consultant for Andersen Consulting in San Francisco.*

This case was previously published in the *Annals of Cases on Information Technology*, Volume 6/ 2004, pp. 312-329, © 2004.

Chapter VIII

The Planned and Materialized Implementation of an Information System

Pekka Reijonen, University of Turku/Laboris, Finland

Jukka Heikkilä, Helsinki School of Economics, Finland

EXECUTIVE SUMMARY

The object of this case study is a marketing and sales information system in two local offices of a regional telephone company. A unified, advanced client/server system was needed due to the merging of three companies into a bigger regional company, keener competition, and the growing complexity of the services provided. The system is tailor-made to meet the needs of the industry and it was developed by a software vendor in close cooperation with the nationwide alliance of regional telephone companies. This study illustrates the difficulties in simultaneously aligning an organization and implementing a new information system. Views on the skills and competence needed in using the system vary, and lead to the negligence of education and training. The consequent lack of skills and knowledge of some users, especially of those not using the system regularly, create profound problems in the whole work process and in productivity as the first, obvious work practices become the dominant mode of operation bypassing the desired integrated workflow. The findings are discussed and reflected to concepts of institutionalization, positive reinforcement, and productivity paradox. This case emphasizes the importance of the organizational implementation and adaptation process which ought to begin after the implementation of the technical system.

BACKGROUND

It was only in the beginning of the 1990s when the telecommunications sector was deregulated in Finland. Long distance lines (i.e., crossing the local telecommunication areas), which were previously operated by a state-owned company, opened to competition. Deregulation took place simultaneously with the emergence of radio-based telecommunications, most notably with the introduction of the wireless analog NMT and later digital GSM and DCS networks. Radio-based telecommunication is also open to competition. The mobile phone has grasped a significant share of the phone traffic (there are more than 2.3 million mobile phone subscribers in a country of 5 million inhabitants and 2.8 million subscriber lines). As a consequence, the rates of hard-wired long distance calls have fallen by 80% and the rates of local calls by 50% since deregulation. Deregulation has affected the regional teleoperators the least, but as the former licensing of telecommunication areas is deregulated, too, local operators have started to merge.

Currently there are two main telecommunications operators in Finland: Sonera plc (formerly known as the state-owned Telecom Finland) and the Finnet consortium, which is owned by 46 regional telephone companies. There is a third player, owned by Swedish Telia, but as its market share is less than 2%, long-distance and radio-based telecommunication markets have practically a duopoly. Our case company belongs to the Finnet consortium, the market share of which is about 50% of the total telecommunications turnover, 16 billion FIM (about 3 billion USD), and the market is expected to grow to 22 billion FIM by the year 2000.

The nationwide Finnet consortium was established by the regional telephone operators in order to provide seamless long-distance calls for their subscribers and to keep up with the pace of fast rapidly-developing technology. The consortium developed a digital SDH-based backbone network, primarily on optical cabling, for long-distance and mobile-voice calls and data communication.

The regional telephone company, called here Areal Phone Ltd. (AP), implemented a new information system (IS) about one year before this case study. The implementation was not, actually, a free choice, rather the company was driven to changes. The reason for the new IS was simple: AP is the result of a merger of three mutual, local telephone

Table 1. Breakdown of the teleoperators' market in Finland (Telecommunications Statistics, 1998)

Regional phone calls	3.3 billion FIM
Long distance calls	4.1 billion FIM
International calls	1.3 billion FIM
Data communications	1.8 billion FIM
Equipment	4.7 billion FIM

Table 2. Areal Phone Ltd. in figures (1997)

Turnover	145 million FIM (28 million USD)
Investments	69 million FIM (13 million USD)
Gross margin	43 %
Revenue	11 million FIM (2 million USD)
Balance sheet	320 million FIM (62 million USD)
Telephone subscribers (wired)	75 293
Cable television subscribers	25 827
Employees	240
Inhabitants in the territory	152 500
Area of the territory	3236 square kilometers (2000 mile^2)
Municipalities in the territory	23

companies. The three companies each had their own tailor-made operative information systems, so there was a need to unify the operations and ISs.

The reasons for the merger stem from the previously mentioned changes in the telephone operators' business environment. On the one hand, deregulation has driven small local companies to seek economies of scale by merging. On the other hand, it is considered difficult for small companies to keep up with the rapid development of digital telephone technology, where different data services are becoming more and more important (for example, the percentage of Internet users in Finland is one of the highest in the world).

At present AP serves about 74 000 wire-based telephone subscribers and 26,000 cable television (CATV) subscribers (see Table 2). The company functions in 23 municipalities, where there are 152,000 inhabitants. The private customer base consists mainly of urban households but a significant number of customers live in the countryside. This means that maintaining the infrastructure is a demanding and expensive responsibility of the telephone company. The digitalization of the telephone network that started from switches has continued at the subscriber end and the number of ISDN installations has been growing constantly. For example, in 1997 the turnover of AP data services grew over 30%. Data services are especially important in enterprises, which is the other important customer sector of AP. The possibilities of AP offering data and other add-on services are excellent because their network is totally digitized.

The main business of AP is still to maintain the telephone network and operating calls. The future trend is to increase the complexity of value-added services both for the private customers (e.g., CATV-network renewal for data communications and xDSL connections) and enterprises (e.g., running outsourced switchboards and high-speed data communications). However, as the hype for an information society continues, telephone operators are seen as vital players; hence, the interest for mergers continues: in 1997, AP acquired the telephone company of a town situated nearby. In 1997, AP also introduced their own radio-based DCS/GSM 1800 city phone network.

SETTING THE STAGE

The new information system at AP was seen as a means to cope with the competition and to improve marketing, sales, installation, and maintenance of the increasingly complex value-added services. As we browsed through the documentation, it became clear, however, that the project had concentrated only on defining data structures and little attention was paid to actual business concerns and work process (re)organization.

The selection process of the new IS for AP had taken about one year. The turn-key information system chosen had not been used earlier in any of the three separate companies. The coalition worked in close cooperation with a Finnish representative of a large international software vendor. The new IS for operations control was to cover practically all the functions of the company (see Figure 1).

The broad domain was not news to any of the companies, because their former ISs covered also the majority of operations. As the new IS was different from any of the three former ISs, the data conversion from the old ISs was anticipated to become one of the most demanding tasks of the implementation process.

The new IS was implemented on an Oracle database using Oracle Forms and Report Writer development tools. The most computing intensive routines were programmed in C++. The system runs on a UNIX-server and the workstations (PCs) are connected to the server via an ethernet-connection. The OC system is used with a terminal program, so the architecture actually corresponds to a mainframe system. Office programs, however, are used directly from the server.

The initial problems with the capacity of the server were due to the greater-than-expected number of concurrent connections. The number of connections grew high as the IS did not support incomplete tasks (i.e., if a task was interrupted all the previous input was lost and the task had to be restarted from the very beginning). Hence the users

Figure 1. Subsystems of the OC system (Operations Control) and their interconnections (Subsystem #1 ["Marketing and Sales"] was the main object of the deployment evaluation project)

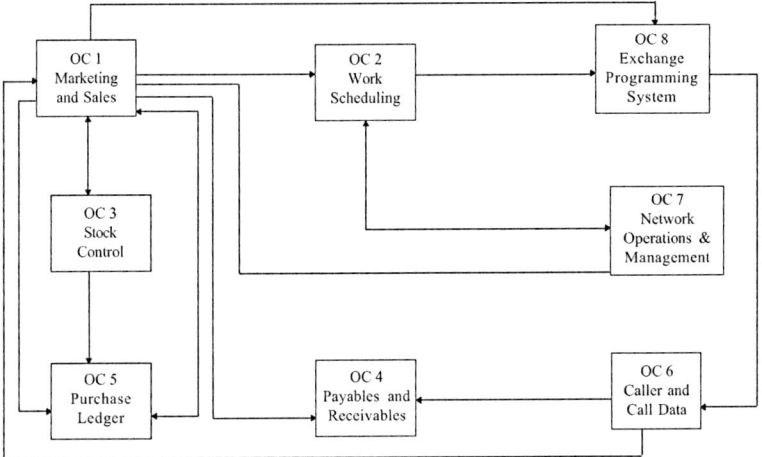

opened several concurrent terminal sessions in order to manage their work fluently in a customer service situation. At the time of the study, this capacity bottleneck had been removed and the response times of the system reached an acceptable level.

However, there arose additional, "fuzzy" problems that led the CIO of AP to contact our research unit at the university to evaluate the situation at the marketing and sales subsystem. The perception of the problem by management was vague; it was based on casual user complaints about the poor usability of the system. According to management, the problems concerned mainly the private customer sector and were caused primarily by deficient features of the IS: the IS did not support business processes anymore because it had been designed for selling hard-wired connections instead of add-on services. Therefore, management decided to set up a project for streamlining the interface in order to improve the functionality of the IS by "reducing the number of screens by half." Our evaluation was planned to complement the requirements analysis of the forthcoming renewal project. In addition, management wanted to compare the IS usage and usability problems to earlier evaluations made by our research unit.

The subsystem studied (see Figure 1) was labeled "Marketing and Sales," but actually "Customer and Product" would better describe the system. This is because the marketing part of the subsystem was actually abandoned—mostly for its lack of usability and functionality — after a short trial period. The subsystem is linked to all other subsystems (see Figure 1) because it is used for maintaining customer records (e.g., names and addresses, and long-term contracts).

The goal of the evaluation was to describe end users' work tasks, work processes, and the use situation of the IS, so that more informed decisions about further development of the organization and the IS could be made. The study was carried out at all three regional offices of AP, but we use two offices of comparable size and departments to illustrate our case.

CASE DESCRIPTION

In both offices sales function of AP is organized according to their clientele, into Private and Enterprise departments. There were 15 end users in the Private Subscribers Department and 11 in the Enterprise Department. Management named these two groups as the target groups of the evaluation. It soon turned out, however, that about an equal number of end users are using the same subsystems outside the Marketing and Sales Department, and their IS usage is directly related to the work processes of the personnel in marketing and sales.

The average age of the users is about 37 years in both departments. In the Private Department, nearly all users are women, but in the Enterprise Department the proportion of women is only about 20%. The end users are competent long-term employees and are accustomed to the organizational culture: mean employment time is nine years in the Private Department and 14 years in the Enterprise Department.

The users are also experienced computer users (on average, 8-10 years of usage), even though most users have experience only in the systems used in the current work place. In the Private Department, the IS is the main tool of its end users — on average, 87% of the total computer usage, which is equivalent to about 74% of the working hours spent using the new information system. In the Enterprise Department, the average usage time is 48% (about 25% of the working hours) and the variance in usage time is considerably higher. The situation in which the IS is used is also different: in the Private Department, most of the usage (40% of the working hours) took place in customer service situations, whereas computer usage in the Enterprise Department is more back-office-like; in only 4% of the working hours was the computer used while dealing with customers.

Before the IS implementation, the program vendor first trained and educated a few principal users who in turn trained their peers. The average training time was eight days in the Private Department and five days in the Enterprise Department — both figures can be considered impressive in light of our earlier research (Laboris, 1997) — despite the fact that in the Enterprise Department the training is unevenly distributed and over half of the users had received less than four days of training. The reported reason for missing the training courses was urgent duties but, on the basis of interviews, a lack of motivation seemed to be the major reason.

Technological Concerns

The new IS has a text-based interface managed with the keyboard. This solution was decided on in the cooperative planning association long before the implementation of the system. The decision was justified by the requirement that "the appearance of the interface must be the same in all kinds of workstations like PCs, X-terminals, and ASCII-terminals." Even though consistency and uniformity are important aspects in interface design, the grounds for this decision are not relevant to AP because all the workstations are PCs. Contrary to its initial idea, the decision eventually limited the possible range of interface options (e.g., it eliminated the possibility to implement a graphical user interface) and the development tools.

As a result, the system is somewhat difficult to interact with — or, actually, there is little interaction. The system consists of 90 separate screens, and the coupling of the individual screens to the process of completing a specific task is weak. For the end users this means that they have to memorize the screens and in which order they are used for each specific work task. This makes the system hard to learn and much practice and repetition is required in order to master the system fluently. This is confirmed by the fact that users' estimates of the time required to learn to use the system to the extent that one can work efficiently vary from two months to two years (the average was nine months). The estimates of the average learning time differ in the two departments: the average estimate of the learning time is four months in the Private and 13 months in the Enterprise Department. As most of the learning takes place during system usage, it is obvious that the perceived learning time is correlated with the amount of system usage ($r = 0.59$, $p = 0.01$, $n = 18$). At the time of the study (one year after its implementation), users rated their current ability at 7.7 (on a scale ranging from 4 to 10) in the Private Department and at 6.8 in the Enterprise Department.

Users are unanimous that the system has been, and is, difficult to learn and to use. There are too many screens and there is little help built into the system to find the right screens in the right order. Obviously, the IS supports occasional users poorly, but with a sufficient amount of practice, the user eventually can learn it by heart. One clerk who has been using the system for five to six hours a day for one year expressed this by saying:

Figure 2. End user computer satisfaction (EUCS) scores of the two groups (the differences in the means between the groups are statistically significant [t-test, p < 0.05] except for the factor "Timeliness")

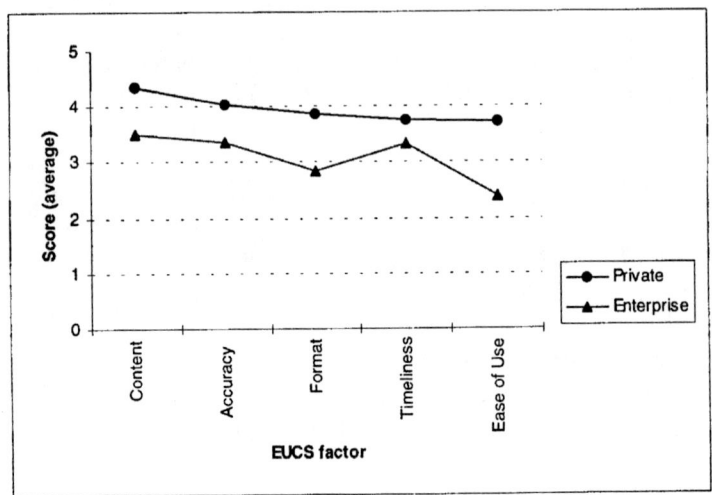

That task is so routine nowadays, we have done them so much, that one does not need to think much anymore.

But the system usage is obligatory for less often or irregularly performed tasks, too. In these cases, it has become a common practice to write the necessary information on a piece of paper and key the data in the IS later (with help from co-workers or support personnel).

The overall evaluation of the system by an EUCS-measure demonstrates that all features of the system are rated lower in the Enterprise Department than in the Private Department (Figure 2).

The user opinions give grounds for two possible conclusions: the system really is not well suited to the work in the Enterprise Department, and/or the computer self-efficacy of the users is low, i.e., the users feel that they cannot manage their computer supported work tasks properly (c.f. Compeau & Higgins, 1995).

Organizational Concerns

The consequence of the poor alignment of the system functions and the actual work flow is that less-frequent users did not learn to utilize the system properly. Because some user groups (e.g., supervisors of the installation function, technicians and some of the sales managers in the Enterprise Department) received little education in the system usage and on the ideal planned work flows, they developed apparently successful, but manual routines, to work around the system's limitations.

Management launched a redesign project to streamline the process in the Private Department, but actually the problems were more severe in the Enterprise Department. In the private sector, the service procedure was rather simple: customers were either face-to-face or on the telephone with the clerk who used the IS to carry out the service (see Figure 2). However, the system did not fully meet the requirements of marketing, selling, installing, and maintaining more complex services such as corporate ISDN-subscriptions. The willingness and skills to learn and use the system were also lower in the Enterprise Department.

In the Private Department, the implementation was carried out according to the plan: at the time of the study, most end users have reached such a skill level that the speed of the work process was acceptable in both offices. The work is interrupted by breakdowns only occasionally (about half of the users have problems weekly, and 57% of the problems are due to the lack of skills and knowledge, mostly a problem for newly enrolled clerks). In other words, there are few severe problems in the use of the IS. Also the work processes closely follow the planned workflow. Although the system is clumsy for some individual tasks, and some operations are not institutionalized yet, the users have learned to manage their work with the IS and to resolve the few problems with assistance from their fellow workers. In other words, private customer service either in direct contact face-to-face or by phone proceeds rather fluently. Direct customer contact also increases the clerks' motivation for developing their skills in system usage because, by being able to use the IS fluently, they could avoid embarrassing social situations which might arise if they had to say to a customer: "Well, actually, I cannot perform my work tasks."

In the Enterprise Department, there has emerged an unplanned supplemental work role (substitute IS user in Figure 3) because the sales managers were unwilling to learn and use the IS. Instead of inputting the order information into the IS, they deliver orders as handwritten notes to "the substitute IS user" who then inputs the orders into the IS. The sales managers' notes rarely include product numbers, only the name of the product or service. It is not unusual that while inputting data, extra contacts between the sales manager and "the substitute IS user" are needed because the product name the sales manager uses is not found in the IS.

In the other office, technicians and their supervisors also ceased to use the IS after a few weeks of initial use. Actually, they ceased to use the IS as they were told to stop using it. This is because they had received neither enough training nor enough practice to master the IS as occasional users. For example, when the technicians were to record in the IS the working hours spent and materials consumed for a certain order, they often could not find the right order nor the right customer number. Instead, they keyed in the data in the first found occurrence of this specific customer — or created a new customer number. This led to a situation where some orders were not invoiced, the incorrect department of an enterprise was invoiced, and the database of the IS was eroding because of the duplicates. Except for extra work, this also caused customer complaints and in the present competitive environment, flawless customer service, quick response, and a good reputation are critical.

In order to correct the situation, the IT usage of these user groups was replaced with pencil and paper: after inputting the work order, the substitute IS user printed out the order, and data of the later phases of the work process were documented on it. The notes made on the paper document were then inputted into the IS by the substitute IS user as

Figure 3. Use of the IS in the sales processes (from order to invoicing): end-user roles, IS usage, and interactions in the work processes (the solid lines represent the planned interactions and work flows; the dashed lines represent the situation which was institutionalized in the other office)

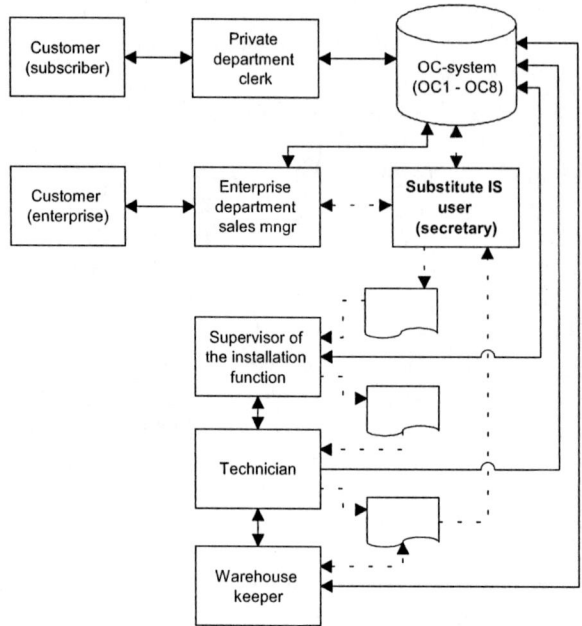

the final step of the process after the installation was completed. This meant new procedures and a lot of extra work in finding missing data, since the data on the paper-based documents were only occasionally complete. As a consequence, the invoicing process was delayed dramatically, in some cases up to several months, despite the constant overtime work of the substitute IS user.

A closer look at the episodes and encounters of the process that led to the ban of usage in the Enterprise Department revealed that the key problem was the end users' inability to cope with the IS. The incapability of technicians and supervisors at the other office resulted in chaos, temporarily solved by the use of paper instead of the IS. However, this arrangement became permanent. For the sales managers, it was also a question of their power in the organization: a substitute user was assigned to the "less valuable" task of learning to use the IS and keying in the data.

Current Challenges

Even though the situation might have been avoided by managing the organizational implementation process "better," there are two interrelated facts, which have contributed to the evolvement of the problems.

First, the IS is not user-friendly with its character-based windowed user interface, and it does not have any support for learning or remembering the flow of operations. Its functionality for some tasks is questionable as it was originally developed for a different business environment. The learning time is long (but not extraordinary) and as a consequence, some user groups are not willing to adopt the system. And we are afraid that there will be more problems of this kind. This is because the way the IS was built seems to have become an industry standard: a shared relational database is normalized to atomistic tables, which are updated by using screens and procedures (half) automatically created from the columns of the table with an application generator. This easily leads to a great number of separate screens. But, as the starting point for development is a semantic data model, the design is seldom compatible with the work procedures of the organization. It is also difficult to learn to navigate in a bunch of separate, seemingly unrelated screens.

The second problem is the inertia caused by the IS in a rapidly changing environment. For example, the IS has been designed so that there can only be one subscriber telephone number per customer ID. This means problems with the rest of the possible numbers of an ISDN line. Or, when a customer enterprise wants to regroup invoices according to its new organizational structure, there are no mechanisms for handling the update. It is inevitable that information systems freeze the current mode of operation and all changes in operations almost always call for reprogramming of ISs. And in the case of the IS, the advantage of sharing the costs of development in the first place has become an obstacle for further development. The software is owned by the consortium, the members of which have to be convinced and agree upon the need for changing the software.

To us the case clearly illustrates the common dilemma in IS implementation: an information system that is somewhat awkward to use, slightly incompatible in work processes, and difficult to change (but otherwise technically functional) can be implemented in one organization whereas in another, almost similar organization, the IS is abandoned.

DISCUSSION

The design, production, and implementation phases of a computer-based information system are commonly considered the key activities in its life cycle — in money, time and other resources. Skillful software development is about the technical maneuvers in conducting the phases. However, the value of the IS, how technologically progressive it might be, can not be realized until it is effectively implemented and deployed. As pointed out by Kling and Allen (1996), effective use of information technology does not depend on technology alone: organizational issues often play an important role. In order to emphasize the importance of non-technical factors, they use the term "organizational implementation": "Organizational implementation means making a computer system accessible to those who could or should use it, and integrating its use into the routine work practices" (ibid, p. 269). In the world of outsourced development, this is more important than ever.

Organizational implementation necessitates, for example, that users are willing to adopt the IS and that they have the necessary skills and knowledge (both domain and technology/tools) needed in the deployment of the CBIS. In research, the need for education and training in computer-related skills is commonly recognized and its importance emphasized (Nelson et al., 1995; Clement, 1994). In practice, however, the training issues are often neglected and it has even been argued that one of the reasons for less successful CBIS implementations is the lack of user training (Mitev, 1996).

One reason for the relative ignorance of training and education issues is that the deficiency of competence usually becomes apparent only in the beginning of the usage phase of a new CBIS. If the usage of the CBIS is obligatory, users learn — sooner or later — to deploy the system while doing their daily work. When users have learned to manage their work to the extent that there are no apparent problems, the problems disappear. And, as there are no problems, there is no need to evaluate the efficiency and effectiveness of the actual implementation and work practices! Or, only the technical functionality, such as response times, is checked.

The time required to learn to use a new CBIS depends on many factors, for example, the structure and appearance of the CBIS, the amount of use (i.e., the practice obtained), and the individual learning capacity. Learning times of operative ISs, such as the OC-system, are rather long with huge variation; in most cases the learning period is several months. The learning process can be characterized as a trial-and-error diffusion process where a single user or a group of users solve the problems encountered when integrating their work practices into the features of the CBIS. It is not a uniform process, rather there are multiple simultaneous processes (Heikkilä, 1995). In this kind of learning environment people tend to accept the first functional solution and stick to it as long as it works. Its effectiveness is never questioned, nor are alternative solutions actively sought: the functional solution will do. A typical feature in this learning period is that users search for help mainly from their peers (in 80-90% of the problems) whereas manuals, help-desks and other types of support arrangements play only a secondary role (Heikkilä, 1995; Laboris, 1997). It is worth noting that managers are seldom in advising roles because they rarely can give support in CBIS-related problems. As a consequence, managers often only have a vague idea of the real CBIS deployment situation. Therefore, the effectiveness of the applied procedures depends mainly on the initially invented workable solution which becomes the best practice in that environment.

Implementation of a new CBIS generally also changes the division of labor and incentive structure which cause changes in the work tasks and motivation of individuals. After implementation users must learn this new division of labor and potentially some new work tasks as well as the functions of the new CBIS. The division of labor is planned and institutionalized rather carefully, therefore, some kind of "role book" is thus already available when a CBIS is designed. The process of dealing out the roles to the actors is also part of the management of the implementation. However, management, control, and evaluation after implementation are often shallow and based more on beliefs, attitudes, and opinions than technological or economical data (Kumar, 1990; Kling & Iacono, 1984). Iivari (1986) has concluded the following about the intra-organizational implementation of multi-user ISs: there are actually two coexisting sets of activities in implementing an IS, or innovation in general; the technically-oriented rational-constructive tasks and a political process of resolving conflicts of interests and changing the organization.

In practice, many of the problems, which are caused by the changes in the division of labor and encountered when deploying a CBIS, are solved with the same kind of ad hoc principle as other work-related problems. As management's concerns are in restructuring the industry, the most important thing is to get the work done by the doers, no matter how it is done. And, in the same way as an individual repeats an action after a positive reward, an organization gives legitimate status rather quickly to an arrangement which seems to be functional (see positive reinforcement trap; Argyris, 1985). Finally, the division of labor becomes institutionalized and legitimated (Berger & Luckman, 1966). At this stage, the way of acting feels to be the right one and the actor becomes confident: "This is how things are done." In his book on organizational culture, Schein (1992) has described the process as follows: "When a solution to a problem works repeatedly, it comes to be taken for granted. What was once a hypothesis, supported only by a hunch or a value, comes gradually to be treated as a reality" (ibid, p. 21).

The insufficient training and the consequent emerging ad hoc work practices that get institutionalized with the IS can easily lead to inefficient and unchallenged operations, as our case study clearly demonstrates. There is growing evidence of the high hidden costs of using computers in the workplace: about two-thirds of the total costs seem to come from the invisible opportunity costs in the usage phase of the computer technology (Heikkilä, 1995; Gartner Group, 1995; van Hillegersberg & Korthals Altes, 1998). In other words, the out-of-pocket costs from the IS Department budget represent only one-third of the total costs. Two-thirds of the costs arise from the users' working hours spent in tackling computer-related problems, advising other users, learning new computer skills, and taking part in various CBIS planning, development, and coordinating taskforces. And the number of the hours of education, training and peer-hands-on-learning on applying ISs in work settings has not kept up with the increase in IS-based activities.

How much of this hassle can be removed by training is still unclear, but the fact is that it is one of the key actions to diminish the time wasted. And what has been found so far, is that the traditional teaching courses are deficient in meeting the needs of users applying ISs in their work practices. More innovative ways of learning are needed, such as peer tutoring, support activities, and workshops for the users of ISs (Reijonen, 1998).

The problems highlighted in this case are not typical in just one kind of business sector or type of information system, but a more general trend which we have noticed in several organizations. Hence, the emphasis is on the level of the general findings and the case is used to illustrate how problems can be manifested in a particular situation.

ACKNOWLEDGMENTS

This study was financed partially by the grants from The Finnish Work Environment Fund and The Academy of Finland.

REFERENCES

Argyris, C. (1985). *Strategy, change and defensive routines*. Boston: Pitman.

Berger, P. L., & Luckman, T. (1966). *The social construction of reality*. Middlesex: Penguin Books.

Brynjolfsson, E., & Yang, S. (1997). Information technology and productivity: A review of the literature. *Advances in Computers, 43*, 179-214.

Clement, A. (1994). Computing at work: Empowering action by 'low level users'. *Communications of the ACM, 37*(1), 53-63.

Compeau, D. R., & Higgins, C. A. (1995). Computer self-efficacy: Development of a measure and initial test. *MIS Quarterly, 19*(2), 189-211.

Doll, W. J., & Torkzadeh, G. (1988). The measurement of end-user computing satisfaction. *MIS Quarterly, 12*(2), 259-274.

Gartner Group. (1995). *Desktop computing: Management strategies to control the rapidly escalating cost of ownership.* Gartner Group.

Heikkilä, J. (1995). *Diffusion of a learning intensive technology into organisations: The case of PC-technology.* Doctoral Thesis, Helsinki School of Economics and Business Administration A-104.

Iivari, J. (1986). An innovation research perspective on information system implementation. *International Journal of Information Management, 6*, 123-144.

Kling, R., & Allen, J. P. (1996). Can computer science solve organizational problems? The case for organizational informatics. In R. Kling (Ed.), *Computerization and controversy* (2nd ed., pp. 261-276). New York: Academic Press.

Kling, R., & Iacono, S. (1984). The control of information systems developments after implementation. *Communications of the ACM, 27*(12), 1218-1226.

Kumar, K. (1990). Post implementation evaluation of computer-based information systems: Current practices. *Communications of the ACM, 33*(2), 203-212.

Laboris. (1997). *A collection of customer reports on the evaluation of the deployment of information technology in organizations* (1993-1997). University of Turku, Laboris, Laboratory for Information Systems Research, Finland (in Finnish).

Mitev, N. N. (1996). More than a failure? The computerized reservation systems at French Railways. *Information Technology & People, 9*(4), 8-19.

Nelson, R. R., Whitener, E. M., & Philcox, H. H. (1995). The assessment of end-user training needs. *Communications of the ACM, 38*(7), 27-39.

Reijonen, P. (1998, June 4-6). End-user training and support: A comparison of two approaches. In W. R. J. Baets (Ed.), *Proceedings of the 6th European Conference on Information Systems*, Aix-en-Provence, France (pp. 660-672).

Schein, E. H. (1992). *Organizational culture and leadership* (2nd ed.). San Francisco: Jossey-Bass.

Telecommunications Statistics. (1998). The Finnish Ministry of Transport and Communications. Retrieved from http://www.vn.fi/lm/telecom/stats/index.htm

van Hillegersberg, J., & Korthals Altes, P. (1998, June 4-6). Managing IT-infrastructures: A search for hidden costs. In W. R. J. Baets (Ed.), *Proceedings of the 6th European Conference on Information Systems*, Aix-en-Provence, France (pp. 1655-1662).

ENDNOTES

[1]   At a first glance the mutual telephone companies were geographical (natural) monopolies. But as the mutual companies were owned by the subscribers, who also elected representatives to the board of directors, the monopoly power was never actually exploited to its full extent. Nevertheless, this led to a price discrimination between leased lines and "owned" lines, which is currently banned by the

European Union regulation.

2 The subscribers of the conventional mobile phones (NMT and GSM) are served via a separate, nation-wide company, Radiolinja, owned by the Finnet consortium.

3 As Finnet operates GSM networks via Radiolinja, the regional telephone companies are most interested in getting their share of the radio-based telecommunications by providing geographically limited mobile services for private and business customers at favorable prices.

4 Preliminary discussions about the development of the system had actually started several years earlier at Finnet. One of the companies that merged with AP took part in the requirements analysis. The first installations took place only a couple of years before this study, and nobody at AP was experienced with the system.

5 UNIX environment and client/server architecture were selected because of its openness (it is possible to select between different software vendors) and versatility (technically easy to extend).

6 The data were gathered with a survey questionnaire (26 end users), by interviewing end users, managers and IS-personnel (16 interviews), from documentation (e.g., documents of the CBIS development project, end-user manuals), by observing the use situation, and by using the training version of the CBIS.

7 EUCS is End User Computing Satisfaction (see Doll & Torkzadeh, 1988).

8 For example, it takes almost half an hour to register and schedule installation for a new telephone subscriber.

9 For instance, some Private Department clerks preferred to call the warehouse keeper for the price of a product instead of checking it themselves from the database.

10 Both the product numbers and product names were changed when the new IS was implemented. The new numbering/naming system has its benefits because it is becoming a standard in the Finnet-group.

11 The design of the IS did not make the situation easier because there were no practical means for managing enterprise level customer data, so, for example, the departments of a large enterprise occurred as separate customers with unique customer numbers in the database.

12 The use of paper-based documents also meant that the IS no longer was a real time system, so answering customer inquiries of the state of an order always required contacting supervisors or technicians.

Pekka Reijonen earned his MSocSc in psychology from the University of Turku. He is a researcher at the Information Systems Laboratory Laboris, University of Turku. His research interests are usability and deployment issues of information systems in organizations and he currently is working on a project defining end users' skills and knowledge requirements and their evaluation. His earlier work has appeared in the British Journal of Psychology, Education & Computing, Interacting with Computers, *and* information systems conferences.

Jukka Heikkilä holds a PhD in information systems from the Helsinki School of Economics. He is a junior research fellow of the Academy of Finland at the Helsinki

School of Economics. His current research focuses on the problem of adopting, implementing, and integrating innovative technologies to support business processes. He is also a member of the Information Systems Laboratory Laboris, University of Turku specializing in the evaluation of the utilization problems of implemented IS in organizations. He has published in the Information & Management, European Journal of Information Systems, Journal of Systems Management, Journal of Global Information Management, *and* Scandinavian Journal of Information Systems.

This case was previously published in the *Annals of Cases on Information Technology Applications and Management in Organizations*, Volume 1/1999, pp. 48-59, © 1999.

Chapter IX

The T1-Auto Inc. Production Part Testing (PPT) Process:
A Workflow Automation Success Story

Charles T. Caine, Oakland University, USA

Thomas W. Lauer, Oakland University, USA

Eileen Peacock, Oakland University, USA

EXECUTIVE SUMMARY

This case describes the development, design, and implementation of a workflow automation system at a tier one automotive supplier, T1-Auto. T1 is a developer and manufacturer of anti-lock brake systems. In 1991, T1-Auto had outsourced its IT Department. They retained a management core consisting of the CIO and five managers, but transitioned approximately 80 other members of the department to the outsourcing firm. In 1994, Lotus Notes™ was installed as the corporate standard e-mail and workflow platform. A team of four Notes™ developers wrote workflow-based and knowledge management-based applications. Another team of three administrators managed the Notes™ infrastructure. The first workflow application written at T1-Auto was developed for the Finance Department. The finance team quickly realized the workflow benefit of streamlining and tracking the capital expense request process. The Notes™ development team and the project sponsor, the Controller, worked closely to develop the application. Following this initial success, the power and value of workflow technology caught on quickly at T1-Auto. One of the most successful projects was the Electronic Lab Testing Process described in this paper. The Electronics Lab and Testing System (ELTS) was identified as a Transaction Workflow problem by the IT Lotus Notes™ team. Because the ELTS involved policies and procedures that crossed many groups and divisions within T1-Auto, and since the process was consistent across the organization, the solution lent itself very well to Lotus Notes™. However, while T1-Auto was experiencing rapid growth and the number of tests was increasing, the testing process was prone to communication and coordination errors. As part of their production and product development processes, their electronics laboratory was required to test electronic components that were part of the brake systems. Clearly the testing process was critical to T1 since delays or errors could adversely affect both product development and production. The case goes on to describe the design and development of the Lotus Notes™ workflow management system. The design description includes process maps for the as-is and the new system. In addition, descriptions of the testing phase, the pilot, and the roll out are included. The case concludes with a discussion of project success factors and planned future enhancements.

T1-AUTO BACKGROUND

T1-Auto Inc. is a leading producer of brake components for passenger cars and light trucks. The most significant automotive products manufactured and marketed by T1-Auto are anti-lock braking systems ("ABS"), disc and drum brakes, disc brake rotors, hubs and drums for passenger cars and light trucks. T1-Auto is one of the leaders in the production of ABS, supplying both two-wheel and four-wheel systems, and was the leading manufacturer of two-wheel ABS in North America for light trucks. In order to meet increased ABS demand, T1-Auto built new plants in Michigan and Europe. T1-Auto is also a leader in the production of foundation (conventional) brakes, and benefits from its strategic position as a major supplier of ABS and foundation brakes for light trucks, vans and sport utility vehicles. T1-Auto also produced electronic door and truck lock actuators. T1-Auto operated six plants in the Michigan and Ohio areas and had one plant and engineering facility in Europe. The company operated a central engineering and testing facility near its corporate headquarters outside Detroit, Michigan.

In 1991, T1-Auto had outsourced its IT Department. They retained a management core consisting of the CIO and five managers, but transitioned approximately 80 other members of the department to the outsourcing firm. In 1994, Lotus Notes™ was installed as the corporate standard e-mail and workflow platform. A team of four Notes™ developers wrote workflow-based and knowledge management-based applications. Another team of three administrators managed the Notes™ infrastructure. The first workflow application written at T1-Auto was developed for the Finance Department. The finance team quickly realized the workflow benefit of streamlining and tracking the capital expense request process. The Notes™ development team and the project sponsor, the Controller, worked closely to develop the application. Following this initial success, the power and value of workflow technology caught on quickly at T1-Auto. One of the most successful projects was the Electronic Lab Testing Process described in this paper.

Since the incursion of foreign automobiles in the 1970s, the automotive industry has been characterized by intense rivalry of the participants. This affects the OEMs, the U.S. big three, and cascades through the supply chain. Because T1 is dependent on a few customers, they are subject to their demands for price, quality, and delivery conditions (Porter, 1985). The most severe pressures concern time to market. The OEM continually searches for ways to shorten the design cycle for a new vehicle platform. When a tier 1 supplier such as T1 signs a contract to produce a part, they are given hard deadlines to meet. The means for improving speed to market is most often found by changing the long entrenched inefficient processes that add costs to the supply chain (AIAG, 1997; Lauer, 2000). In the words of David E. Cole, director of the Center for Automotive Research at the University of Michigan, "The business model that is emerging is one that is extraordinarily fast and with a minimum amount of paperwork. Automakers need to get the product out to the customer quicker" (Verespej, 2001).

SETTING THE STAGE

Wayne Russell, Electronics Lab Supervisor at T1-Auto Inc., grabs his cup of coffee each morning and sits down at his PC to check his Lotus Notes™ e-mail. Using the same Lotus Notes™ software, he switches over to another database and checks the status of the various tests being performed in the electronics lab. Wayne opens the "Electronics Lab Test System" and tracks the tests that are being performed on parts from the various plants at T1-Auto. Finally, all of the electronic tests were being submitted and tracked in a consistent and efficient manner.

Our company was growing, Wayne reflects. This caused communication gaps because of new people not knowing the testing process. Before the Notes™ system, we had not defined the testing process. In the midst of a test, people might be making changes on the fly to the procedures that were being carried out. With all these changes going on, it was difficult to know who the key players, who needed to know what.

The new Electronics Lab Test System (ELTS) created and enforced a consistent process for submitting parts to the electronics test lab. The system also handled the various events and sub-processes that occur within the testing process. By leveraging the recently installed Lotus Notes™ infrastructure at T1-Auto, the ELTS greatly improved the testing process and quickly resulted in a positive payback for a relatively small investment. This workflow system achieved all of the expected benefits of implementing a workflow system, including: less paper, higher throughput, standardization of procedures, improved customer services, and an improved ability to audit the process.

CASE DESCRIPTION

Production Part Testing Process

The electronics test lab received primarily two types of test requests, production tests and ad-hoc engineering tests. The bulk of the tests were production tests. Ad-hoc engineering tests were performed to check production parts against new requirements or to ensure new products would meet customer and engineering specifications. Production testing was performed to ensure part quality for the life of its manufacturing production.

Production tests consisted of randomly testing finished parts as they came off the production line. Each product that was produced by T1-Auto had test specifications and a test schedule created for it by the engineering team. Products typically had periodic tests defined on a weekly, monthly, quarterly, or yearly basis. Each plant had a quality team assigned to ensure that the test schedules for all of the parts were carried out. Plant quality personnel would pull finished parts from the line and place them in containers for shipment to T1-Auto's test lab.

Product engineers in the company could request the ad-hoc engineering tests at any time. One or more parts would be sent to the lab with specific test instructions including the type of test, number of tests, and specific test parameters.

The test lab consisted of various test machines capable of stress testing parts causing them to run through their particular actions thousands of times in a short period of time. The test lab could also simulate the working environment of a part using special environmental chambers capable of simulating severe weather conditions such as heat and humidity and subjecting the parts to elements such as a road salt.

Once the tests were received at the electronics test lab, the individual test had to be assigned to a lab technician and scheduled. The schedule had to coordinate the technicians, test equipment, and test requirements such as the required completion date.

After the parts had been tested, test reports were written and returned to the plant quality team and product engineer for production tests, or the submitting engineer for ad-hoc tests. If an issue occurred during testing, the engineer would be contacted for instructions. If the issue was related to the testing process or procedure, the engineer could ask the technician to continue the test. If the issue was related to a part or set of parts falling below expectations, the engineer could then ask that the test be stopped and that the parts be sent to his office for evaluation.

When a production test is stopped due to issues uncovered during the testing process, a process designed to inform all of the necessary participants must be followed. A form called the Unsatisfactory Test Results Notification (UTRN) would be filled out and sent to key participants, including the product engineer, sales representative, plant quality team, and product manager. Sales representatives for the particular product are contacted so they can notify the customers about the testing status of the products as soon as possible. In many instances, T1-Auto was contractually bound to notify its customers within twenty-four or forty-eight hours of the time that testing irregularities are discovered. These deadlines created substantial problems for T1, often disrupting previously planned testing which led to further problems. Product engineers would then start a standard ISO/QS 9000-quality process called the "Corrective Action." (See Exhibit 1 for a workflow chart depicting the original pre-ELTS process).

The Problem Set

The ELTS was quickly identified as a transaction workflow problem by the IT Lotus Notes™ team. Since the ELTS involved policies and procedures that crossed many groups and divisions within T1-Auto, and because the process was consistent across the organization, the solution lent itself very well to Lotus Notes™. However, since T1-Auto was experiencing rapid growth and the number of tests was increasing, the testing process was prone to communication and coordination errors.

Process

Given that T1-Auto was experiencing significant growth and the testing process involved so many different groups, consistency in the process became a concern. Ensuring that all the forms were complete and that they were filled out properly was a problem. Since the participants in the testing process had a number of other responsibilities, the likelihood of delays in time-sensitive events was high. To further complicate matters, the test lab had no advanced notification of required tests. This led to planning inefficiencies in scheduling staff and test equipment. Tests were scheduled once the truck arrived each day with the required tests from each plant. Management became aware that advance test schedule notification would improve utilization of test lab personnel and equipment.

Communication

Another problem with the testing process was that test lab technicians often did not know who were the current product or quality engineers for each product. This could result in an e-mail notification of unsatisfactory test results being sent to the wrong individuals. This sort of incorrect communication alone could add two to three days to the entire process. In general, communication speed is critical in the auto industry where contractual conditions often include delivery requirements for new product development. Failure on the part of T1 to provide timely notification of test issues to its customers could cascade through the systems causing significant delays.

Another communication problem could occur when a customer or engineer called the plant or test labs to inquire about test status. Because the process was entirely paper-based, typically the best the plant quality personnel could do was to report on the portion of the process they controlled. The plant could tell if the parts have been pulled and shipped to the lab which indicated that testing had begun, or if the test report had been returned and the test completed. For any more detail, the plant personnel would have to call the lab for an update. Again, the paper-based system used at the lab was inefficient, since a status inquiry would require tracking down the physical test submittal paperwork and locating the assigned technician. Simple status inquiries added excessive human interaction and non-value-added overhead to the process. Furthermore, because of the amount of effort required to handle all the testing paperwork, the entire process was error prone.

Knowledge Management

As with all paper-based processes, historical information was difficult to locate. Test reports were copied and sent to the required participants. Historical reporting or analysis meant rifling through file cabinets and transcribing data to Excel™ spreadsheets creating another non-value-added and time consuming process constraint. Communication pertaining to specific tests could also be lost since the communication (e-mail, telephone calls, memo's, and hallway conversations) was not collected and stored with the testing documentation. It was possible at times for a duplicate process error or non value-added event to take place on a test because the necessary historical information was unavailable.

Summary

Inefficiencies in the testing process adversely affected part production. The testing process was contractually mandated. Therefore, it had to be done well. At the same time, due to time to market pressures, it had to be done expeditiously. The existing process was replete with inefficiencies that resulted in delays, communication problems with the customer, and the lack of a knowledge base for analyzing previous tests. Many of these problems stemmed from difficulties in handling all the documentation associated with the testing process. These included document tracking and document sign-offs.

Automated Workflow Management

Workflow is concerned with the automation of procedures wherein documents, information or tasks are passed between participants according to a defined set of rules to achieve, or contribute to, an overall business goal. While workflow may be manually organized, in practice most workflow is normally organized within the context of an IT system to provide computerized support for the procedural automation and it is to this area that the work of the Coalition is directed (Workflow Management Coalition, 1995).

The ELTS software took only two months to develop by a single Notes™ developer. This rapid development time was achieved using the Lotus Notes™ development environment. All of the processing and development was accomplished with the core Notes™ application and the Lotus Notes™ macro development language. Electronic versions of the Lab Work Request, Unsatisfactory Test Results Notification, and Test Results Notification forms were duplicated in the Notes™ ELTS database. Additional forms were added for systems management and to allow the system users to carry out electronic discussions using a comment form within the ELTS system. The system was developed, tested, and deployed using proven software deployment methodologies.

Lotus Notes™

The ELTS was initially developed using version 3.0 of Lotus Notes™ that was released in May of 1993. The first release of Notes™ shipped in 1989 and had become widely deployed and accepted as the leading e-mail and workflow application on the PC desktop in major corporations. With Notes™, users could create and share information using personal computers and networks (Florio, n.d.). Today, Lotus Notes™ is on version 5.0 and claims over 50 million users. Lotus Notes™ was being tested at T1-Auto for a specific workflow application in 1994. Due to its power and workflow capabilities, it quickly spread to other applications and became the corporate e-mail platform. Prior to Lotus Notes™, T1-Auto had no corporate e-mail package. Version 4.5 of Notes™, released in 1996, tightly integrated the power of Notes™ with the World Wide Web and called the new server side of the product Lotus Domino™. With a small amount of tweaking, Notes™ applications can become Web-enabled and accessible via a Web browser.

Powerful workflow applications can be developed in Lotus Notes™ by designing forms and views. Forms are electronic versions of paper-based forms and contain fields. Fields store the information that is either entered on the form or automatically generated, such as the current date. Besides typical field formats such as numbers, dates, and text, Lotus Notes™ contains a signature field that allows for electronic signing of documents. Once a form has been created and filled in, it is called a document. Views within Notes™ display all of a specific set of documents that belong to an application. These views display in a list, using rows and columns, all of the desired fields from the documents. Multiple views for an application are typically created sorting the documents by different fields. The forms, documents, and views for a specific application are stored in a Lotus Notes™ database. A typical workflow developed in this manner application is stored in only one Notes™ database. (At the time of this development project, Notes™ was the best option for developing workflow systems. It should be noted that at present, there are a number of other options available, e.g., Microsoft Exchange™).

The "Electronic Lab Test System" (ELTS)

The Lotus Notes™ development team at T1-Auto followed a standard development methodology for the development of the ELTS application. The project was broken down into the following phases: design, development and testing, pilot, full roll-out, and ongoing operations. Prototyping was employed during the design phase to ensure that requirements were well understood and that they could be met.

Design

The initial design simply created electronic versions of the Lab Submittal, Unsatisfactory Test Results Notification (UTRN), and Test Completion Notification (TCN) forms. The intent was to document and automate the existing process. It became clear that process efficiencies could be removed and value could be created if the test schedule was put online by including a Lab Submittal form with minimal information. A process map was developed using Visio™, a graphical charting program. The process map was used to document the existing process and develop requirements for the new system. Exhibit 1 contains a flowchart of the original process. The standard Lotus Notes™ development environment lacks the ability to graphically design and test workflow scenarios. It does, however, offer an add-on product called Lotus Workflow™ that provides this capability. The ELTS system was designed and created without the use of a graphical design and testing tool.

The development team quickly (less than a week) developed prototypes of the electronic forms and demonstrated them to the plant and lab personnel. Once the forms were approved, requirements were documented with the users showing how the new application would work and how the completed forms would appear in Notes™ views. Once the requirements were approved, development began immediately.

Development and Testing

The completion of the Design phase yielded a clear understanding of the requirements for the ELTS application. Since the majority of the work for developing the forms was completed in the prototype phase, the remaining work consisted of:

- Complete the forms,
- Design and build views,
- Code the form and view logic,
- Create the agents,
- Set and test security,
- Write "About" and "Using" help pages.

Forms were all checked for consistency and proper operation. Fields were checked for proper function and usability. The design team had set standards to ensure a consistent "look and feel" and ease of use from one application to the next.

Notes™ views were created to display current and future tests status, tests by plant, and test by month and day. Views were developed from the perspective of each process participant to make applications familiar and easy to learn and use. For instance, a view was developed for the plant personnel that listed each test in progress by plant, status and date. This made it easier for the plants to find their specific tests. For the lab users, views were developed that listed the tests by status, date and plant. Lab users were more interested in tests from all plants by date. Views were also created to develop specific reports for management that provided test status information.

Once the forms and views were completed, the developers wrote the code that automated the workings of the applications. Fields that would need to be automatically set by computation as opposed to user entry had to be coded. Views could also contain fields that would be computed for display such as number of days until a test was required.[1] Notes™ program agents are small programs that can be triggered to run on a fixed schedule or when an event occurs, such as saving a form. One of the agents developed for the ELTS system was run each morning and checked all forms to see if they are being processed in a timely fashion. For instance, if a UTRN was not acknowledged by the test engineer, an automatic e-mail notification would be resent to the engineer as well as the engineer's manager.

While security is very important for workflow applications in general, it is even more critical for the ELTS system given the sensitive nature of a potential test problem. Security is defined on a very granular basis for the system from who can access the system to who can enter data into a specific form field. Access levels range from permission to view a specific form (reader) to permission to edit data on a form (manager). Typically, the Notes™ team uses roles for each application and manages security based on putting users into roles. This lessened the overall security management overhead necessary for each application. A role was created for the ETLS systems called "Test Engineer." Every engineer using the system was put into this role and all were all granted exactly same security privileges within the ELTS system.

Once the system had been coded and tested, the last step prior to releasing the application was to write the "About" and "Using" Notes™ documents. The "About" document is accessed from the Notes™ menu and contains information about the intent and ownership of each application within the system. The "About" document also contains the application version and release date. The "Using" document is also accessed from the main menu and provides information on the specific usage of each application. The development team typically put a description for each fields on the forms. The "Using" document provided online help for the system.

Once user documentation was written and the system tested, the ELTS system was ready for the pilot test.

The Pilot

Prior to full implementation, a pilot version of the new application was installed. A cross-functional team of users from the engineering group, lab, and one plant was formed for pilot testing of the ELTS. The piloting of the project lasted only three weeks. A number of small improvements were made to the application to incorporate the necessary changes identified during the pilot. Most of the change centered on the usability of the application and not the core functionality. Most of the changes were made to the application in near real-time. Members of the development team worked with users in person and over the phone while viewing the pilot version. As the user talked, the developer made and tested many of the changes. If the changes were accepted, they were installed and tested in the production copy of the database.

The pilot was performed in parallel with the existing test process in order to minimize the impact to the organization and to individual workers. This parallel effort doubled the required work for the participants but lowered the overall risk of the project. The pilot group had to perform their typical duties as well as perform the testing process with the new system. This double-duty caused some heartache with an already busy group, but since the pilot only lasted three weeks, the members gracefully agreed to the extra effort. After the three-week pilot was completed, the application was deemed ready for full implementation.

Another by-product of the pilot was a list of additional functionality to extend the application into other processes. One such recommendation was to extend the UTRN sub-process to include the corrective action process by automatically creating and populating an electronic version of the corrective action form. This functionality was added after the initial version of the application was released.

Roll-Out

Training for the new system was minimal since the user interface was familiar to most users. Wayne remembers:

Since the company used Lotus Notes™ as its e-mail system, it was a familiar interface for our people. Basically, anyone who was familiar with Lotus Notes™ and the product being tested could easily become a user of the system. Once they were on the system, they could easily navigate around the ELTS and receive information.

ELTS was deployed to about 50 users dispersed over seven sites. The users included plant quality personnel, test lab managers and technicians, product engineers, quality engineers, and sales people. Many of the participants needed only view access into the new system. The system provided automatic notification via e-mail for many users when their participation in the process was necessary.

Ongoing Maintenance

T1-Auto kept a small development team intact that maintained existing applications as well as wrote new applications. This team was augmented by an offsite team of Notes™ developers from T1-Auto's outsourcing partner. Given the ability to quickly and easily make changes to these systems, the development team had to balance rapid turn-around with stream of change requests. The team tried to collect requests for each application and make monthly changes. This system provided a reasonable way to quickly update each system while maintaining version control and quality standards. The T1 development team was responsible for developing and maintaining over 100 Notes™ databases. Their costs were comparable to other development groups at the time, but T1 management felt that their productivity and impact was superior.

Project Success Factors

Abdallah Shanti, The CIO at T1-Auto, recognized the benefit the ELTS system brought to the organization.

The ELTS project resulted in a benefit of approximately $900,000 annually through the elimination of redundant work. However, the biggest benefit came from more efficient and traceable process that was introduced at the labs. Parts were no longer lost in the middle of a test; testers stuck to their assigned tasks; and plant managers were able to view a test's progress in real-time. A key fact integral to the effective management of the testing process was, when there were irregularities in the test, the owner of the project was instantly notified. All this added up to better service to the customer.

The estimated software development cost of $18,000 over two months generated an estimated yearly saving of $900,000 in hard benefits, and potentially even more in non-quantifiable benefits.

Another valuable benefit was that the ELTS enabled T1 to begin mapping knowledge about the testing process. A knowledge store was developed that tracked all the information on each test. Included with this information was the unstructured data such as discussions and workflow notifications. No longer would there be conflicts regarding when a part was sent and received by the participants. No longer were problems encountered when production engineers did not respond to failed test. (If they didn't, they knew their supervisor would automatically be notified as a result of the inaction).

For the Plants

The plant quality technicians now had one system for scheduling and tracking the PPT process. In addition, they were given real-time access to test status and were able to quickly look up and answer questions related to tests. Many of the reports they were asked to generate were made obsolete by incorporating the reports into standard Lotus Notes™ views that were accessible to all participants. With the use of color in views and workflow technologies, process irregularities were identified by the system and action was quickly taken to correct problems. Finally, the plant had to spend less time filling out reports and copying data from one system to another. The quality staff in each plant consisted of a very few people, so no staff reductions were required when the system was implemented. Since cutting cost had not been a primary objective, headcount reduction was not an issue. Finally, the task of updating test status and receiving the Test Completion Notification form was eliminated.

For the Lab

The test labs were given access to the PPT schedule and were better able to schedule their operations. The automatic generation and copying of data from the PPT form to the Test Completion and Test Failure Notification forms reduced errors and time. Workflow notification reduced the time it took for test lab technicians to notify product engineers of testing problems. Test lab supervisors were given a system to track all PPT test status and monitor communication with test technicians. By updating the test status regularly, the test technicians had to spend less time answering questions from engineers and plants particularly when inquiries were status checks.

For the Engineers

Product engineers were immediately notified of problems, eliminating one entire day from the process. Engineers were also given ready access to historical test information for each of their products. This information proved to be invaluable when analyzing failure trends over time. For instance, engineers could identify if a failure occurred each year during hot months in the production process. The capability to access and analyze this type of information led to a much quicker resolution for many problems.

Summary

Overall, there were a number of factors that led to the success of the project. Within T1 there had already been some successful workflow projects. Thus, PPT users were open to the ELTS project. Once it proved itself, that it made their work easier and made them more productive, they became supporters of the system. The use of prototyping during the design phase ensured that requirements would be met and provided a vehicle for user involvement. The use of the pilot solidified the project's success. Not only did it provide a proof of concept within the testing environment, but it also enabled fine-tuning of the ELTS requirements.

CURRENT PROBLEMS/CHALLENGES

Although Wayne Russell was very pleased with the new PPT process, he realized that T1-Auto was on a never-ending path to change. There was continual pressure for T1-Auto and other tier one suppliers to improves their processes. They were expected to squeeze out cost while at the same time make their processes more responsive to their customers. That could mean anything from quick response to new OEM product line requirements to testing in the face of warranty claims.

T1 were considering enhancing the ELTS system by adding Web enablement using Lotus Domino™. By completing this enhancement, it would be possible to create a portal allowing direct involvement to T1-Auto's customers. Another possible enhancement was to integrate the ELTS system with the manufacturing scheduling system. This would allow the ELTS system to be updated by automatically canceling tests for parts if they are not scheduled for production. Integration could also allow the initial test documents to be automatically generated to eliminate manual loading of this information. A final planned enhancement is to integrate the testing system with the Product Data Management (PDM) system. This would tie all of the parts design and specification data together with the parts testing data.

REFERENCES

Automotive Information Action Group. (1997). *Manufacturing Assembly Pilot (MAP) Project Final Report*. Southfield, MI: AIAG.

Florio, S. (n.d.). The history of Notes and Domino. Retrieved January 2, 2004, from http://www.notes.net/history.nsf/

Lauer, T. W. (2000). Side effects of mandatory EDI order processing in the automotive supply chain. *Business Process Management Journal, 6*(5), 366-375.

Porter, M. E. (1985). *Competitive advantage: Creating and sustaining superior performance*. New York: The Free Press.

Verespej, M. (2001). Automakers put wheels on supply chains. *Industry Week, 15.*

Workflow Management Coalition. (1995). *The Workflow Reference Model version 1.1* (Document Number TC001-1003). Retrieved October 25, 2002, from http://www.wfmc.org/standards/docs/tc003v11.pdf

APPENDIX

Exhibit 1. Old PPT process

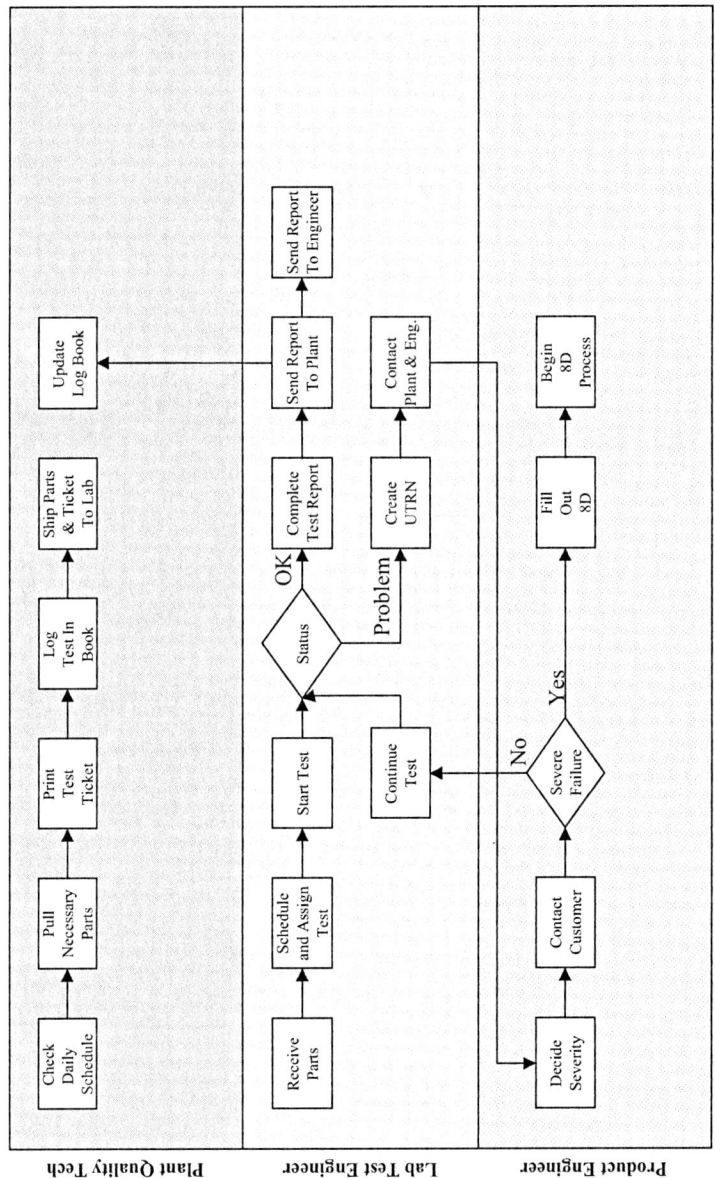

Exhibit 2. Workflow-enabled PPT process

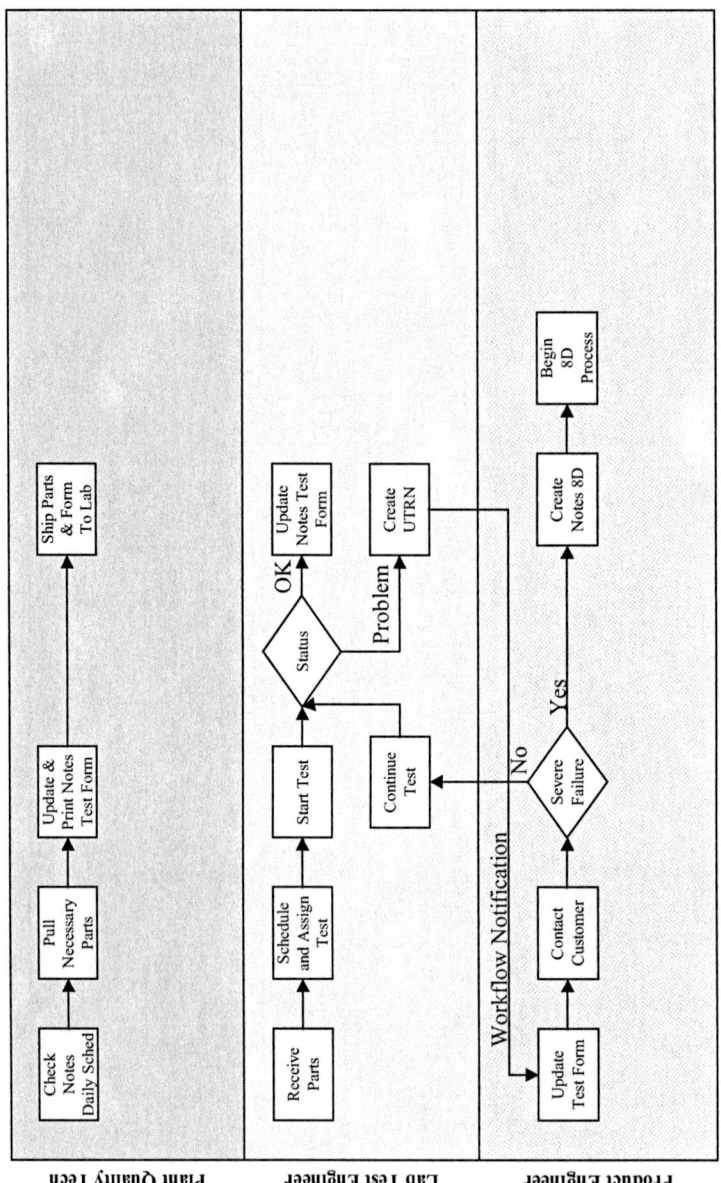

Charles T. Caine is presently VP of IS for Resourcing Services Company LLC in Troy, Michigan. He has 18 years of systems experience from software development, infrastructure design and support, IT consulting, and IT strategy development. He has worked for EDS and Perot Systems supporting clients in the automotive, manufacturing, transportation, and consumer products industries. He received his BS in computer science from the University of North Texas in Denton, and his MBA from Oakland University in Rochester, MI.

Thomas W. Lauer holds a BA in mathematics, an MBA, and a PhD in management information systems, all from Indiana University. He also holds an MA in public administration from the University of New Mexico. He is currently a professor of MIS at Oakland University in Rochester. Professor Lauer has more than 30 publications and has received a number of grants for his research in MIS.

Eileen Peacock is currently associate dean and professor of accounting in the School of Business Administration, Oakland University. She obtained her PhD from the University of Birmingham, UK. Her research interests lie in managerial accounting including activity-based costing, behavioral aspects of accounting, and AIS curriculum development. She has published in a variety of journals and magazines including The British Accounting Review, Review of Information Systems, Internal Auditing, Journal of Cost Management, Journal of International Accounting, Auditing and Taxation, Management Accounting, *and* Review of Accounting Information Systems.

This case was previously published in the *Annals of Cases on Information Technology*, Volume 5/2003, pp. 74-87, © 2003.

Chapter X

Implementing an Integrated Software Product at Northern Steel

Annie Guénette, École des Hautes Études Commerciales, Canada

Nadine LeBlanc, École des Hautes Études Commerciales, Canada

Henri Barki, École des Hautes Études Commerciales, Canada

EXECUTIVE SUMMARY

This case describes the implementation of the payroll and human resources modules of an integrated software product in a large manufacturing organization. The firm is located in a large metropolitan city and system implementation took place following a major organizational restructuring (from a public to a private enterprise) and downsizing (from 10,000 to 2,000 employees) effort. The extensive maintenance required by the existing legacy systems and the high cost of modifying them to address the year 2000 problem motivated the company to acquire an integrated software product from a vendor, and adapt it to the organization. Implementing the software took longer than scheduled and was 35% over budget. Some of the problems encountered include conflicts between the Accounting and Human Resources departments, technical difficulties in building interfaces to existing systems, inadequate staffing of the project team, the IT director who left during the project, and a poorly functioning steering committee.

BACKGROUND

The Steel Industry

In 1997, the North American steel industry consisted of 96 companies (Standard & Poor's, 1997). Reflecting the particularities of the steel making process (see Figure 1), this industry is principally made up of two types of producers: integrated mills and minimills. Integrated producers such as USX, Bethlehem, and Dofasco make use of expensive plants and equipment to produce from two to four million tons of various steel products per year. In contrast, minimills such as Nucor, Birmingham Steel, and Co-Steel Lasco produce 400,000 to over two millions tons per year by melting recycled ferrous scrap in electric arc furnaces to make a limited number of commodity carbon steel products. Originally small-scale plants serving local markets for structural steel products, North American minimills have become major players in the 1990s and compete with integrated producers in most product areas (Standard & Poor's, 1997).

The trends experienced by all players of the industry throughout the 1990s include the globalization of major customers (e.g., automotive, appliances), continued pressure from offshore excess capacity (e.g., Russia, Asia), continued pressure for environmental improvement, rapid and accelerating technological change, and the challenges of alternate materials in traditional markets (e.g., autos, bridge construction). Many firms have responded to these pressures by reinventing steel making through corporate repositioning (e.g., increased customer focus, cost reduction, organizational restructuring), by making large investments in technology and employee skills, and through increased collaborative research efforts (e.g., thin strip casting, process modeling, electronic sensors) (Bain, 1992).

Northern Steel

Established in the northeast as a public company over 30 years ago, Northern Steel was a money losing operation throughout the latter part of the 1980s and in the early

Exhibit 1.

A wide gap in labor productivity remains between minimills and integrated producers. For example, in 1996 Nucor's 6,600 workers produced 8.4 million tons of raw steel, or 1,273 tons per employee. LTV, the United States' third-largest integrated steel maker, produced 8.8 million tons of raw steel with a work force of 14,000; this equaled 629 tons per worker.

Minimills have a labor cost advantage because they typically employ nonunion labor, whose compensation is often directly linked to production and profits. Integrated companies employ union labor which is more expensive over the course of the business cycle.

To survive in the long term, integrated producers need a technological breakthrough that will make them more competitive with minimills by reducing their labor and capital costs. Despite aggressive cost reduction and more flexible labor practices, integrated steel making remains more capital- and labor-intensive than minimill steel making (Standard & Poor's, 1997).

Figure 1. Steel making process (Bain, 1992)

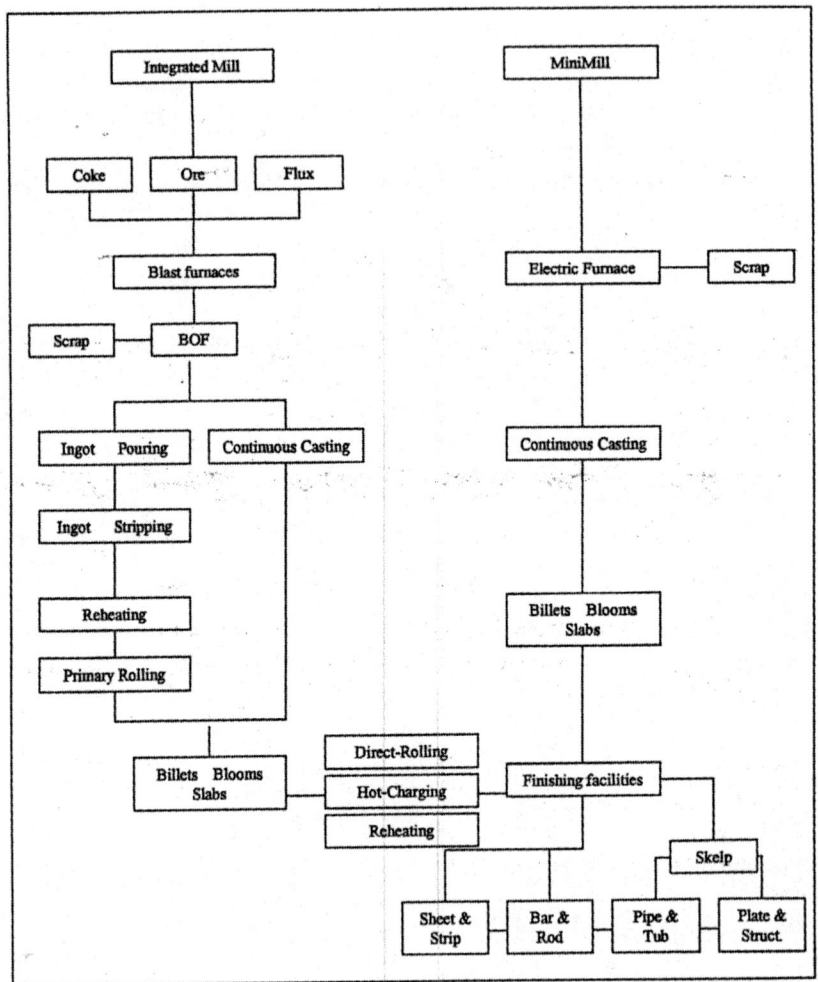

1990s. Employing a workforce of over 10,000 employees, it was sold to an international conglomerate and became a private enterprise in 1994. The new ownership immediately started a process of downsizing the company to improve productivity and increase its market share. Over the next couple of years, numerous facilities and divisions were either sold or closed, and the size of the workforce was reduced to around 2,000. By 1997, a much leaner and customer-oriented Northern Steel had only six facilities that manufactured and distributed around 1.5 million tons of steel products in North America (including sheets, slabs, weld pipes, billets, bars, rods, wires, and pellets). In 1997, Northern Steel's annual revenues were $900M, and more importantly, it had become profitable once again. However, the organizational culture developed over the past decades as a heavily unionized public company proved more difficult to change. The apathy of a slow moving bureaucracy and a nine to five mentality could still be seen rearing its head relatively frequently despite the privatization and the downsizing, and the strong pressures they entailed to improve productivity and to cut costs in all areas.

SETTING THE STAGE

IT at Northern Steel

Since the early 1970s Northern Steel had computerized its main transaction systems including its payroll and general ledger with most programs written in COBOL. With the introduction of microcomputing in the 1980s, many user departments started developing their own applications. Some of these were quite substantial. For example, one accounts payable application which handled over $50M had been developed by a user in the Accounting Department.

During the mainframe years, the IT Department had a budget of around $3M and had a staff of 30. By early 1990s, problems in finding and hiring personnel having the requisite skills to maintain the legacy systems, and the constant difficulties experienced in servicing the old mainframe, convinced Northern Steel to scrap its outdated hardware and to outsource the operation and maintenance of its mainframe systems to a large consulting company. The widespread practice of end-user development had also resulted in a decentralized IT structure. Thus, by 1990, Northern Steel's IT Department was reduced to a staff of only five professionals who were responsible for telecommunications and network management, establishing company-wide software and hardware standards, and for approving their purchase. The remaining IT responsibilities were decentralized to the different manufacturing facilities and functional areas, with many former IT personnel working in, and reporting to, user departments. For example, the staff responsible for payroll systems worked in the Accounting Department and reported directly to the company comptroller. At Northern Steel, IT was frequently seen as a cost item. Many senior executives felt that, in general, the less Northern Steel spent on IT the better off the company would be.

The Payroll System

In 1994, an IT strategic plan was prepared and it budgeted $25M to the updating of all legacy systems and their migration towards newer and more accessible information technologies. Among these, revamping the legacy payroll systems was given top priority in view of several considerations. The Accounting Department, responsible for Northern Steel's payroll, was quite anxious to update the legacy payroll system and applied continuous pressure to prioritize its implementation. The extensive modifications the payroll system needed in order to deal with the Y2K problem added considerable urgency to the problem. Moreover, the existing payroll programs mostly contained spaghetti code written by programmers many of whom had long left the organization. As a result, the payroll system was already costly to maintain and difficult to manage. Frequently, changing one of its parts introduced bugs to its other modules. Thus, making all the Y2K changes to the legacy system was, at best, a scary proposition. Finally, the fact that the employees of Northern Steel were represented by several unions each with a different collective bargaining agreement meant that the payroll system was already complex, and had to be modified every time Northern Steel signed a new collective bargaining agreement.

The idea to replace the existing payroll system was initiated in 1980, and a proposal was obtained from integrated business solutions (IBS), a software development/consulting company. However, Northern Steel's lack of resources at that time had prevented the then public company from pursuing the issue further. Following the privatization of Northern Steel, and according to the 1994 strategic plan, alternative solutions to replacing the legacy payroll system were examined. Initially, outsourcing the payroll was considered. The available products promised improved payroll accuracy and up to 80% reduction in payroll preparation time and offered a wide range of options including tax filing and deposit services, automatic check signing and insertion into envelopes, unemployment compensation management features, and a variety of management reports. While such an alternative appeared satisfactory from a purely payroll perspective, it would do little to improve the existing human resource (HR) systems, many of which were also tightly coupled to the payroll system.

At this time, the director of the HR Department suggested the purchase of an integrated software product to replace the existing payroll and HR systems. In his dealings with a large bank, he had observed integrated business solutions' (IBS) software product being used, and had been quite favorably impressed by the bank's positive evaluation. Since Northern Steel lacked the necessary resources for in-house development, adopting an integrated software product to replace both the payroll and HR systems simultaneously, while initially more expensive and challenging, appeared as a better long-term solution. Given the general consensus about the need to scrap the legacy payroll system and adopt an integrated software product, it was decided to prepare a request for proposal (RFP) and to send it to suppliers.

In June 1995 the IT Department prepared a relatively short RFP document that consisted of 14 points. These provided broadly described requirements (e.g., the modules needed such as payroll, workman's security, employee skills, medical, etc.; the need for compatibility with the planned organizational platforms such as Unix and Oracle, and a client/server architecture; the possibility of modifying the product so as to fit Northern Steel's business processes) and guidelines concerning what information the suppliers needed to provide (e.g., the estimated cost to Northern Steel of modifications that would be needed, the cost of maintenance services, the availability of support from the supplier). Suppliers of integrated software products were contacted and asked to submit offers. The evaluation of the submitted proposals was completed by the end of September 1995.

Whereas other competitors offered software that integrated several business processes, including HR and payroll, IBS sold a "best of breed" product that consisted only of an integrated payroll and HR system but its offer included considerable customer support and service. With strong backing from the HR Department, the Selection Committee chose IBS's offer as being the most suitable for Northern Steel. However, this was not a unanimous decision since the Accounting Department did not agree that IBS's software was the one that most suited their needs. Despite opposition from accounting, the Selection Committee voted to go ahead with IBS and an estimated project schedule and budget was prepared. The budgeted amount included the necessary funds to purchase IBS's payroll and HR modules, including its employee benefits and compensation modules, and to finance a project for their implementation. IBS's pension module was excluded from the purchase in view of the fact that Northern Steel had relatively recently implemented a new pension system. The necessary interfaces between the existing pension system and IBS would have to be built during the project. The project schedule and budget were submitted to senior management for approval and the allocation of the requisite funds.

CASE DESCRIPTION

Senior management approved the Selection Committee's decision, and allocated the requested funds in January 1996. Based on the Committee's recommendation, Steve Collins was assigned as project leader. Working in the Accounting Department and reporting to the comptroller, Steve had been with Northern Steel since 1980 and was well-known across the company. Over the years, he had developed several accounting applications as an end user and had gained experience as a project leader by managing several projects in accounting and other departments. Steve was quite excited about the IBS project. While this was going to be Northern Steel's first experience with the implementation of an integrated software product and Steve's first attempt at managing a large project team, he did not foresee too many problems: "With a package, all we have to do is implement it, and it should not take too long."

To oversee the project, a Steering Committee was formed. Its members included the IT Director Mark Owens, an experienced and well-respected professional, and senior executives both from the Accounting and HR departments. In addition, a five-member project team including two IT Department staff and one representative each from accounting and HR was also created. The selection of each was left to their respective departments.

The next two months were spent in a period of "fit analysis." During this time, a User Committee consisting of 10 HR employees, two accounting employees and one IT staff was also created. This committee's mandate was to examine the IBS software and to recommend a list of modifications to be made to IBS so as to satisfy Northern Steel's payroll and HR information requirements. The modifications and additional functionalities requested by the User Committee were communicated to the project team in an informal manner, and at times, verbally. Examples of the type of changes requested from the IBS software included changing the paycheck format, incorporating the different overtime rate calculations, calculating payroll based on hourly salaries, and preparation of new reports. Based on the requests they received from the User Committee, the project team prepared a written work requisition and subsequently evaluated their feasibility and decided which IBS functionalities to keep and which ones to modify according to the work requisitions.

All too frequently, the modifications and supplemental functionalities requested by the User Committee appeared more like a wish list to the project team. According to Steve, users had difficulty understanding that certain hard to reconcile differences existed between the way things were done at Northern Steel and the way the IBS software was designed. As a result, the project team ended up discarding approximately half of the requests received from the User Committee.

Figure 2.

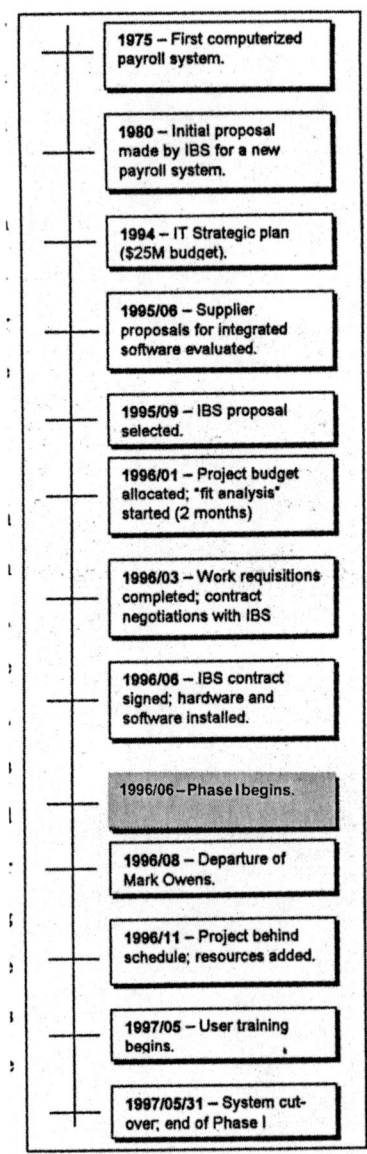

By March 1996 a list of work requisitions was finally arrived at and submitted to the Legal Department for approval. These were approved in June 1996 and a contract for a two-phase project signed with IBS. Phase I of the project, scheduled for completion in February 1997, was started immediately following the signing of the contract. The objective of this phase was to implement the payroll module of IBS as well as parts of its HR system such as its insurance, accidents, and HR administration modules. IBS started to install the hardware and software environment (the network, the server, work stations, Unix and Oracle) and to train Northern Steel's IT staff in their use.

During the first six months of the project, little progress was made. All user members of the project team continued to work full time in their original jobs and so, could not devote much time to the IBS project. Moreover, the prevalent organizational culture meant that many users had little motivation to stay after hours for the sake of the project. In addition, replacing the old systems with IBS meant that a number of jobs would disappear from the HR Department, creating some anxiety and lack of cooperation among the HR personnel. Another blow came in August 1996 when Mark Owens left Northern Steel. Up until then he had been one of the project champions and had been instrumental in resolving several conflicts that had arisen between accounting and HR. His departure also created a certain void in the Steering Committee. Its meetings were gradually held less frequently and eventually disappeared.

An added difficulty which slowed progress stemmed from the technical challenges encountered in building interfaces between IBS and the existing pension system. Moreover, during this period Northern Steel was in the process of negotiating a collective bargaining agreement with one of its large unions. Preoccupied with this process, many senior executives were unwilling to spend too much time on the IBS project. For example, when Steve tried to discuss the project with the director of the HR Department, the response he got was "Listen, I am trying to negotiate salaries for the next five years and we're talking over a billion dollars here, so don't bug me."

By November 1996 the project had fallen behind schedule by several months and Steve was quite concerned with the slow rate of progress. He finally convinced senior management that remedial action was needed and this resulted in the addition of two more members to the project team, one from accounting and one from HR. In addition, all team members were now assigned to the project on a full time basis. These changes improved the situation considerably, and the project started to advance at a more rapid pace.

The next six months were spent with making the necessary modifications to the IBS software and developing its database. While this activity was IBS's responsibility, all data being converted had to be validated by the users concerned. Their input was necessary since they knew all the compensation and benefits rules and procedures that were being incorporated into hundreds of tables in the new database. However, this process took longer than anticipated. The old payroll and HR systems consisted of numerous poorly structured files some of which contained over 200 fields. Cleaning them up and making sure all the data they contained would be conserved in Oracle's relational tables was in itself a challenging task. According to Steve, an added difficulty came from the user representatives in the project team who had problems grasping elementary database concepts and needed a lot of help with the design.

Moreover, the user members of the team were frequently unable to decide without consulting other users. During these consultations it became clear that many of the work requisitions discarded earlier in the project were actually needed. For example, at Northern Steel, salaries were calculated on a daily basis but the IBS software determined paycheck amounts by first converting these into hourly rates which were then summed up over two-week periods. This process sometimes resulted in rounding errors of one cent. However, during consultations it was discovered that not only any such deviations were totally unacceptable to the users (and to the unions) but that paychecks had to be exact to three decimal points. This meant that there was no choice but to make the necessary changes to the IBS software to satisfy this particular requirement. Numerous such requirements resurfaced as Northern Steel's business processes were once more passed under scrutiny. As a result, many previously discarded work requisitions had to be reintegrated into the modifications already made to the IBS software. During this period, Steve frequently felt frustrated by the lack of flexibility exhibited by the user departments who approached the whole process with a "Today we are at this point and we won't go back" attitude. In addition, the project was being delayed further and since all budgeted funds had been spent, additional funding had to be requested from senior management. Steve felt the pressure mounting.

Nailing down all the requirements, completing the database design, writing and testing more than the thirty data conversion programs needed to populate the IBS tables took the project well into May 1997. At this time, preparations to cut-over from the old systems to IBS were undertaken. The accounting and HR personnel were trained over the first three weeks in May through a 3-day training program that was repeated to groups of users. Those who participated during the project were trained first, and in turn, they trained the other users.

While running the old and the new system in parallel was considered as an option, it was discarded due to the approximately 20 additional personnel and the associated expenditures it would have required. All the data from the old systems had to be converted and transferred to the IBS databases rapidly and without error. This requirement was mainly due to the unions who were quite inflexible with regards to potential delays or errors in the issuing of worker paychecks. Despite firm promises that all errors would be eventually corrected, they refused to accept any delays or mistakes in the paychecks. This meant that the all conversion and cut-over activities would have to be completed over a weekend, putting additional pressure on the project team.

The cut-over took place during the last weekend in May. Tests were run first with one employee, then with 50, and finally with all employees. During these tests, both the old system and IBS had to be run in parallel and this turned out to be more complicated than anticipated. In addition, some of the conversion programs were found to contain several errors, all of which had to be corrected on the fly. It all made for a very stressful Sunday. However, by late Sunday night all modules were working properly and IBS was up and running.

After a four-month delay from its scheduled completion date and at a cost that was 35% over budget, the IBS system was put into use at the end of May 1997. During the year that followed it has functioned without any major problems. The users have no major complaints regarding functionality or ease of use.

CURRENT CHALLENGES

In August 1998, Steve was wondering whether or not to start the second phase of the IBS project. During Phase II, scheduled to take six months, the remaining HR modules of IBS which include the medical, workman's security, and employee skills systems were to be implemented. These modules were considerably more complicated than those installed in Phase I and Steve was not yet sure that enough time had elapsed to allow everyone to catch their breath and to get used to the IBS software.

REFERENCES

Bain, T. (1992). *Banking the furnace: Restructuring of the steel industry in eight countries.* W.E. Upjohn Institute for Employment Research.

Standard & Poor's. (1997, July 24). *Metal: Industrial survey.*

ENDNOTE

[1]   In order to maintain confidentiality, the name and geographical location of the companies concerned as well as the names of the individuals involved in the case have been disguised.

Annie Guénette has an undergraduate degree in information systems from the École des Hautes Études Commerciales in Montréal where she is also completing her MSc in information systems.

Nadine LeBlanc has an undergraduate degree in information systems from the École des Hautes Études Commerciales in Montréal where she is also completing her MSc in information systems.

Henri Barki is a professor in the Information Technologies Department and director of research at the École des Hautes Études Commerciales, Montréal. His publications have appeared in Information Systems Research, Information & Management, INFOR, Management Science, *and* MIS Quarterly.

This case was previously published in the *Annals of Cases on Information Technology Applications and Management in Organizations*, Volume 1/1999, pp. 60-67, © 1999.

Chapter XI

The Relation Between BPR and ERP Systems:
A Failed Project

David Paper, Utah State University, USA

Kenneth B. Tingey, Utah State University, USA

Wai Mok, University of Alabama in Huntsville, USA

EXECUTIVE SUMMARY

Vicro Communications (we use a pseudonym to mask the identity of the organization) sought to reengineer its basic business processes with the aid of data-centric enterprise software. Vicro management however made the mistake of relying completely on the software to improve the performance of its business processes. It was hoped that the software would increase information sharing, process efficiency, standardization of IT platforms, and data mining/warehousing capabilities. Management, however, made no attempt to rethink existing processes before embarking on a very expensive implementation of the software. Moreover, management made no attempt to obtain feedback or opinions from employees familiar with existing business or legacy systems prior to investing the software. Unfortunately for Vicro, the reengineering effort failed miserably even after investing hundreds of millions of dollars in software implementation. As a result, performance was not improved and the software is currently being phased out.

BACKGROUND

Vicro Communications is an international provider of products and services that help companies communicate through print and digital technologies. As a leading supplier of document formatted information, print outsourcing and data based marketing, Vicro designs, manufactures and delivers business communication products, services and solutions to customers.

Vicro operates in complementary marketplaces: Forms, Print Management and Related Products, which includes Label Systems and Integrated Business Solutions including personalized direct marketing, statement printing and database management. With more than a century of service, Vicro owns and operates over 100 manufacturing and distribution/warehousing facilities worldwide. With approximately 14,000 employees serving 47 countries, it provides leading edge, high-tech solutions that enable companies to adapt to the dynamics of change. Vicro is a large company with approximately 2.45 billion in 1999 and 2.26 billion dollars in 2000 revenue. The appendix contains additional financial information.

Vicro provides consulting, project management, reengineering and distribution of high volume, customized communications to its clients. It delivers personalized, easy-to-read documents intended to facilitate a positive impression on an organization's customers. Its reengineering and redesign services intend to ensure that an organization's business communications have high quality and clarity.

Equipped with the latest print and digital technologies, Vicro has become a market leader in managing critical business communications. It offers products and services that include statement/billing, cards, government noticing, policyholder and plan member communication, and database marketing.

SETTING THE STAGE

Vicro is a conservative organization in that (it purports that) it doesn't embrace "bleeding edge" technology to obtain a competitive advantage. It has been in existence for many years and depends on a good reputation with its clients and positive "word-of-mouth" to attract and maintain its client base. Hence, Vicro wants to deploy proven technology that will help satisfy and exceed customer requests and expectations. The major technologies utilized include mainframe systems to store centralized production data and serve the core applications of the business and client-server technologies for development and daily operations such as e-mail, file transfer, Web access, etc.

Vicro Communications was chosen as a case study because the authors knew that it had experimented with business process reengineering (BPR) to streamline its operations and that information technology (IT) was intended as a key facilitator. Since we were interested in why BPR efforts (facilitated by IT) succeed or fail, and had contacts at Vicro, we initiated this research project. We chose the case study approach to gain a rich understanding of what really happened and why events unfolded as they did.

Business process reengineering (BPR) was used as a literature base to frame the study. The BPR literature reveals that many BPR efforts are unsuccessful. Based on this premise, it seemed a good research undertaking to explore why this is the case.

A synopsis of salient BPR literature is included as a resource for the reader. In the early 1990s, business process reengineering (BPR) came blazing onto the business stage as a savior of under performing organizations. Early advocates of BPR (Davenport, 1993; Hammer & Champy, 1993; Harrington, 1991) touted BPR as the next revolution in obtaining breakthrough performance via process improvement and process change. However, BPR has failed to live up to expectations in many organizations (Davenport, 1993; Hammer & Champy, 1993; Kotter, 1995; Bergey et al., 1999). Some of the reasons include adoption of a flawed BPR strategy, inappropriate use of consultants, a workforce tied to old technologies, failure to invest in training, a legacy system out of control, IT architecture misaligned with BPR objectives, an inflexible management team, and a lack of long-term commitment (Bergey et al., 1999). As one can see from this list, it seems obvious that many organizations failed to realize the scope and resource requirements of BPR.

Patience is another key aspect. BPR initiatives can lose momentum as managers face limited resources, slow pay-off, diminished employee enthusiasm, and increased resistance to change (Harkness et al., 1996). When short-term BPR results are not obtained, management tends to lose interest and top management is less willing to allocate new resources to the project (Paper, 1998a). One solution to this problem is targeting a BPR initiative that is "manageable" and that will garner quick results (Paper, 1998a). Another solution is for top management to be actively involved in the effort (Kettinger et al., 1997).

Assuming that the organization understands the scope of BPR and is patient, the project still may not succeed without careful consideration of the type of process initiative. Paper (1998a) argues that the BPR initiative should be driven by a focus on the customer, strategic business issues or senior management directives. Failure to do so greatly reduces the chances for success.

IT has been touted as one of the key enablers of BPR (Davenport, 1993). However, IT can be one of the biggest obstacles if not properly aligned with business objectives (Broadbent et al., 1999). The heritage of a legacy system can contribute greatly to BPR failure (Bergey et al., 1999). Many legacy systems are not under control because they lack proper documentation, historical measurements, and change control processes (Bergey et al., 1999; Paper, 1998b). Due to the scope and complexities inherent to a typical legacy system infrastructure, it should be treated with the same priority as the cultural and organizational structures when undergoing process change (Broadbent et al., 1999; Clark et al., 1997; Cross et al., 1997).

Although the proliferation of research articles has been abundant, research findings have provided limited explanatory power concerning the underlying reasons behind BPR failure. To address this problem, several recent in-depth case studies have appeared in the IS literature to add explanatory power to this issue (Broadbent et al., 1999; Clark et al., 1997; Cooper, 2000; Cross et al., 1997; Harkness et al., 1996; Paper, 1999). However, much more work of this type needs to be undertaken. Hence, we embarked on a case study to gain insights into the IT-enabled BPR phenomenon.

CASE DESCRIPTION

Vicro Communications was under-performing according to its board of directors and major stockholders. That is, its market share was declining, its revenues were not growing as expected, and its share price was plummeting. The stakeholders agreed that drastic improvements were needed. It thereby decided that reengineering of its basic business processes was the correct path to undertake. In addition, it was agreed that the BPR efforts would be facilitated by data-centric enterprise software. The stakeholders believed that the power of IT would complement BPR efforts. We mask the name of the vendor by calling the software high profile technology (HPT). Since Vicro is conservative in terms of IT investments, it chose enterprise software that had been in existence for over 30 years with worldwide name recognition. It was hoped that this software would facilitate automation of redesigned processes while improving overall system performance in terms of increased information sharing, process efficiency, standardization of IT platforms, and data mining/warehousing capabilities.

Although the software investment was very significant, top management felt that it was a good decision. Top management based the software investment decision solely on vendor promises, market share of the software in its market niche, name recognition, and CEO endorsement. No effort was made to obtain opinions and/or feedback from employees at the process level or those engaged in existing systems development and maintenance. Moreover, the state of legacy systems and processes were never considered as a factor in the decision.

In short, Vicro management attempted to solve its performance problems with expensive enterprise software. There appeared to be a communication breakdown between what the stakeholders wanted and the decided course of action because the original intention of the stakeholders was to complement BPR with IT, not to depend solely on an IT solution to solve the problem. The communication breakdown extended even further. Top management mandated the plan to use the enterprise software without interactions with other managers and process workers. Furthermore, Vicro made no attempt to align business process changes with business objectives or IT objectives. The BPR literature agrees that this is one of the biggest reasons for failure.

The remainder of the case description attempts to illuminate for the reader what happened to the reengineering effort. We describe in detail how we gained and analyzed the data for the case. We adhered to a phenomenological approach. The strength of this approach is that it allows themes to emerge from the case over time. These emergent themes then become the basis for classification of the data so that it can be analyzed in an organized fashion.

Data Analysis

The phenomenological approach called for us to identify a set of patterns or themes that emerge from the data. To accomplish this we set up a series of interviews that were subsequently transcribed and analyzed. The formal interviews were conducted with our main contact. Informal interviews were conducted with several site employees over time. Each researcher iteratively combed through the transcripts for several hours to allow a set of common patterns or themes to emerge from the data. Each theme was color coded to facilitate easy identification in the transcript. The colors were negotiated and agreed upon after the first iteration. After each iteration, (we did three complete iterations), the researchers met to compare themes. After the final iteration, the researchers negotiated a set of *themes* or categories that are laid out later in this section †technology usage, process improvement, HPT adoption, CEO mandate, enterprise integration, and resistance to change. Using these themes (which actually became categories to facilitate organization of transcribed data), the researchers were able to more easily analyze the case as pieces of the transcript naturally fell into one of the categories. Placement of the data from the transcripts into one of the themes was based on the experience and judgment of the researchers and verified by member checks with the main respondent.

The general theme of the formal interviews was to obtain information about the use of breakthrough technology in the BPR process. The interviews began with general questions concerning the use of computers to manage processes. The interviewee was then encouraged to divulge his opinions and ideas related to process redesign and the use of high profile technology (HPT) to facilitate or inhibit such initiatives (we use HPT to mask the name of the software vendor). The interviews were audio taped to maintain the integrity of the data and to allow proper analysis of the transcripts produced. The interviews were framed within the context of technology usage and process improvement to provide an easily understandable context for the interviewee and guide the discussion. Informal interviews were conducted with several employees onsite. These interviews were done on an ad-hoc basis, that is, we conducted them when we saw an opportunity to do so. Since we have close ties with Vicro, we were able to informally speak with several employees about the BPR effort and its relationship with HPT.

The initial contact that enabled entry into Vicro Communications was Ron Dickerson (the name has been masked to protect the respondent from reprisal). Ron is the National Manufacturing Systems Project Manager. He is also one of the key facilitators of reengineering and new technology initiatives at Vicro. Ron was the major direct player and contact in the case, but we did meet and speak with several users, project managers, and business managers on an informal basis (informal interviews). Ron is located at the headquarters of the technological communications division of Vicro. As such, this site is responsible for streamlining business systems and facilitating process reengineering across the organization.

The first contact with Ron was a phone interview on December 4, 2000 to garner preliminary information and discuss the merits of the research. Another phone interview was conducted on February 2, 2001 to discuss the fundamentals of BPR at Vicro and some of the major obstacles to success. Preliminary phone interviews were conducted to set up a time for a formal 'sit down' interview and acclimate the interviewee to the 'essence' of the topics to be discussed. The first formal interview with Ron was conducted on April 16, 2001. We administered two additional formal interviews on May 29, 2001, and July 1, 2001.

In the remainder of this section, we summarize (and synthesize) the analyzed data by classification theme. The information garnered from the data is based on in-depth analysis of the recorded transcripts and other data collected from telephone and e-mail interactions. Each section is a negotiated theme wherein the data summarized from the transcripts is presented and discussed. Included are pertinent segments of the respondent's comments (comments are indented) followed by an explanation of how these comments relate to the theme.

Technology Usage

This theme relates to the importance of computers at work. That is, are computers important for accomplishing tasks, activities, and objectives within delineated processes?

In my job, [computers] are essential. I live and die with my computer. If used correctly ... [computers] can help. Used incorrectly, they can be a burden. We have ... personal computers ... I have a laptop because I travel a lot and take it with me everywhere I go ... The plants have mainframes where all of their data is fed to clients ... a lot of servers that store and work our data, but the facilities in particular use mainframes ... so we have kind of both worlds [personal computers connected to servers and mainframes].

In his job, Ron cannot get by without computers. He uses them for data collection, analysis, reporting, communicating, and documenting. The nature of the business Ron manages is communication services. Hence, the business makes demands on technology in order to facilitate data collection, storage, migration, reporting, communication, etc. Computers are critical in facilitating day-to-day operations as well as data-centric problem solving. Computers are important because they store the business data of the organization. Discussions with other employees on an informal basis were consistent with Ron's assessment. Everyone we spoke with believes that computers are vital part to accomplishing daily routines and assignments.

Process Improvement

This theme relates to the relationship between BPR and data-centric enterprise technology. That is, does technology facilitate or inhibit BPR and, if so, how does it do this?

The idea [of] enterprise system[s] is that what someone does helps someone else down the line and that information is fed to them. This is not always the case. Our biggest problem was that we were not willing to change our processes ... [When we got HPT] we ended up trying to modify the software to fit our processes which was a horrible approach ... we didn't change anything and in the end we ended up bolting on hundreds of different systems ... because no one was willing to change and they wanted to keep doing the same process.

The relationship between BPR and technology was essentially nonexistent. There was no consideration for the existing process prior to implementation of HPT. No attempt was made to design new processes or redesign existing ones to match the procedures of HPT. That is, there were no synergies between the business objectives and the technology. In addition, Vicro Communications programmers were not allowed to alter the enterprise system (HPT) to match existing processes. This is not uncommon when purchasing enterprise software. Looking at a common desktop operating system can draw an analogy. In most cases, source code is proprietary and is therefore not open or shared. Vendors of these types of products often force compliance to their rules as terms of the purchase. In this case, HPT did not enable information sharing across the enterprise.

A lot of people sent up tons of red flags, but when it came down to crunch time they had a lot of pressure from up above to get this in ... [existing processes] were working reasonably well, but there was a lot of room fro improvement. They could have eliminated a lot of redundant steps. There were a lot of manual processes. Even just automating it would have helped to some degree, obviously it wouldn't have changed the process, but it would have taken maybe the manual labor out of it.

HPT has what it calls "best practices," but its best practices are many times too generic. Vicro Communications is in the communications business and its best practices should be based on the best performing organizations in its industry, not on those dictated by HPT. Top management decided to implement HPT without consideration for: (1) the existing processes, (2) how existing processes could be redesigned, or (3) the match between the enterprise software's view of best practices and the best practices of the industry within which Vicro Communications operates. Process workers knew that there was a mismatch between their processes and the ones dictated by HPT. However, they were powerless to resist the mandates from top management.

HPT Adoption

This theme relates to the adoption of HPT (and the issues surrounding the adoption of the technology itself). That is, why was HPT adopted and for what purpose?

We do mailings for our clients and we bring in millions of dollars ... to cover the cost of mailing ... We bring that in and we just sit on it ...We love to have the cash flow and the float on the cash ... but ... its client funds and it's a liability. There [is] no way to handle [this] in HPT ... we had a system that was written 12-13 years ago to track those funds, and then they had to be manually keyed into the GL, and system is still in place ... we modernized [the program], but it is still the same system and we automated it somewhat so that instead of hand-writing journal vouchers to be keyed into HPT, now the system will automatically kick them out. Someone still has to take them now from this home-grown system and key them into HPT, because there [is] no interface.

HPT was adopted to streamline processes and standardize databases on one platform. However, the "best practices" built into the software did not match the existing processes. The result is that legacy systems are still in place to handle many processes that are unique to Vicro Communications. HPT is not flexible enough to handle customization of processes. Instead of improving process flow, HPT actually doubled activity because it is running and the legacy systems still have to operate as they have always done. The bottom-line is that HPT didn't work as planned and the postage funds example (mailings) shows that business could not have been conducted using HPT for this process.

We got into time issues ... do we want to spend all of this time investigating it and coming up with a new ... process ... or is it easier just to keep the old process and try to ... bolt on to HPT [and] dump the raw numbers in? ... or instead of actually doing it with HPT, keep doing it the way we were doing it and just dump it in there ... the training expense with HPT was unreal.

The time commitments to learn how to use the software and apply it to existing processes are prohibitive. HPT is not user friendly. It also has a tremendous learning curve. Further, "best practices" built into HPT do not align well with Vicro Communications existing processes. HPT is not a flexible software tool.

You almost have to become an expert in one part of the finance module ... not only in just [one] module, but one part of a module. The human resources for that are unreal.

There are over 20 modules in HPT. Just one part of one module takes tremendous time and practice to master and gain expertise. Further, mastery does not guarantee that the best practice will work for a given process. That is, once a module is mastered, it may not be useful for automation of a process that doesn't fit the rules of the HPT best practices. In short, HPT is not easily customizable and it forces its idea of best practices on an organization regardless of industry or business.

CEO Mandate

This theme relates to the HPT mandate called for by the CEO to implement HPT. That is, why was the mandate forced upon the organization?

The reason why HPT is so marketable is because it says "we will force everyone."

HPT appeared to be an ideal solution to BPR problems since the HPT vendors advocated its ability to standardize all processes on one platform. Assuming that HPT is completely implemented, it effectively forces everyone to use a standard. Thus, a non-technical CEO can easily be tempted to opt for solutions like HPT. Of course top management found out that the HPT solution was not as effective as promised.

There were time constraints from above, they had spent a lot of money on [HPT], and the board of directors was saying, we want to see some results, so just get it out there, which didn't leave time to investigate and change processes. Instead, we took the old processes and the old systems in a lot of cases [and] just took the numbers from them to dump them into HPT. There was some [tension] ... between divisions. Our division does it this way and we don't want to change, we like it and the other division says, we do it this way, so there was obviously some battle of wills.

Since the boards of directors (including the CEO) have invested hundreds of millions of dollars in HPT over the past several years, they wanted results. However, top management had no real experience with BPR or data-centric enterprise software. The CEO and board may have had a good understanding of the business and industry within which Vicro Communications competes, but it did not understand the fundamentals of IT-enabled process flow redesign.

In our division, we were working on our own ERP [enterprise resource planning] system ... we had spent a number of years developing a data collection system that collected data from the production floor, employee hours, machine hours, pieces, feed of paper ... actually it was a pretty good system ... when HPT came along, they [top management] just put it on hold ... because HPT is going to replace it ... now that HPT out of the hundreds and hundreds of people that were employed just for that [HPT development], they are down to four or five people. Guess what we have been doing for the last year? Updating the old system again for our division, we are updating it [legacy systems], we are back to it ... we are now putting it into one of our plants.

Although the management reporting, accounting, and production systems were working pretty well, HPT was purchased and implemented to replace them. Hundreds of HPT consultants were brought onsite to implement HPT. Over time, it was found that HPT wasn't working as promised. Hence, the number of HPT consultants was drastically reduced and Vicro is actually moving back to their legacy systems and upgrading them. That is, they are trying to phase out HPT.

One major problem facing Vicro Communications over the past few years has been change of leadership. The current CEO is the third one hired in the past few years. He was therefore confronted with the HPT problems when he assumed office. He had two choices; he could continue to support HPT or he could phase it out. Considering the lack of effectiveness even with massive amounts of budgeted resources, his choice to phase it out was not unexpected by the organization.

[The CEO] would just assign someone to go [look at quality of processes] ... what ended up happening is that 70%-75% of our company is our forms division ... the HPT team ... dominated ... our forms division ... we ... got told, these are the practices you will use ... they never did address our issues and the differences from our division vs. the other divisions ... [after failure of HPT] ... the stop got put on and it got stopped and we went back to our home-grown systems ... accounts payable [is] ... pretty much the same process in all divisions, but other components ... manufacturing, our sales force are different ... so much is different once you start looking down into the different divisions, down to those finite levels, it never got that far.

The CEO never looked at the state of the existing processes. There was never an assessment made concerning which processes were working well and which were not. Further, there was never any analysis of processes across divisions. That is, there was never any concern for differences in processing from one division to the next. Generic processes like accounts payable are pretty much the same, but most processes are very different.

The new CEO ... who came in December [2000] – his goal was $100 million in savings [and cost reductions] this year ... when you consider that we spent $280 [million] ... on HPT over about a three-year period ... [we could have already met the goal in savings] ... the CEO at the time [prior CEO] ... threw a bunch of money into HPT ... halfway through the HPT project, he was sent packing. And they brought in a new CEO. He was there for less than a year, but he just kept dumping money into HPT as if it were still the answer ... and now, he's gone and they sent the second CEO packing and brought in their own guy – who is now slashing and cutting and chopping [with no regard for process quality].

The CEO who originally bought in HPT was fired because the software was draining money from the organization with no visible results. The next CEO was an advocate of HPT and promised that results would be forthcoming soon. However, he was fired in less than a year. The most recent CEO has embarked on a cost-cutting strategy. He doesn't seem to be concerned with process quality, that is, the cost cutting is 10% across-the-board regardless of productivity. Both formal and informal interviews revealed that this has done little to improve overall performance, but has drastically decreased employee morale.

Enterprise Integration

This theme relates to the efforts at Vicro Communications to promote enterprise information sharing, standardization of processes, increased efficiencies, and process improvement. That is, how is the environment being changed to promote these efforts?

HPT to me is a technology solution to a business problem rather than a business solution to a business problem ... [The top management] solution was, lets throw a bunch of money into IT — that'll solve it. [To handle integration with so many home-grown systems] you need to pass through work ... the only thing we have is at the month-end close, we just feed up the final numbers ... the only integration between divisions is accounting — we just roll the numbers up at the month end and quarter ends and year ends ... the idea was initially we need to have more integration ... if we can have everyone in a centralized shared service of purchasing ... we can have more purchasing power [and more information sharing].

HPT was brought into Vicro to facilitate enterprise integration. Top management however failed to grasp the essence of the problem. HPT is a technology solution that may be appropriate for some organizations, but it is not flexible enough to allow customization. Further, business processes should be engineered based on business objectives prior or at least in conjunction with IT implementation. It appears that HPT did nothing to facilitate enterprise integration. It actually worsened the situation because it drained $280 million dollars in cash from the organization that could have been put to better use.

The company was having legitimate problems. The forms division just wasn't profitable and it was a huge chunk of our business and we just didn't keep up with the computer age ... People just don't need pads of forms anymore to write sales and pricing ... so when the business was going bad, their [top management] solution wasn't [to] reevaluate what we are doing with our forms division, it [was] lets throw a bunch of money [at the problem].

Vicro Communications produces forms for a variety of organizations. However, the organization failed to realize that 'paper and pencil' forms would eventually be replaced by computer or Web-based forms. When profit margins for the forms division began plummeting, the solution was to invest in HPT rather than looking at rethinking the business.

As far as integration between divisions like HPT was going to give us, I would say that is completely dead. I think we'll just keep rolling up division number to get the final numbers and go from there.

Ron' division has effectively abandoned HPT and gone back to the legacy systems they used before the technology was adopted. It appears that HPT did absolutely nothing to increase enterprise integration.

Resistance to Change

This theme relates to internal resistance to change brought on by the HPT process reengineering initiative. That is, how are people reacting to the implementation of HPT as a process improvement tool?

Our biggest problem in our division [is that] ... we have all of these little plants [that] ... have been allowed to do a lot of things for themselves ... So now we try to bring a system in [HPT] and we're forcing them to change some process — if we use Logan [and] we have decided that they are doing it best and try to tell the plants in Chicago and out in Maryland and Connecticut that they need to do it this way, huge resistance [occurs] and the hardest thing in our group [Logan, UT plant] is [that] we haven't had the support at the top.

HPT was forced upon all divisions without input from workers. Further, best practices are defined from a plant that is doing well, like Logan, UT. Other divisions are then told to use these "so-called" best practices, but no real support is given from top management to enforce them. Resistance to change is therefore very strong because there is no real punishment for failing to adhere to the best practices and there is no employee involvement in the effort to obtain buy-in.

No one's had the balls to tell them [other divisions] that this is how you are going to do it. They [other divisions] can stall for months and not implement it. We will get a system implemented and they [other divisions] will kind of halfway use it because no one has said this is how it is, it is not optional ... It has been very frustrating from our standpoint even within the division

Actually, no one had the power to tell other divisions what to do. Top management endorsed HPT, but did not actively pursue implementation on an enterprise-wide basis.

[When each CEO] actually really did stress it [HPT} ... there was the least amount of resistance. It was said, this is it, [and] it is coming in.

When one of the three CEOs was actively pontificating that HPT was going to be the standard, resistance was less profound. The problem was that each CEO would stress HPT for a short time and then get distracted by other business. Once the CEO pressure was off, resistance to change increased dramatically.

Sometimes our division [Logan] is kind of looked on as the maverick division, but no one comes down on it because it has also been very profitable ... It is still frustrating ... because we will [build] a really nice system and one plant will use it because it's the greatest thing since sliced bread and the other plant—they might just keep doing their manual process or whatever they are doing because they have been given that leeway.

Highly effective processes can be either used or mirrored in other divisions with active top management support. We can see that in this case top management was not actively involved in identifying effective processes. They just bought into HPT with the hopes that it would solve all of the organization's problems. Hence, effective pockets of well-designed processes and systems in support of these processes could never really impact the entire enterprise.

Even if you do change [processes] ... with HPT it [is] hard for [people] to let go of their old processes ... [people] didn't want to let go of the old numbers from the old legacy system ... [people] had worked there for 18 years and the customer's account number was this. Well, that format didn't fit HPT so there's a new number, so what was happening was that ... in some of the text fields in HPT they would still type in the old number, so that they could run analysis by the old number ... You're pretty stuck, we weren't able to use the same account numbers that the customers had had forever ... it was very stressful to the people ... very stressful.

Long-time veterans of the company were used to using specific customer numbers for clients. HPT was so inflexible that it would not allow the existing numbers to be input in its databases. Resistance to change was thereby greatly increased because people had internalized these numbers and couldn't understand why they had to change them. The original effort was effectively doubled because people were forced to use HPT, but actually used the old legacy numbers for reporting purposes. Hence, HPT did not really add any value to the process.

[HPT] cost a lot of money, and I said, for our division, we have the world's most expensive AP [accounts payable] system ... for our division ... we went back to our home grown system.

The sarcasm here is obvious. From the $280 million dollars spent on HPT, Ron' division is only using it for accounts payable (AP) and this is only because they must show top management that they are using the product. It is well known throughout the organization about failure of HPT and this has done little to quell resistance to change. In fact, it appears the top management cares little for employee opinion about ongoing efforts.

CURRENT CHALLENGES/PROBLEMS FACING THE ORGANIZATION

This section describes current issues facing the organization. It also develops a useful set of factors that should be useful to other organizations facing similar problems.

Current Issues

The BPR effort did not produce significant performance improvements and HPT did not live up to vendor promises. From our discussions with Ron and informal interactions with other employees, it appears that the BPR effort was not well designed or planned by top management (we don't think that there was a plan). The CEO and the board seemed to embrace BPR as something that they had to do to keep competitive rather than as a holistic method to transform the organization. That is, BPR was fashionable rather than substantive.

Ron was a key player in the BPR effort, but was not able to convince the CEO of the scope required for successful transformation. Effective BPR requires people to understand the process and the role technology plays in the transformation effort. It also requires a lot of capital (which will make a big dent in the operating budget). In addition, it requires a participatory commitment from top management. Top management cannot delegate BPR. It must be actively involved in its planning, design, and deployment. In Vicro's case, BPR was delegated to people like Ron. Ron is a capable executive, but only has clout within his domain. He did not have the tremendous political clout required to change organizational processes and battle resistance to change on an enterprise basis. The only parties with enough power to implement real enterprise-wide change are top managers.

In terms of the enterprise software, Ron was never consulted about the investment in HPT nor was he ever a part of the organization-wide plan to implement the software. This was counterproductive considering that Ron is responsible for many of the enterprise systems in his area. Top management trusted the software vendor to plan and implement HPT to improve performance (even though the vendor has no experience or understanding of the Vicro business model). Ron knew that this plan was flawed, but was not consulted. Moreover, many other key people at the process level were left out of the software decision. By not involving employees at all levels of the organization, resistance to change is bound to increase. At Vicro, people resisted HPT because they didn't understand its complexities and they were never consulted for their opinions about how existing processes and systems work. Hence, there was a mismatch between what the existing processes are really doing and what the encapsulated processes (best practices) within the software itself do.

HPT has a set of business processes built into the logic of the enterprise system it touts. The vendor calls these processes "best practices." There were never any discussions (at any level within the organization) with the vendors to see if its "best practices" fit with what Vicro really wanted to accomplish. As such, HPT attempted to force its "best practices" onto the Vicro business processes it was supposed to support. Since Vicro could not practically change its customized business processes to fit HPT, performance wasn't improved and it had to resort to using its legacy systems to keep the business going. In the end, Vicro spent 280 million dollars on software that did not work and thereby didn't improve business performance. As a result, enterprise integration wasn't improved.

Currently, Vicro is only using HPT for accounts payable (AP). This means that Vicro has maybe the most expensive AP system in the world. This debacle has caused very few organizational changes except for a revolving set of CEOs (at least most of the blame was put in the right place). Since the budget for this software is approximately 10% of total revenue for year 2000, it may put Vicro's future viability in jeopardy. We got a sense from Ron that this might be the case. He (as well as many other Vicro employees) believes that they face a precarious future with the company as a result of the BPR/HPT failure.

The reasoning behind adopting HPT in the first place was to better integrate the enterprise in terms of information sharing, reporting, standardization, and effective processes (BPR). From the case we saw that HPT was a complete failure. However, from the themes we were able to garner a set of factors that can vary from one organization to the next. Hence, we believe that analyzing these factors can help other organizations better deal with enterprise BPR and system adoption success.

Contributions to BPR Literature

Further analysis of the data by classification theme enabled us to generate three "super themes" — immersion, fluidity, and top management support or mandate. One of the principles of phenomenology is to continue classification until it cannot go any further and simplify as much as possible. This process of simplification also establishes the basis for new theory. Although we do not have many themes, we wanted to see if the negotiated themes are actually part of a simpler set of even fewer themes. As such, we were able to synthesize even further down to only three themes. Technology usage and HPT adoption reveal that Vicro Communications is immersed in technology, that is, they depend on technology to do their work. Process improvement reveals that Vicro is attempting to become a more fluid organization, that is, they want information to flow freely so that it can be shared across seamless processes that effectively support business activities, and to delight their customers. CEO mandates reveal that top management was concerned with fluidity and immersion issues and wanted to do something about it. Although their choice of HPT appears to be misguided, they realized that change must be supported from the top. Resistance to change reveals that fluidity and CEO mandates are inextricably tied to how people perceive change. Process improvement is accomplished through people at the process level who do the work. They therefore need the resources and support of management to engineer and redesign processes.

In short, Vicro is immersed in technology because people need them do to their work, that is, technology is critical and important. Technology also enables people to comply to work demands. For instance, if an ad hoc report is required within two hours, the use of technology (databases, networks, computer terminals, and printers) allows people to comply with demands. Fluidity relates to responsiveness, effectiveness, knowledge sharing, and knowledge capture. The objective of process improvement is to improve responsiveness to customers by designing and redesigning effective processes. To improve responsiveness to customers processes must enable effective knowledge sharing and capture, reduce unnecessary costs, and save time. In addition, enterprise systems must work in alignment with business processes. Database technology, networks, software, operating systems, and desktops are the main technology components in a business. Each of these components need to be streamlined in a seamless manner, that is, desktops should be able to talk to databases through networks without concern for hardware, software, and operating system platforms.

Analysis of the Vicro case enabled the researchers to generate a set of themes and, from these themes, a set of "super themes." The BPR literature is thus enhanced because we now have evidence to support the importance of immersion, fluidity, and CEO mandate. Although the idea of CEO mandate has appeared in the literature, immersion and fluidity are new concepts at least with respect to IT-enabled BPR. Moreover, the BPR literature does not really consider the role of enterprise software in BPR (with few exceptions). Since enterprise software issues are more relevant today than in the past as many organizations buy-into HPT and other vendors, we believe that this case adds significantly to this area of research.

Summary

Vicro Communications made no attempt to analyze existing processes and systems to see if they were fluid. That is, they failed to obtain feedback and opinions from people along the process path and legacy systems experts. From the data, it was apparent that many people at Vicro were uncomfortable with HPT and BPR. Many of these same people tried to communicate to management what they thought was going wrong with planning and implementation, but were never listened to or asked for their opinions. We believe that this is the main reason for failure. Every organization has experts with legacy systems and business processes. Both systems and processes must be understood on an enterprise level if change is going to be successful. Hence, people that know the business and systems should be carefully consulted before enterprise change and software is undertaken. Vicro is immersed in technology, but fluidity is a major obstacle and probably the major reason for the failure of HPT. Business processes are not streamlined and efficient at this point so automating them with software will only speed up bad processes. Vicro went one step further. They didn't try to automate existing processes; they forced HPT processes onto a business that is too customized to client needs to work.

Through what we learned from this case study, we hope to shed light on what can happen when decision makers rely on outside vendor promises to improve performance without regard for their employees' knowledge and a comprehensive understanding of the existing state of their business processes. In the Vicro case, analysis of the data suggests that the software investment decision was paramount to the failure of the effort. Vendor promises were not kept and Vicro was stuck "holding the bag." In short, the reengineering effort failed miserably even after investing hundreds of millions of dollars in software implementation. As a result, performance was not improved and the software is being phased out.

REFERENCES

Bergey, J., Smith, D., Tiley, S., Weiderman, N., & Woods, S. (1999). Why reengineering projects fail. In *Carnegie Mellon Software Engineering Institute — Product Line Practice Initiative* (pp. 1-30).

Broadbent, M., Weill, P., & St. Claire, D. (1999). The implications of information technology infrastructure for business process redesign. *MIS Quarterly, 23*(2), 159-182.

Clark, C. E., Cavanaugh, N. C., Brown, C. V., & Sambamurthy, V. (1997). Building change-readiness capabilities in the IS organization: Insights from the Bell Atlantic experience. *MIS Quarterly, 21*(4), 425-454.

Cooper, R. (2000). Information technology development creativity: A case study of attempted radical change. *MIS Quarterly, 24*(2), 245-276.

Cross, J., Earl, M. J., & Sampler, J. L. (1997). Transformation of the IT function at British Petroleum. *MIS Quarterly, 21*(4), 401-423.

Davenport, T. H. (1993). *Process innovation: Reengineering work through information technology*. Boston: Harvard Business School Press.

Hammer, M., & Champy, J. (1993). *Reengineering the corporation*. New York: Harper Collins.

Harkness, W. L., Kettinger, W. J., & Segars, A. H. (1996). Sustaining process improvement and innovation in the information services function: Lessons learned at the Bose Corporation. *MIS Quarterly, 20*(3), 349-368.

Harrington, H. J. (1991). *Business process improvement: The breakthrough strategy for total quality, productivity, and competitiveness.* New York: McGraw-Hill.

Kettinger, W. J., Teng, J. T. C., & Guha, S. (1997). Business process change: A study of methodologies, techniques, and tools. *MIS Quarterly, 21*(1), 55-81.

Kotter, J. P. (1995, March-April). Leading change: Why transformation efforts fail. *Harvard Business Review, 73*, 59-67.

Paper, D. (1998a). BPR: Creating the conditions for success. *Long Range Planning, 31*(3), 426-435.

Paper, D. (1998b). Identifying critical factors for successful BPR: An episode at Barnett Bank. *Failure & Lessons Learned in Information Technology Management, 2*(3), 107-115.

Paper, D. (1999). The enterprise transformation paradigm: The case of Honeywell's Industrial Automation and Control Unit. *Journal of Information Technology Cases and Applications, 1*(1), 4-23.

FURTHER READING

Amabile, T. M. (1997). Motivating creativity in organizations: On doing what you love and loving what you do. *California Management Review, 1*, 39-58.

Ambrosini, V., & Bowman, C. (2001). Tacit knowledge: Some suggestions for operationalization. *Journal of Management Studies, 38*(6), 811-829.

Ballou, R. H. (1995). Reengineering at American Express: The Travel Services Group's work in process. *Interfaces, 25*(3), 22-29.

Bartlett, C. A., & Ghoshal, S. (1994). Changing the role of top management: Beyond strategy to purpose. *Harvard Business Review, 72*(6), 79-88.

Beer, M., Eisenstat, R. A., & Spector, B. (1990). Why change programs don't produce change. *Harvard Business Review, 68*(6), 158-166.

Bergey, J., Smith, D., Tiley, S., Weiderman, N., & Woods, S. (1999). Why reengineering projects fail. In *Carnegie Mellon Software Engineering Institute — Product Line Practice Initiative* (pp. 1-30).

Guha, S., Kettinger, W. J., & Teng, J. T. C. (1993, Summer). Business process reengineering: Building a comprehensive methodology. *Information Systems Management, 10*, 13-22.

Khalil, O. E. M. (1997). Implications for the role of information systems in a business process reengineering environment. *Information Resources Management Journal, 10*(1), 36-43.

Kim, C. (1996, Fall). A comprehensive methodology for business process reengineering. *Journal of Computer Information Systems, 36*, 53-57.

Nemeth, C. J. (1997). Managing innovation: When less is more. *California Management Review, 40*(1), 59-74.

Orlikowski, W. J., & Hofman, J. D. (1997) Improvisational model for change management: The case of Groupware Technologies. *Sloan Management Review, 38*(2), 11-22.

Pfeffer, J. (1998). Seven principles of successful organizations. *California Management Review, 2*(40), 96-124.

APPENDIX

Vicro Financials

Table 1. Vicro Communications five year summary

Income Statistics	2000	1999	1998	1997	1996
Sales	2,258,418	2,425,116	2,717,702	2,631,014	2,517,673
Income loss from operations	(46,234)	141,681	(630,500)	49,411	142,608
Per Dollar of Sales	$(0.02)	$ 0.058	$(0.232)	$ 0.019	$ 0.057
Income tax expense (recovery)	(17,377)	35,286	(94,330)	49,171	48,570
Percent of pre-tax earnings	21.3%	27.4%	14.7%	47.2%	24.4%
Net earnings (loss)	(66,372)	92,599	(547,866)	55,099	149,923
Per Dollar of sales	$(0.029)	$ 0.038	$(.202)	$ 0.021	$ 0.06
Per common share	$(0.75)	$ 1.05	$(6.19)	$ 0.59	$ 1.50
Dividends	17,594	17,692	34,057	85,830	94,183
Per common share	$ 0.20	$ 0.20	$.385	$ 0.94	$ 0.94
Earnings retained in (losses and dividends funded by) the business	(83,966)	74,907	(581,923)	(30,731)	55,740

Balance sheet and other statistics	2000	1999	1998	1997	1996
Current Assets	$699,641	$ 750,860	$894,343	$ 965,078	$1,369,579
Current liabilities	468,247	622,464	941,034	790,454	485,739
Working Capital	231,394	128,396	(46,691)	174,624	883,840
Ratio of current assets to current liabilities	1.5:1	1.2:1	1.0:1	1.2:1	2.8:1
Property, plant, and equipment (net)	409,099	458,808	466,198	635,770	603,750
Long-term debt	272,465	201,686	4,841	49,109	53,811
Ratio of debt to equity	0.4:1	0.3:1	0.0:1	0.0:1	0.0:1
Shareholders' equity	624,685	672,674	610,145	1,185,612	1,549,819
Per common share	$ 7.06	$ 7.60	$6.90	$ 13.40	$ 15.49
Total assets	1,868,426	1,630,293	1,726,135	2,174,572	2,224,040
Average number of shares outstanding	88,457	88,457	88,456	93,200	99,967
Number of shareholders of record at year-end	4,455	5,074	5,506	6,482	6,901
Number of employees	16,166	15,812	17,135	20,084	18,849

Table 2. First call earnings estimates summary, Vicro Communications

Fiscal Year Ending Dec							Last Changed: 15-Jan-2002	
Year Ending	Q1 Mar	Q2 Jun	Q3 Sep	Q4 Dec	Fisc Yr Annual	Num Brok	Cal Yr Annual	Num Brok
2003					0.64	4	0.64	4
2002	0.07	0.08	0.10	0.14	0.40	6	0.40	6
2001	-0.08A	-0.06A	0.02A	0.05	-0.08	6	-0.08	6
2000	-0.09A	-0.14A	0.01A	-0.15A	-0.37A	4	-0.37	4
1999	0.11A	0.02A	0.10A	0.20A	0.43A	7	0.43	6
1998	0.06A	-0.24A	-0.03A	0.17A	-0.04A	6	-0.04	6

Consensus Recommendation: 2.8

David Paper is an associate professor at Utah State University in the Business Information Systems department, USA. He has several refereed publications appearing in journals such as Journal of Information Technology Cases and Applications, Communications of the AIS, Long Range Planning, Creativity and Innovation, Accounting Management and Information Technologies, Journal of Managerial Issues, Business Process Management Journal, Journal of Computer Information Systems, Journal of Database Management, *and* Information Strategy: The Executive's Journal. *He has worked for Texas Instruments, DLS, Inc., and the Phoenix Small Business Administration. He has consulted with the Utah Department of Transportation and is currently consulting with the Space Dynamics Laboratory in Logan, Utah, USA. His teaching and research interests include process management, database management, e-commerce, business process reengineering, and organizational transformation.*

Kenneth B. Tingey is a doctoral student at Utah State University in the Business Information Systems Department, USA. He has more than 25 years of experience in industry, working as a venture capital fund founder and general partner, entrepreneur, general and line manager, and executive staff assistant. He is founder, chairman, and CEO of OpenNet Corporation, an enterprise software developer. His academic credentials include a master's degree in Pacific international affairs from the University of California, San Diego, a Master of Business Administration from Brigham Young University, a Bachelor of Arts in music education from Utah State University, and a Baccalaureate Major in accounting from Brigham Young University. His professional affiliations include Strategic Information Division of Ziff-Davis Publishing Company, the Ventana Growth Fund, and Sunrider International. In addition, he has conducted many business consulting and systems development projects on contract with direct selling companies, software development companies, and government contractors. Mr. Tingey has engaged in many enterprise-level systems development projects with special emphasis on requirements of supporting the mission of institutions by means of information processing models and information technology tools. Mr. Tingey is the author of Dual Control, *a book on the need to support top-down policies and horizontal processes in a unified system environment.*

Wai Yin Mok is an assistant professor of information systems at the University of Alabama in Huntsville, USA. From 1999 to 2001, he was an assistant professor of information systems at Utah State University. From 1996 to 1999, he was an assistant professor of computer science at the University of Akron in Ohio. He was an assistant lecturer of computing at Hong Kong Polytechnic from October 1992 to August 1993. His papers appear in journals such as ACM Transactions on Database Systems, IEEE Transactions on Knowledge & Data Engineering, Journal of Database Management, *and* Data & Knowledge Engineering, *and* Information Processing Letters. *He serves on the editorial review board of the* Journal of Database Management. *He received a BS, an MS, and a PhD in computer science from Brigham Young University in 1990, 1992, and 1996, respectively.*

This case was previously published in the *Annals of Cases on Information Technology*, Volume 5/2003, pp. 45-62, © 2003.

Chapter XII

Introducing Expert Systems at The Corporation*

Jay Liebowitz, George Washington University, USA

EXECUTIVE SUMMARY

This case study highlights the concept that the "management" of the technology is usually the limiting factor causing the demise of a project rather than the "technology" itself. This real case study involves creating an awareness of a new technology within the company (hereafter named "The Corporation") and trying to start a (much needed) project using this technology. The technology in question here is "expert systems." An expert system is a computer program that emulates the behavior of an expert in a well-defined domain of knowledge. At The Corporation, a few key top management executives thought that an expert system could be used to help The Corporation in configuring its minicomputer systems. The Corporation enlisted the help of a consultant to develop a feasibility study of using expert systems for configuration management at The Corporation. In doing so, an awareness of expert systems technology was created throughout the company in almost all divisions — customer service, sales, marketing, finance, information systems, manufacturing, etc. The hidden agenda of the consultant was to start an expert systems project for configuration management at The Corporation, if the feasibility study deemed it worthwhile. The case study describes many of the hurdles that had to be jumped, and shows the importance of understanding the corporate culture of the organization, especially in the difficult times of mergers and acquisitions, economic downturns, and tough competition. Let's now see how the case unfolds.

BACKGROUND ON THE HISTORY OF THE ORGANIZATION

The Corporation (a pseudonym) is a manufacturer of minicomputer systems who acquired a larger company in the same line of business. The Corporation manufactures a variety of computer families for use in real time simulation, software development, computer architecture research, and a host of other applications. The Corporation, a $280 million company, has about 900 employees at its corporate headquarters and about 200 employees at its manufacturing plant.

Over recent years, downsizing and outsourcing have been dominant strategies used in The Corporation. A sense of uneasiness plagued many of the employees as they were unsure of whether their jobs were protected, as many of their colleagues were receiving pink slips. Those whose jobs were valued remained, but were troubled with large workloads due to the reduction in force.

A critical component of The Corporation's operations is configuration management. Configuration management, or more appropriately configuration control, refers to configuring the hardware (and software) correctly for a customer's order. The Corporation wanted to improve the number of times it took to correctly configure an order. About 90% of the computer configurations were done incorrectly the first time, and it would normally take about 12 times to correctly configure the order. The vice president of development (who was quickly let go after the acquisition) and the vice president/general manager of manufacturing and customer service wanted to explore the feasibility of developing a Configurator using expert systems technology. The Configurator would configure an order and provide a quotation correctly the first time.

By having such a system, it was thought that the amount of time spent in contracts/ configuration control could be reduced, thereby freeing up time for manufacturing to build the customer's system and ship within the delivery date. Additionally, an automated Configurator could facilitate the creation of forecasting reports, improve customer relations, and provide timely and accurate configuration information out into the field to the analysts and sales representatives. It was also felt that an expert system may be a good vehicle for building up the corporate memory of the firm so that valuable knowledge and experiential learning would not be lost.

In the following sections, a discussion of how an awareness of expert systems technology was created within The Corporation will be made. However, due to mainly organizational reasons, the development and implementation of an expert configuration system were never realized.

SETTING THE STAGE

Getting Started

The first step in getting an expert systems project started is to create an awareness of expert systems technology. Fortunately, at The Corporation, the vice president of development and the vice president/general manager of manufacturing and customer service had some familiarization with expert systems and they thought that it would be

Figure 1. Top-level organizational chart for The Corporation

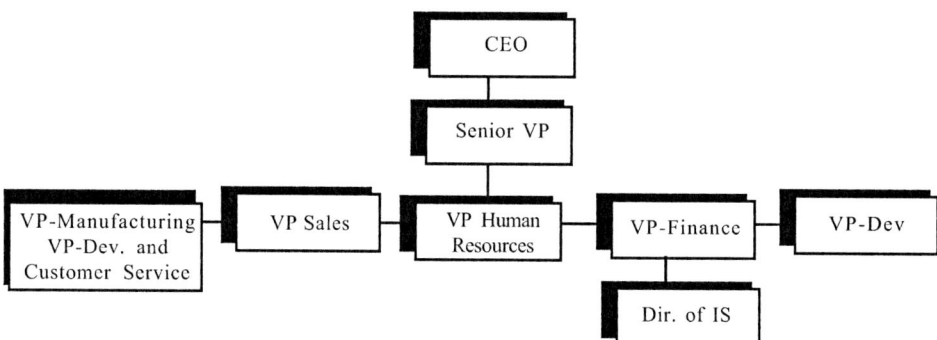

useful to see how expert systems technology could help improve The Corporation's operations. They then hired a consultant specializing in expert systems to write a feasibility study on if and how expert systems could be used in configuration management at The Corporation.

The initial step of creating an awareness of expert systems technology was already made easy because some top management executives were backing the study. This top level, champion support helped to pave the way for this study. Unfortunately, however, the vice president of development (the main champion) left the company about a week after the consultant arrived. This was due to the acquisition, and all the former top management officials were now replaced with top management from the acquiring company. This event was not a major hindrance because the other sponsor of the project (i.e., the vice president/general manager of manufacturing and customer service) was a supporter of this feasibility study effort. Although, the main champion of the effort (i.e., the former vice president of development) was lost, and the new vice president of development did not seem to be interested in the value of expert systems. An awareness and understanding of expert systems still had to be created within the organization at all operating levels.

In order to permeate this notion of expert systems throughout the company, it was necessary to use a variety of methods to achieve this goal. These methods included: a top-down approach, a bottom-up approach, and an introductory seminar approach.

PROJECT/CASE DESCRIPTION

Top-Down Approach

In order to develop the feasibility study, the company's operations had to be understood and this involved speaking with top management at The Corporation on down through the various departments (customer service, manufacturing, engineering and development, sales, marketing, information systems, finance, etc.). In studying the organizational chart of The Corporation, the consultant realized that there was not a vice president of information systems, which is a rarity for a computer-oriented, high technology firm. Instead, the traditional, conservative hierarchy was used whereby the director of information systems reported to the vice president of finance. This realization would later come back to haunt the effort.

By having the support of some top management executives, it was rather easy to gain access to other individuals in the company in order to make them aware of this study and expert systems. In speaking with the various individuals throughout The Corporation, the thrusts of the conversations were to determine what their needs were if an expert configuration system were built and what practices and policies at The Corporation would have to change to successfully build and use an expert configuration system. The 35 individuals who were interviewed felt comfortable in that their comments would be incorporated into the feasibility study which would be sent to all top management officials. In essence, they felt ownership in the study.

An essential player in this effort was the proposed expert if an expert configuration system would be built. It was also important to obtain the input and advice from this expert in regards to the need in having such an expert system. Additionally, it was important to gain the expert's support, so that this project wasn't forced upon her. There really was only one expert who had been around long enough to understand the configurations of the different product lines. She was severely overworked, and would welcome an aid to assist herself and her staff in the configuration management/control area. The expert was kept abreast of the work during this feasibility study effort, so that there would be no surprises at the end of this study.

Another innovative idea to create a better awareness of expert systems at the top management level was to circulate a pad to each top management executive with each page of the notepad embossed with the saying "Artificial Intelligence/Expert Systems are for Real at The Corporation." This approach was borrowed from a similar technique applied by the expert systems manager at 3M Corporation. As top management would use these pads to write short notes, they would constantly be reminded of expert systems.

Bottom-Up Approach

Not only was a top-down approach used in gaining user support, but also a bottom-up method was utilized. This bottom-up approach involved working with the actual users of the proposed expert configuration system in order to obtain their views and requirements in having such a system. The consultant and a member of the Information Systems Department at The Corporation took the three day real-time sales/analysts training course. Most of the configuration analysts and sales representatives at The Corporation in the real-time market were gathered during this course. We were able to have a round table discussion with about 16 analysts (the primary users of the expert configuration system) and also speak with several sales representatives (secondary users). The analysts were very interested in the project, and offered valuable comments and insights. Additionally, as an outgrowth of our meeting, four analysts from throughout the country were designated to assemble a list of their requirements for having an expert configuration system. This information was very helpful in formulating the feasibility study.

Almost everyone in the company was very excited about the prospects of having a better way of configuring orders. The consultant and members of the Information

Systems Department were particularly cautious to make sure that expectations were kept under control.

INTRODUCTORY SEMINAR

Another way which was used to create an awareness of this project and expert systems throughout the company was to offer an introductory seminar on artificial intelligence/expert systems to key individuals at The Corporation. This hour presentation served to boost the interests of those who attended, and helped to better familiarize themselves with expert systems technology and applications. Descriptions of expert configuration systems used by competitors stimulated increased interest in keeping up with the competition via expert systems. Even though the attendance at this seminar was less than expected, we were able to gain further support for this project.

TECHNOLOGY, MANAGEMENT, AND ORGANIZATIONAL CONCERNS

The Feasibility Study and Requirements Document

After interviewing the many individuals throughout The Corporation and performing an analysis of the results, the feasibility study was written and sent to all top management executives. The feasibility study included the following sections:

Executive Summary
1.0 Does a Need Exist for Developing an Aid for Facilitating The Corporation's Configuration Management Function
2.0 Can Expert Systems Technology Solve the Configuration Need
    2.1 Survey of Expert Systems for Configuration Management
    2.2 Can Expert Systems Technology be Used at The Corporation
3.0 Alternatives for Using Expert Systems Technology for The Corporation's Configuration Management
4.0 Cost/Benefit Analysis and Risk Assessment
5.0 Building an Expert System Prototype for Configuring a Specific Product Line at The Corporation
6.0 Recommendations and Next Steps
7.0 References and Appendices

As part of the feasibility study, a thorough survey of existing expert configuration systems worldwide was included. It showed that major companies had been successfully using expert systems for configuration management. This gave confidence to The Corporation's officials to see that expert systems have been proven technology for successful use in configuration management activities. A separate summary of the recommendations based upon this feasibility study was also circulated to top management.

Other groundwork was laid to "set" the stage. A functional requirements document for an expert configuration system was prepared and sent to top management. This document included:

1.0   Purpose of Expert System
2.0   Reference Documents
3.0   User Information
4.0   Functional Requirements
      4.1 Database Access
      4.2 User Interface
      4.3 Input/Output Content
      4.4 Control Structure (Inference Engine)
      4.5 Knowledge Representation
      4.6 Hardware
5.0   Documentation
6.0   Training
7.0   Maintenance Requirements

By preparing this document, it showed top management that we had gone one step further by generating requirements for an expert configuration system, instead of merely stating that it is feasible to build an expert configuration system at The Corporation. Also, we lined up a local company who had expertise in developing expert configuration systems for developing the proposed expert configuration system at The Corporation and providing technology transfer in order to better acquaint the information systems staff on expert systems. We also spoke with a major, local university to provide some education/courses on expert systems onsite at The Corporation to designated individuals. We also had the support of the expert to go ahead with this project.

Final Approval from Top Management

The stage was set and a meeting had been scheduled with top management in order to get the go-ahead on developing the expert system. We quickly learned that "timing" is a critical part in making any business decision made. Unfortunately, the meeting was cancelled due to emergency budget planning sessions, and it became apparent that it would be difficult to reschedule this meeting with top management. We even tried to hold a videoconference between top management in two locations, but only half of the management attendees could be available. Additionally, the project's approval became clouded with other timing issues, such as a transition in information systems management and then a new company-wide hiring freeze. During this time, the (then current) director of information systems said that he had certain signing authority and he would sign for the funds to develop the expert systems prototype. Four days before signing off, a new chief information officer (CIO) was hired to ultimately replace the previous director of information systems. The consultant and information systems staff briefed the new CIO on the proposed expert configuration system, but the CIO felt that other priorities were perhaps higher. Even the vice president of manufacturing (one of the original advocates of the study) said that although the company was doing configuration poorly, at least it could be done whereas other important areas were not being done at all. Coupled with these events, The Corporation was trying to cut back new projects in order to help ensure profitability for the company.

"WHAT ACTUALLY HAPPENED" WITH THE PROJECT

What did these events mean to the expert configuration system project? Principally, the decision to go-ahead with the project was delayed until new management felt comfortable in funding new projects and aligning their priorities. Management said that they would revisit and reconsider this expert configuration system project in four to six months. This never happened due to other perceived priorities by top management. Additionally, staff turnover and firings had increased in order to reengineer the company and control costs.

Poor business decisions continued to plague the company. The former director of information systems of The Corporation had heard from colleagues still at The Corporation that the new CIO agreed to buy a million dollar software package, only to find out later that it was incompatible with The Corporation's hardware.

SUCCESSES AND FAILURES

What can be learned from this case study? First, obtaining top management and user support are critical elements to the success of a project, especially one involving a new technology. Integral to gaining this support is the ability to create a thorough awareness of the technology (e.g., expert systems) within the company. This awareness, especially at the upper levels of management in the company, should help in gaining the financial and moral support from top management in order to go ahead with the project. Recognizing the internal politics, organizational culture, and external climate of The Corporation are essential elements where the development team must have strong sensitivity.

Another important lesson learned is to strongly involve the expert(s) and users in helping to gain support for the expert system project. In this case, the expert (who was the only person who knew the different product lines for configuration) should have been used in a more vocal capacity in order to urge top management that the company needs this expert configuration system for productivity, training, and longevity purposes. If the expert were to leave the company, the configuration task would be extremely difficult to perform due to the lack of expertise in The Corporation in knowing the various product lines and associated configurations. This fact should have been emphasized more to top management in order to further convince them of the need for such a configuration system.

A potential major flaw in introducing expert systems to The Corporation may have been the chaotic state of the company. With firings, staff turnover, reengineering, and cost control measures rampant within the company, it would be difficult to introduce a new technology, like expert systems, within the firm. Even though the use of an expert system for configuration management was being marketed by the development team as a good business decision, there were too many other priorities that needed immediate attention by top management. Perhaps it would have been better to have stimulated interest in the project after the initial chaos had settled. There were too many high priority items on the platter for top management that needed attention.

EPILOGUE

The birth of a new technology at The Corporation (namely, expert systems) is slowly emerging. It took about three months from the first time the words "expert systems" were uttered at The Corporation to the time it took to saturate top management with the notion of expert systems. The best outcome of this project was creating an awareness of expert systems throughout the company. The Corporation took this important first step in bringing expert systems technology to the company. The hybrid approaches used to get the project underway at The Corporation were very successful in terms of introducing expert systems technology to The Corporation's employees and management.

Knowing the corporate climate and culture was an important lesson learned from this experience. Having an appreciation for the organizational structure, internal politics, organizational barriers, and possible resistance to change were key concepts that should have been appreciated more than just understanding the technology. Also, trying to implement a new technology in a chaotic environment (due to the recent acquisition and resulting restructuring of The Corporation) was a difficult task indeed.

This case study hopefully illuminates some useful techniques that other companies may use to create an awareness of expert or intelligent systems within the organization for eventual expert/intelligent system funding and support. A critical concept is a thorough appreciation for understanding the "management" of the technology versus just the technology itself. After all, without careful attention to these matters, the project may be a technical success but a technology transfer failure.

QUESTIONS FOR DISCUSSION

1.   Since The Corporation was in a chaotic environment due to the acquisition of a large company, was it worthwhile to try to introduce a new technology into The Corporation under these trying times?
2.   How could the consultant gain better support from the new vice president of development and the vice president of manufacturing?
3.   What could have been done differently to have been able to get the expert system prototype project funded?
4.   In what ways could the domain expert in configuration control have helped more in trying to get the expert system funded?
5.   How could the users of the eventual expert system (i.e., sales representatives and configuration analysts) have been more active in order to get the expert configuration system project started?

REFERENCES

DeSalvo, D., & Liebowitz, J. (Eds.). (1990). *Managing artificial intelligence and expert systems.* NJ: Prentice Hall/Yourdon.

Kerr, R. M. (1992). Expert systems in production scheduling: Lessons from a failed implementation. *Journal of Systems Software, 19.*

Lee, J. K., Liebowitz, J., & Chae, J. M. (Eds.). (1996). *Proceedings of the 3ʳᵈ World Congress on Expert Systems.* New York: Cognizant Communication Corp.

Liebowitz, J. (Ed.). (1994). *Worldwide expert system activities and trends.* New York: Cognizant Communication Corp.

Turban, E., & Liebowitz, J. (Eds.). (1992). *Managing expert systems.* Harrisburg, PA: Idea Group Publishing.

SELECTED BIBLIOGRAPHY

International Society for Intelligent Systems/James Madison University. (1995). *Developing your first expert system* [CD ROM].

Lee, J. K., Liebowitz, J., & Chae, Y. M. (Eds.). (1996). *Proceedings of the 3ʳᵈ World Congress on Expert Systems.* New York: Cognizant Communication Corp.

Liebowitz, J. (Ed.). (1990). *Expert systems for business and management.* Englewood Cliffs, NJ: Prentice Hall.

Liebowitz, J. (1992). *Institutionalizing expert systems: A handbook for managers.* Englewood Cliffs, NJ: Prentice Hall.

Liebowitz, J. (Ed.). (1994). *Worldwide expert system activities and trends.* New York: Cognizant Communication Corporation, Elmsford.

Turban, E., & Liebowitz, J. (Eds.). (1992). *Managing expert systems.* Harrisburg, PA: Idea Group Publishing.

ENDNOTE

[*] "The Corporation" is being used to protect the name of the organization. Please use the following for a Web-based version of this case: http://cac.psu.edu/~gjy1/case

Dr. Jay Liebowitz held the chaired professorship in artifical intelligence at the U.S. Army War College, and is also professor of management science at George Washington University. He has published 18 books and more than 200 journal articles, mostly in expert systems and information technology. He is the editor-in-chief of the international journal Expert Systems With Applications *and the associate editor of the international journal* Telematics and Informatics. *He is the founder of The World Congress on Expert Systems and he has lectured in more than 20 countries.*

This case was previously published in J. Liebowitz and M. Khosrow-Pour (Eds.), *Cases on Information Technology Management in Modern Organizations*, pp. 1-8, © 1997.

Chapter XIII

Leveraging IT and a Business Network by a Small Medical Practice

Simpson Poon, Charles Stuart University, Australia

Daniel May, Monash University, Australia

EXECUTIVE SUMMARY

Although many medical information technologies require significant financial investment and are often out of reach of small medical practices, it is possible through careful alignment of IT and customer strategy, together with a network of strategic alliances to exploit IT effectively. In this case we present a small cardiology consultancy that has engaged in strategic planning in its attempt to leverage IT expertise to attain competitive advantage. We propose that through a network of alliances, a relatively small medical enterprise can benefit from its limited IT investment. The case study indicates the importance of a team of champions with both IT and medical knowledge and the notion of mutual benefit. We also discuss some of the issues faced by all participants in this alliance relationship. The objectives of this case are to provide readers the opportunity to:

1. *Discuss how a small medical practice can leverage skills, expertise and opportunities within a strategic alliance to enhance its competitive advantage without heavy up-front financial investments.*

2.   *Explore how small businesses in the professional and knowledge-based industry can gain strategic advantage through IT.*

3.   *Understand the pros and cons of strategic alliances and potential issues related to building trust, consolidating relationships among members and risk management of such alliances on an ongoing basis.*

4.   *Think about the plausibility of business transformation by moving from one industry (specialised cardiology services) to another (medical informatics).*

BACKGROUND

Although specialised cardiologist was often considered a fairly exclusive profession, it was not without competition. In specialised medical consultancies, medical technology was playing an increasingly critical role in determining the future success of their operations. Across all dimensions of healthcare, technology held the promise of enabling clinicians to provide more accurate diagnosis and engage more effectively in the clinical process (Achrol, 1997; Slack, 1997; van Bemmel & Musen, 1997). In addition to medical technology, information technology (IT) had also become an important support for medical professionals to be more effective. IT applications in the medical field varied greatly in their diversity, with the most commonly seen being office information systems (Collen, 1995). Such systems helped to achieve effective information management in hospitals, clinics and other medical establishments. Based on the principles of management information systems, office information systems could support effective retrieval and cross-referencing of medical records including patient histories, past treatments, prescriptions among other functions. Although playing an important role, these systems often worked in isolation and were not effectively integrated with other medical and clinical technologies to support better decision making (Slack, 1997). Additionally, significant investment in cost and effort must also be made in the purchase/ development of IT infrastructure, its deployment, maintenance and administration (Charvet-Protat & Thoral, 1998; Lock, 1996; Staggers & Repko, 1996).

In general, a small medical practice[1] faces the challenge of many small-and-medium enterprises (SMEs): it did not have the resources or financing of a large enterprise to enable it to enjoy expansion and compete at will (Eden et al., 1997). Thus, while IT was a powerful tool, it could also be a significant burden to invest in IT and related technology when there was insufficient business volume to justify the investment. However, despite its cost, IT could be the fulcrum upon which the SME's competitive advantage was leveraged. Some examples of IT as leverage: (a) human resource: an alternative to hiring additional administrative staff was to let staff perform their own scheduling supported by commonly accessible group diaries; (b) time management: electronic questionnaires could be used to elicit information from patients prior to consultation, enabling clinicians to focus on key diagnoses and reducing the pressure of increasingly limited consultation times; (c) patient management: electronic systems could reach out to the home of the patient to further improve the clinical management process and empower the patient; and (d) customer service: there could be enhanced service to customers by clinicians, providing more timely and informative diagnoses, and access to records and information.

Figure 1. An integrated framework for IT, medical technology, and patient management

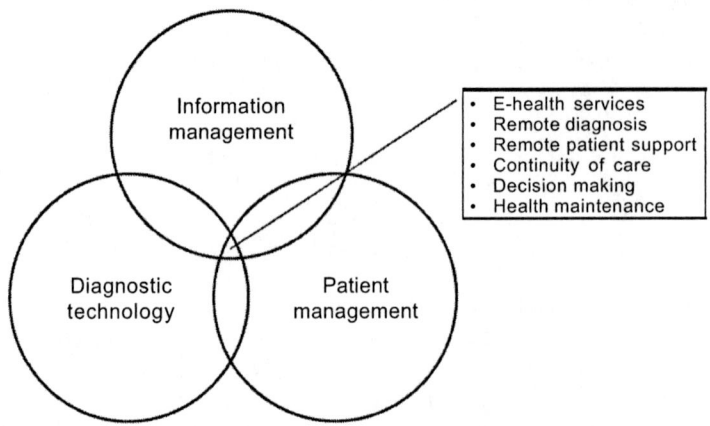

SETTING THE STAGE

Coupled with the advances in medical imaging, bioinformatics and advanced diagnosis techniques (van Bemmel & Musen, 1997), there was vast potential for IT to further enhance the quality of healthcare and deliver medical services to the patient on-demand, independent of location and time (Mitchell, 1999). In order to achieve this vision, it was essential to have a strategy to effectively integrate the various aspects of health informatics such as information management, diagnostic technology and patient management among others. Figure 1 illustrates a conceptual framework of how these components come together.

In the following sections, we present the case of a small medical practice specialising in echo-cardiology that had created a strategic vision that integrated health services, technology, and marketing — fuelled by the use of IT and medical technologies, coupled with an innovative and entrepreneurial mindset. By considering the medical practice as a service entity that was driven by knowledge as its primary resource, this cardiology practice had embraced an IT and also in the future a knowledge management (KM) infrastructure as a competitive tool to achieve its strategic objectives. In doing so, it had encountered challenges as a function of its size — and sought to overcome them while retaining the flexibility and advantages of being a small enterprise.

CASE DESCRIPTION

Eastern Cardiology Services (ECS)[2] was established in 1997 by Dr. Jeff Curtin, a cardiology specialist with expertise in the subspeciality of echocardiography. After working in the U.S. for over a decade in consultancy and research, Dr. Curtin returned to Australia for personal reasons. Having been involved with the local public hospital and a group of cardiologists at a private hospital, Dr. Curtin saw an opportunity for

Figure 2. Organisational chart of ECS

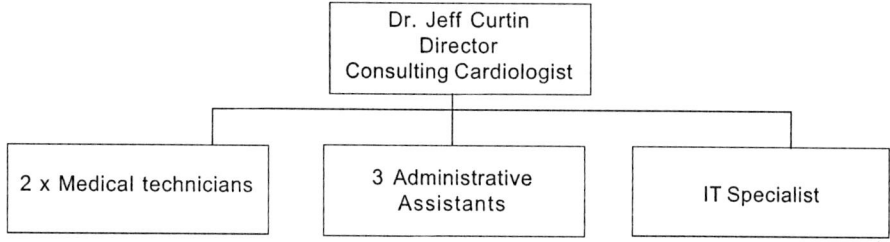

creating a specialist consultancy in cardiology and echocardiography. Although there were existing providers of such services, Dr. Curtin perceived an opportunity to create a niche: by exploiting his expertise in echocardiography, adopting new models of patient management and business practices, and the use of state-of-the-art medical and information technology.

At the time of this write up, Dr. Curtin was the only qualified echo-cardiologist at ECS. He was assisted by two medical technicians who were qualified to operate all the diagnostic equipment. In addition, ECS had a few administrative assistants and an IT specialist, a graduate from a local e-commerce research centre Dr. Curtin collaborated with earlier.

As a small practice entering an existing market, ECS needed to create a competitive advantage that would allow it to be distinguished from its competitors. Curtin saw that this could be achieved through the effective use of medical and information technology in its service delivery — at a breadth and depth not undertaken by the competitors. However, the main constraint of this strategy was the lack of in-house IT expertise at ECS. Curtin addressed this by partnering with an interstate university, leveraging the expertise of an associate professor and a doctoral student, who could use ECS as a subject for industry-based research. ECS jointly with the university had applied for financial support to work on IT projects. In this way, ECS was able to take advantage of IT expertise (in the doctoral student), not only to aid in maintaining the infrastructure but in formulating strategic IT direction as ECS evolved.

Figure 3 illustrates the stakeholder relationships between ECS and its alliance members, and their respective interests. To enable the alliance to work effectively, clear understanding of mutual benefit to the different stakeholders was important. From the perspective of the interstate university, it is important to build a research profile that included industry collaboration. ECS provided such opportunities through the involvement of the doctoral student (working in-residence), funding contributions and participation in research grant applications. Similarly, the local e-commerce research centre was interested in developing industry-based research links. Pursuant to this, an Honours year undergraduate project team has been working with ECS on prototyping an interactive intelligent questionnaire (see Figure 5), an adaptive electronic questionnaire that formed part of ECS's strategic IT initiative. ECS was also exploring other follow-up projects with its alliances.

Figure 3. Stakeholder relationships in the alliance

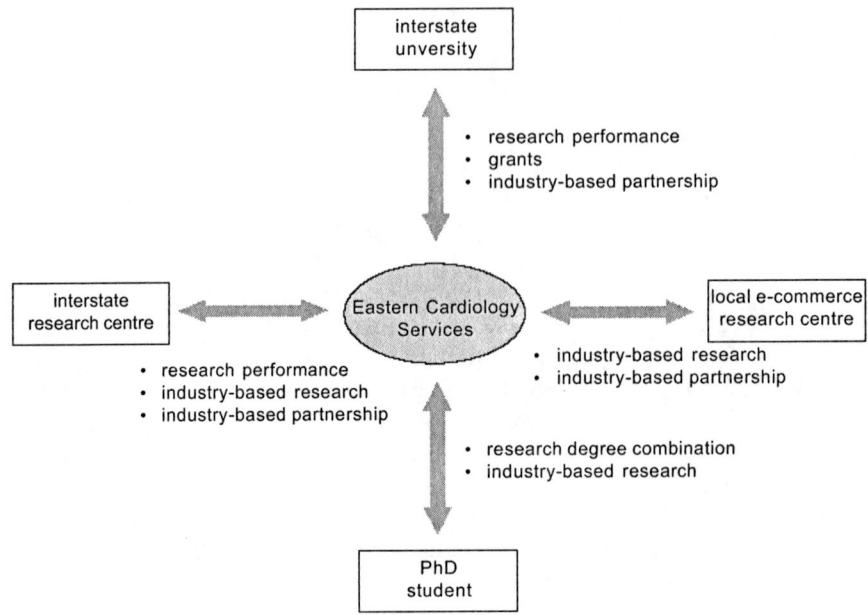

Through this existing connection, the network of relationships had grown: involving a local research centre in electronic commerce (e-commerce) and a large interstate advanced technologies and health research centre. By combining these relationships, ECS had been able to gain access to expertise that would otherwise fall outside the domain of its core competence. More importantly, it was the sharing of funding opportunities and joint publicity which had created a win-win situation. For example, as the Australian Research Council (www.arc.gov.au), a national research funding body, was focusing on linkage research between universities and the commercial sectors, such alliance network would be ready to take advantage of it.

CURRENT CHALLENGES/PROBLEMS FACING THE ORGANISATION

After considering how network alliances could contribute to the success of ECS, it is now important to reflect on how to align IT and strategic orientation to achieve internal success within ECS. After all, it was the objective to achieve ECS's competitive advantage through IT deployment that drove the development of such strategic alliance.

In order to have a focused IT strategy, it was essential to start by examining a firm's strategic orientation. Aligning business strategy with IT strategy had been described as critical to successful IT deployment (Dvorak et al., 1997). Particularly for small practices like ECS, there were scant resources to arrive at an effective strategy through expensive

Figure 4. Strategic vision of ECS

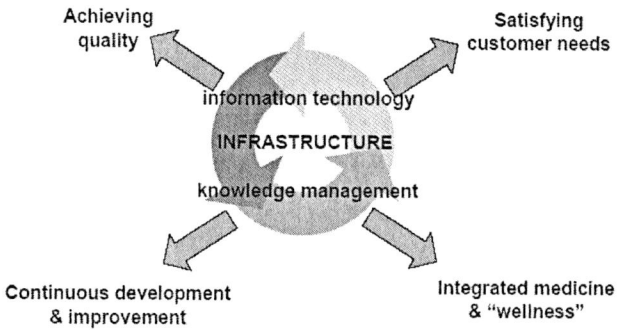

trial and error, often affordable by large corporations. Although an alliance might be in place, the only way that ECS could ensure maximum benefit from the alliance was to play an active role in the strategic alignment process. Such alignment had only been possible through the clear championship of Dr. Curtin, who had been keeping an eye on any valuable partnership both in the medical and IT fields. While input and support had been sought from different stakeholders within ECS, the strategic alignment process had progressed only because it was clearly driven at the decision-maker level.

The strategic orientation of ECS was guided by its mission: that each of their patients would experience the best medical and personal treatment that could be delivered. In achieving this, the strategic vision identified key goals as summarised in Figure 4.

There are four key goals that form the strategic vision of ECS.

Achieving Quality

In the diagnostic and consultation services that it provided, ECS sought to deliver world-class medical care, consistently benchmarking its services against international standards. This goal was fundamental to the existence of ECS: quality service was the key product that ECS saw itself as delivering. In order to achieve this strategic vision, ECS pursued to form strong strategic partnerships with research institutions both locally and nationally to exploit IT to support the delivery of its services. The kinds of quality benchmarks are accuracy, timeliness, customer satisfaction and consistency.

Accuracy was of extreme importance because inaccurate information meant a matter of life-and-death to the patients. So, to stay competitive, ECS not only needed to provide accurate diagnosis, but also the ability to record, retrieve and transfer information accurately. A robust IT system could provide ECS with good information management capability such that medical records were of high accuracy.

Timeliness was another important quality attribute. To provide timely information accuracy, ECS must make sure patients' records were correctly filed and easily retrievable. In a paper-based filing system, this was difficult, sometimes impossible. ECS wanted to ensure that deployment of an IT system, with interface to some of its specialised cardiology equipment, outputs from such equipment could be stored without the need for physical printing and filing.

Customer satisfaction was as important in the medical sector as in other service-oriented sectors. Due to the newly established status of ECS, providing quality services to earn a high customer satisfaction rate was one of the more sustainable competitive advantages. Customer satisfaction in the medical sector could be more complex than in other sectors because patients were often encountering difficult decisions and circumstances due to their illness. ECS was aiming to provide top class patient care by viewing every patient was an unique individual who deserved the best kind of care from all the staff members in ECS — from the cardiologist to the administration staff. Using IT and a customer relationship management (CRM) system, ECS was trying to monitor customer feedback and respond to areas needing attention early.

Delivery of consistent medical services was also a key factor to maintain high customer satisfaction rate. ECS was planning to deploy IT (a patient management and a CRM system) to ensure their services were consistent with their philosophy. Internal guidelines on medical procedures and practices were in place to make sure when ECS grew further, patients might encounter different doctors, but the same consistent services.

Continuous Development and Improvement

As ECS grew, it would need to adapt and evolve accordingly with new business directions, needs and constraints. It would not get everything right initially, despite its efforts to attain the best possible levels of quality and service. Thus, a key goal of ECS was to embody continuous development and improvement across the organisation. This also included the idea of a living corporation (de Geus, 1997) that was able to renew and replenish itself in a sustainable fashion.

To achieve this, ECS saw partnering with specialists in local academic institutions a strategic move. Through leveraging the intellectual support from graduate students, ECS could continuously improve its services via clever deployment of IT. Since IT development could involve high cost and ECS had neither the expertise nor the budget to do it commercially, working with graduate students was a good solution. Graduate students could work on their projects in a focused and relevant manner while producing commercial quality outcomes. As ECS had set up arrangements with local academic institutions such that this relationship would be a continuous one, it had secured a continuous development process.

Integrated Medicine and Wellness

ECS believed that its medical focus should combine traditional and complementary approaches to medical care for a total perspective of medical care — integrated medicine. This also included a view of medical care as maintaining patient wellness, where the patient's health and well-being was maintained (Zollman & Vickers, 1999) rather than disease management, where the focus was on resolving the disease (Hunter & Fairfield, 1997).

The strategic vision was notable in that the key goals are being supported and driven by an infrastructure that combined IT and KM. Like many other medical practices, ECS relies on IT as critical infrastructure. However, ECS took this to the next level by leveraging the use of IT in a "highly relevant manner," wielding it as a critical tool in achieving each of the strategic goals and differentiating it from other cardiology practices in improving service delivery.

In addition, ECS has taken a view of itself as a knowledge enterprise (Drucker, 1957), focusing on the development of a framework that views the key resources of the organisation as its intangible assets and intellectual capital (Stewart, 1997; Sveiby & Lloyd, 1997). While information is distinct from knowledge, it forms an important component of the knowledge generation process (Nonaka, 1991). Thus, ECS is seeking to combine both IT and KM as a tool for competitive advantage and achievement of its mission. Figure 4 also indicated that this core IT/KM infrastructure was iterative, constantly being developed and refined, as its limitations become apparent.

Aligning Corporate and IT Strategy

ECS had taken its first steps in implementing the aligned strategy as described above. An initial set of initiatives was identified pursuant to the strategy.

Interactive Intelligent Questionnaire

This initiative involved the development of an interactive electronic questionnaire that could be deployed across the Internet or in the waiting room of ECS, allowing patients to complete information about their medical condition and summarise the information for the clinician to aid in diagnosis. Deployed over the intranet, the questionnaire could adaptively discard irrelevant questions (based on current answers and known information about the patient) so that the patient only had to answer a relevant number of questions. The underlying IT used to build this interactive intelligent questionnaire was a combination of artificial intelligence (AI) and Web-based technology. The Web was used as a delivery and access system, while the AI technique was used to build the "engine" of the questionnaire. An illustration of how this works is shown in Figure 5. This initiative had been undertaken in collaboration with a local e-commerce research group which provided a team of Honours year Computer Science students to develop a prototype. The motivation behind this project was to "value-add" the increasingly limited time available to the clinician during consultation (Sellu, 1998). The project was recently completed and was nominated as one of the best Honours-year projects in a statewide IT competition. As a result of this initiative, ECS had distilled lessons and concepts from its experience — using them to drive the next set of steps in its strategic vision.

ECS had also started focusing on how it could manage the sources and channels of information that formed part of its business. Key initiatives that were part of the information management strategy included those outlined in the following sections.

Deployment of Digital ECG Management Systems

Digital ECG management systems were medical equipment which allow the storage, editing and retrieval of ECGs diagrams. Compared to the traditional approaches to carry out such diagnosis, this provided the cardiologist the option to file and transfer information electronically, eventually via the Internet. This meant that eventually Dr. Curtin could forward patients' ECGs to other specialists using secured connections

Figure 5. Conceptual diagram of the interactive intelligent questionnaire system

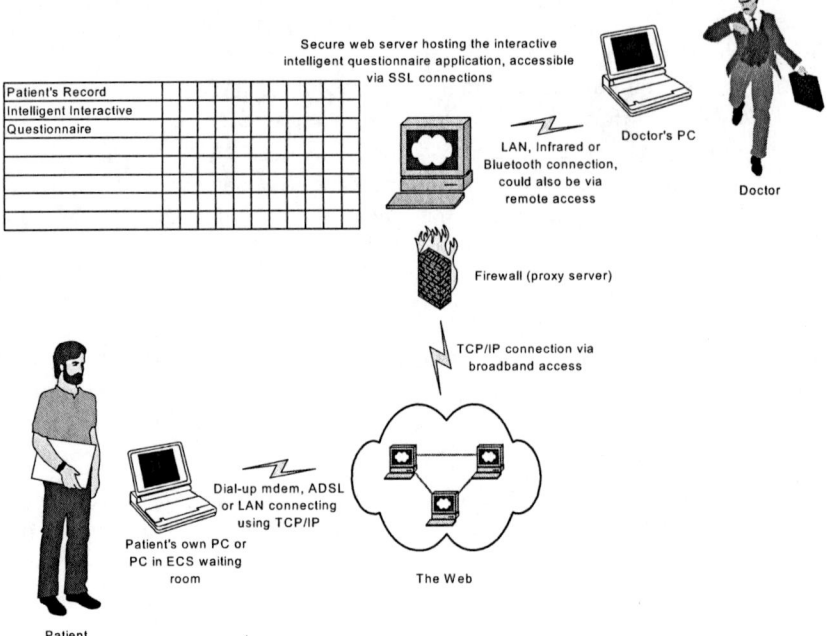

without any paper-based delivery. In addition, the cardiologist could perform analysis of changes on patient's cardiological history and use software tools to do so effectively. This would be a major competitive advantage because cardiologists could be better supported while identifying irregularities in patients' heart activities.

Provides Access to Procedures, Documents, and Resources via the Intranet

As part of the corporate Web site initiative, ECS was aiming to build an intranet using its existing local area network to support internal access and transfer of procedures, documents and resources. As part of the quality assurance strategic vision, making information and documents available on the intranet meant staff members could work more effectively with Dr. Curtin and patients. Traditionally, documents such as operating and administrative procedures were not readily available and when new staff members were recruited, one had to go through a time-consuming induction process by an existing staff member. This decreased the productivity of both the existing and the new staff members. With the intranet and its contents, it was hopeful that much of this relatively static information would be available online for ready access.

Contact Management System

As part of the CRM initiative, it was envisaged that ECS would be implementing a turn-key CRM system which helped to manage and interface with existing databases containing information of vendors, pharmaceutical companies and partnering clinicians. As the business of ECS started to expand, ability to interact with supporting firms and medical practices was considered critical to ECS continuing success. Although at the time of this write-up, ECS already had databases, built using PC-based software to update such information, it was the ability to link these databases up using an integrated approach that would make a difference. Ultimately, ECS hoped to be part of the supply-chain management networks of vendors and pharmaceutical companies to enjoy the fast responses of the delivery services. With the CRM fully implemented, ECS would be able to interact efficiently with the upstream of the value chain (e.g., vendors and partnering clinicians) and downstream stakeholders (e.g., patients) by streamlining information flow in both directions.

Corporate Web Site and Branding

ECS's Web site forms an important part of its strategic vision. The Web site was customer-focused, aiming at providing information in breadth and depth to support patients and clinicians in learning about cardiology, health maintenance and ECS. In particular, the health maintenance dimension of the site was targeted towards focusing on maintaining patient wellness and other aspects of medicine. The Web site will be increasingly used as a primary source of information to support continuing patient education that was location-independent — the objective is to turn the site into a service delivery point (Simsion Bowles & Associates, 1998) rather than limit it to information provision.

Eventually, the Web site would be a one-stop shop for all stakeholders, both of upstream and downstream of the value-chain, as well as internal and external stakeholders of the ECS could interact with the ECS and the necessary systems.

For ECS, a critical part of building the customer relationship was winning the mindshare of customer. As mentioned, achieving quality was one of the strategic visions; ECS had adopted a focused strategy of building a brand in the customer's mind (Ries & Ries, 1998), founded on trusting the quality medical and personal service that ECS delivers, and reinforced through a distinct identity (i.e., targeted design of office, stationery and service culture). This also fit within the knowledge management (KM) perspective where the brand was classed as a key intangible asset in the marketplace that must be managed (Sveiby, 1997).

Although branding in the medical sector, particularly related to service delivery, was different from, say the retail sector, it was word-of-mouth and patients' experiences which counted most. ECS believed quality and branding went hand-in-hand. To achieve quality based on customer satisfaction, IT as a support infrastructure was considered very important. For example, one of the potential initiatives was to post letters of gratitude onto the Web site with RealPlayer clips allowing potential patients to share the experiences of existing patients. Also, providing the latest development echo-cardiology was also considered useful to educate patients and their families about ECS treatment philosophy and approaches. Given ECS's uniqueness in capturing this niche market, consolidating patients' confidence was considered the most critical part of the branding exercise.

The Way Ahead

As the health sector was increasingly dependent on IT, the survival of small specialist firms was often determined by how well they could adopt technology and use them effectively (Chandra et al., 1995). As large health service providers were leveraging heavily off IT to provide better services, competitive pricing and efficient operations, the only way for a small practice to be sustainable was through personalisation and customisation, filling the void created by the "production-line" style customer service characteristics of large medical practices. As such, both leveraging off IT and human resources within its business network were crucial.

ECS embodied the challenges faced by small medical practices and SMEs which were trying to be competitive in an increasingly complex marketplace. This was especially more acute for ECS as it was endeavoring to be innovative, adopting ambitious and unconventional initiatives that distinguished it from similar medical enterprises. While small practices comprised the mainstay of care in primary care models of healthcare (van Bemmel & Musen, 1997), they lacked a breadth and depth of resources that large-scale medical institutions enjoyed. ECS had sought to buttress this weakness by leveraging the skills and resources of external stakeholders and exploiting its small size.

ECS was a work in progress. In just over a year after commencing its strategic vision, it had managed to build a network of partners and a portfolio of projects that it had initiated. From these partners and projects, new nodes in the network were continuing to emerge, as were possibilities for collaboration. As it progressed, the challenges faced by an organisation such as ECS were the following.

Continuing to Maintain the Interests of Stakeholders

As ECS relied on its network of stakeholders, it needed to have an intimate understanding of all the needs, interests and direction of development of its stakeholders (both synergistic and competing interests). In addition, this also included the building and maintenance of the relationship with its most important stakeholder: the customers. The forging of "stakeholder intimacy" was a process of continuous improvement that ECS must maintain in order to position itself for the future.

In terms of the benefits gained among the stakeholders, particularly those by the academic community and ECS were continuous and long term. Given that the Australian Research Council (ARC) had, at the time of this being written up, been emphasising the importance of "linkage" research between academia and the industry, the collaborative relationship was strategically positioning itself for future opportunities. As mentioned by the ARC, one of the guiding principles was:

Encourage and increase partnerships between universities, research institutions, government, business and the wider community at the local, national and international level. (http://www.arc.gov.au/strat_plan/guiding.htm)

Eventually, this relationship formed between the local e-commerce research centre, the interstate university and other research institutes would become a platform to pursue further support from both the public and private sectors.

Collaboration Between Stakeholders

Working jointly with other alliance members had proven to be an effective way to minimise resources and maximise benefits. Ongoing relationships of this kind would evolve: as the relationships between ECS and its stakeholders ripen, further opportunities for collaboration between alliance members would also emerge (augmenting the model where ECS was the sole hub of contact). While this brought more opportunities to bear, it also resulted in additional complexity for ECS in managing and minimising any risk to the relationships. There would be a time in the future that ECS might not be the centre of this alliance relationship, but a worthy follower.

As this case was being written, ECS had been proposing to the participating institutions extension projects based on the interactive intelligent questionnaire. Three projects were proposed:

- Development of a prototype called clinical information management systems customisable interface which helps staff to interact with back-end medical information systems based on a set of preferences.
- Development of a clinical application service provider platform which serves as the middleware between the client software and the back-end processing server so that clinics and specialists working with ECS can share patient records and information securely without the need to exchange paper-based records.
- Exploration of the feasibility of secure document transaction technologies between ECS and its partnering clinics to test out the concept of secure patient data exchange.

The project champion role of ECS was of critical importance if not because of its willingness to participate regardless of the differences in philosophy and timeline, it would have been quite difficult to attain success to date. At the same time, the institutions' willingness to work with a tighter time schedule and real-life projects was also critical to the success of this relationship. To ensure the success of the extension projects, it is likely additional stakeholders — such as those from other segments of the medical field and commercial software developers — might have to be brought in.

Changing the Model as the ECS Grows

ECS's model for collaboration and strategy must be dynamic — rather than static. Complacency and lack of awareness of the needs to adapt could be a fatal error against competitors and a changing regulated healthcare marketplace. Firms like ECS must be prepared to alter their models driving the trajectory of their enterprise along the technological, tactical and strategic dimension. In the case of ECS, it was likely to expand in size, using its own resources and critical mass, while continuing to grow its network of relationships. These changes were likely to force it to question how best to reposition its current strategy and leverage its relationships and increased resources.

If ECS represented the beginning of a trend, then it was likely that small specialist practices would increasingly leverage off its business alliances and technology to gain a competitive edge. This was particularly true among entrepreneurial CEOs of SMEs who span the disciplines of healthcare, commerce and technology as in the case of ECS. The diffusion of IT into the health sector at the small medical practice level would create an interesting balance of power as seen in the business sector. In healthcare, one of the most complex domains, stakeholders increasingly sought to innovate yet were also reliant on each other to bolster core competencies that they lacked. The high ground would belong to those who could manage their core vision and the lattice of networks upon which they would depend.

What'd Happened Since the First-Round Project Was Completed?

The dot-com bubble burst in the year 2000 also had an effect on the direction of ECS. Although a project such as the interactive intelligent questionnaire had demonstrated potential to become a full-blown product with high demand, attracting funding to develop it to such a stage had proven to be difficult. The collaboration with the universities and employing postgraduate students to develop the questionnaire into a functional proto-type was a great success; however, neither Dr. Curtin nor the university partners managed to attract sufficient funding to commercialise the product. At the time this was written, Dr. Curtin was still keen to work with venture capital investors and large medical information systems vendors to breathe life into the product, but given Dr. Curtin's busy schedule (his first appointment usually started at 8.00 am), this had proven to be difficult.

Although one of the Honours-year project students joined ECS after his graduation, the current environment of ECS (average salary, lack of an established IT environment and long working hours) might make retaining staff difficult. For fresh graduates, often they would look for challenges and an environment that could give them maximum exposure to the latest IT systems. To achieve Dr. Curtin's visions, experienced IT staff would be critical to success. As this case was written up, both financial and incentive issues were yet to be resolved.

Due to these constraints, many of ECS's IT projects had difficulties grow beyond the prototype stage. While these prototypes were in workable conditions, they risked being superseded by commercial products if they were not given the opportunity to be further developed into commercial products. The lack of continuing commercial funding might also mean the cooling down of the relationships between ECS and the universities due to the lack of support to engage further postgraduate students. Once this happened, it might be difficult to revitalise them.

Reality Check

As this write-up was completed, Dr. Curtin was spending most of his time on building his medical practice to maintain cash flow. The reality was that there would always be further collaborations but if the medical practice failed, then it would have a major impact on Dr. Curtin. Consequently, Dr. Curtin was putting his medical practice as the priority. After saying so, Dr. Curtin was as keen as ever to maintain an ongoing interest in Honours-year projects of the local university hoping another bright group of students would take one of the prototypes further. He was still keeping close contact with those who had since left the project team (e.g., ex-Honours students) and ex-collaborators (e.g., the director of the local e-commerce centre). It might be one of the best ways to maintain a business network, particularly when mutual trust had already been developed. As with many small businesses, the ultimate survival factor is "cash flow."

REFERENCES

Achrol, R. S. (1997) Changes in the theory of interorganizational relations in marketing: Toward a network paradigm. *Journal of the Academy of Marketing Science, 25*(1), 56-71.

Chandra, R., M. Knickrehm, et al. (1995). Healthcare's IT mistake. *The McKinsey Quarterly,* (3), 91-100.

Charvet-Protat, S., & Thoral, F. (1998). Economic and organizational evaluation of an imaging network (PACS). *Journal de Radiologie, 79*(12), 1453-9.

Collen, M. F. (1995). *A history of medical informatics in the United States, 1950 to 1990.* American Medical Informatics Association.

de Geus, A. (1997). The living company. *Harvard Business Review, 75*(2), 51-59.

Drucker, P. (1957). *Landmarks of tomorrow.* New York: Harper & Row.

Dvorak, R. E., Holen, E., et al. (1997). Six principles of high-performance IT. *The McKinsey Quarterly,* (3), 164-177.

Dyer, J. H., & Singh, H. (1999). The relational view: Cooperative strategy and sources of interorganizational competitive advantage. *Academy of Management Review, 23*(4), 660-679.

Eden, L., Levitas, E., et al. (1997). The production, transfer and spillover of technology: Comparing large and small multinationals as technology producers. *Small Business Economics, 9*(1), 53-66.

Hunter, D., & Fairfield, G. (1997). Disease management. *British Medical Journal, 315*(7099), 50-3.

Lock, C. (1996). What value do computers provide to NHS hospitals? *British Medical Journal, 312*(7043), 1407-10.

Mitchell, J. (1999). *The unstoppable rise of e-health.* Commonwealth of Australia, Department of Communications, Information Technology and the Arts.

Nonaka, I. (1991, November-December). The knowledge-creating company. *Harvard Business Review.*

Packard, D. (1995). *The HP way: How Bill Hewlett and I built our company.* New York: HarperBusiness.

Ries, L., & Ries, A. (1998). *The 22 immutable laws of branding: How to build a product or service into a world-class brand.* HarperCollins.

Sellu, D. (1998). Have we reached crisis management in outpatient clinics? *British Medical Journal,* (316), 635.

Simsion Bowles & Associates. (1998). *Online service delivery: Lessons of experience.* Melbourne, State Government of Victoria.

Slack, W. (1997). *Cybermedicine: How computing empowers doctors and patients for better health care.* Jossey-Bass.

Staggers, N., & Repko, K. (1996). Strategies for successful clinical information system selection. *Computers in Nursing, 14*(3), 146-7, 155.

Stewart, T. A. (1997). *Intellectual capital: The new wealth of organizations*. New York: Doubleday.

Sveiby, K. E. (1997). *The new organizational wealth: Managing and measuring knowledge-based assets*. San Francisco: Berrett-Koehler.

Sveiby, K. E., & Lloyd, T. (1988). *Managing knowhow: Add value ... by valuing creativity*. London: Bloomsbury.

van Bemmel, J. H., & Musen, M. A. (Eds.). (1997). *Handbook of medical informatics*. Springer-Verlag.

Yoffie, D. B., & Cusumano, M. A. (1999) Judo strategy: The competitive dynamics of Internet time. *Harvard Business Review, 77*(1), 70-81.

Zachary, G. P. (1994). *Show-stopper!: The breakneck race to create Windows NT and the next generation at Microsoft*. Free Press.

Zollman, C., & Vickers, A. (1999). Users and practitioners of complementary medicine. *British Medical Journal, 319*(7213), 836-8.

ENDNOTES

[1]   This small medical practice belongs to the micro-business category which in Australia is commonly defined as having less than or equal to 10 full-time members of staff.

[2]   Names of the organisation and the stakeholders have been changed.

Simpson Poon is chair professor of information systems at Charles Sturt University, Australia. He is also a visiting professor at the University of Hong Kong. Dr. Poon earned his PhD in information systems from Monash University, Australia. He was the founding director and honorary fellow of the Centre of E-Commerce and Internet Studies at Murdoch University, Australia. Dr. Poon has been an e-business consultant and had worked with both government and business organisations in Australia and Asia. He has published widely in the area of e-business in both academic and professional journals.

Daniel May is a PhD candidate in the Department of Computer Science and Software Engineering, Monash University, Australia.

This case was previously published in the *Annals of Cases on Information Technology Management*, Volume 4/2002, pp. 512-525, © 2002.

Chapter XIV

Costs and Benefits of Software Engineering in Product Development Environments

Sorel Reisman, California State University, Fullerton

EXECUTIVE SUMMARY

A computer-based cost benefit (CBFM) forecasting model was developed to investigate possible long term effects of improved productivity that might be realized from the use of modern software engineering tools. The model was implemented in the development environment of Company X, a multinational corporation that manufactures embedded processor-based control system products. The primary purpose of the model was to generate comparative data to answer "what-if" questions posed by senior corporate management attempting to understand possible overall effects of introducing the new software development methodologies. The model provided comparative data regarding programmer labor costs, probably this company's most visible yet least understood line item in their monthly status reports. For Company X, the assumptions that were used to develop the CBFM were tailored to senior management's own priorities. Hence the model produced comparative summaries that ultimately allowed Company X to make the decision to begin implementing new software engineering strategies for product development.

BACKGROUND

Information technology plays an important role in the survival of an organization (Drucker, 1988). Coupled with the plummeting costs of computer hardware, it becomes a vital source for deriving efficient and cost-effective solutions to many of today's business problems. A well managed information system enhances a firm's ability to compete favorably and it minimizes the assumptions and guesswork in decision making that could lead to unsatisfactory performance.

In many companies information technology also shapes the process of product development (Abdel-Hamid, 1990). Organizations that are able to adapt new information technologies into their development process have often seen increased productivity and improvements in product quality. Many companies have investigated the utility of such information-based processes as CASE (computer assisted software engineering) methodologies, hoping to realize faster product development cycles, shorter production schedules, higher quality products, and lower overheads.

The cost of software development systems, like any information systems, stems directly from the cost of the resources required to provide and support the functions of the systems. The adoption of software development methodologies can be a serious strategic change. Therefore, "before management can support [software engineering] tool implementation, it must have a realistic understanding of the costs and benefits of the tools" (Smith & Oman, 1990). Cost-benefit analyses usually weigh the relationships between the costs and values of a system (Ein-Dor & Jones, 1985). Like any capital investment, the benefits must exceed the costs to justify the expense.

The economics of software engineering has often focused on software cost estimation, essentially a consideration of the costs related to single development projects. For example, software cost estimation techniques and models have tended to link software development costs to a project's size, its functional complexity, manpower requirements, and ultimately to the duration of the software development. Examples include SLIM (Putnam, 1980), COCOMO (Boehm, 1981), function point analysis (Dreger, 1989), ESTIMACS, and Price-S (Kemerer, 1987).

Because the successful implementation of software development tools requires a critical shift in senior management philosophy, it is essential to be able to justify the upheaval likely to result from these shifts. Within corporate environments such a justification is usually based on economic issues and related benefits, i.e., a sound business case. The economics of justifying these new methodologies to senior management requires a more global view of their costs and benefits than is typically found in the more detailed techniques mentioned above. That view must cut across projects, and focus on the effects of the methodologies on the totality of the development environment. In fact, as software engineering products (e.g., CASE tools) evolve into more integrated systems, they should begin more and more to address the needs of all phases of the system development life cycle.

For companies that may consider software engineering as a process that can improve the development of their own marketable products, a commitment to it should require the cooperative involvement from such diverse areas of the organization as sales, marketing, and customer service. For companies that instead consider that the methodologies may benefit their own internal systems development, the effect on the companies' own internal end user community can be substantial. In either event, the decision to employ these methodologies requires unified agreement within the infrastructure of the organization to support the new practices. (See for example, Cash, 1986; Gibson & Huges, 1994; McClure, 1989; McCleod, 1994; Statland, 1986).

SETTING THE STAGE

Often, because of time limitations, newness of technology, or simply because of a lack of industry knowledge and/or experience, senior management will base strategic decisions on global factors that relate only indirectly to the issue being considered. Consequently, in preparing the justifying business case for implementing new systems, it is important that accurate, appropriate, but most of all relevant factors be presented to senior management. Because relevant factors are primarily concerned with costs and revenues, a justification that can be quantified in that perspective will likely be better received by senior management. While it can be argued that implementing new software engineering methodologies will result in improved product quality (McGrath et al., 1992; Taylor, 1992), it may be very difficult to convince senior management of the appropriateness of those arguments to their own career objectives within the corporate environment.

Nonetheless, arguments based on reduced costs resulting directly from improved developer productivity are more easily understood if the reduced costs are associated with improved productivity. And because the single largest cost factor in most development environments is labor, schemes to reduce that cost without negatively impacting on product quality and schedules will certainly receive significant management attention.

Such was the case with Company X, a U.S.-based subsidiary of a European multinational corporation that develops and markets high technology, embedded microprocessor-based process control products which are distributed throughout the world. The company, which has been in business for more almost 25 years, is obligated to maintain and update the software that is resident in those products for years after they are first shipped. In addition to software maintenance, the company's 100 software engineers continue to develop new products, but without using modern software development tools. In fact, by 1994 almost 40% of all software developers were assigned full time to maintaining existing products while only 60% were developing new ones.

For the previous five years Company X experienced a constant shift of resources into software maintenance, generally an activity that tended to produce almost no incremental revenue. In order to sustain a consistently growing revenue stream, Company X's strategic plan called for the introduction of new products over a five-year period. Resource forecasts for the development of those products indicated that unless significant changes were made in the way that products were developed, the total number of software engineers that would be required to work on new products, and maintain old and new ones would become untenable. Clearly a change was required.

The planning division of Company X decided to investigate the utility of modern integrated software engineering methodologies and tools as a solution to this problem. To accomplish this, they developed a cost benefit forecasting model (CBFM) to allow them to explore, project, and compare the costs and effects of introducing such software engineering tools, tailored to their own development environment, focusing on senior management's particular concerns. The specific objective was to determine whether or not acquiring and implementing new software engineering tools could result in reduced long term development and maintenance costs. Except by implication, the quality of developed products was not addressed in the model. The model was used to analyze alternative scenarios to consider the consequences of Company X's decision to invest in software engineering methodologies as well as CASE products. The CBFM did eventually provide the basis for this company's commitment to new software development tools and practices.

In planning the design of the CBFM, an analysis of Company X's historical product development plans revealed that the cost element that most greatly affected their development budgets was software engineer (developer) headcount. This finding was consistent across all the company's development projects in all its development centers, irrespective of the kinds of products developed as well as the countries in which the centers were located. Consequently, it was decided that since this "variable" was one that was most easily controlled in terms of the company's own recruiting/layoff policies, it would also be a major variable in the cost benefit forecasting considerations.

An examination of the mix of employee titles, seniority, and job responsibilities across Company X's development projects revealed relative uniformity that, for the sake of forecasting, allowed all software engineers to be classified into two categories, those who participated in the development of new code and those who participated in the maintenance of existing code. It was felt that the "sensitivity level" of analyses needed to make the kinds of decisions that were required, precluded the need to differentiate among job categories in greater detail. As a result, developer headcount became the dependent variable upon which other variables were based. These included salary, overhead, training, and workstation (hardware and software) costs.

In keeping with Company X's need for a global view of cost estimating for software engineering justification, the development environment was considered to consist of any number of projects, all of which would eventually utilize the new methodologies and tools. In other words, the CBFM was considered to be a company-wide consolidation of single project cost benefit estimates.

CASE DESCRIPTION

The cost benefit model that was developed was used to compare and contrast two scenarios over a user-definable forecast period. The first was the "no change" scenario. No change implied that current practices would continue and that no drastic changes in the development environment would occur. The second scenario was the "if change" situation. In the context of the model this meant that software engineering products would be introduced into the organization in the first year and would result in a need for additional expenditures; there would also be immediate and long term effects derived from the use of the tools. The extra expenditures would be used for new or additional hardware and software as well as to provide more and ongoing training. Implied benefits that would result from the use of the new tools were considered to be related to improved efficiency of new code development, improved quality of software, and a subsequent improvement in the efficiency of software maintenance.

The CBFM that was developed was independent of any particular software engineering products or processes. Its design was based upon Company X management's

Table 1. Current status of Company X

Current year:	$n = 1995$
Total development personnel headcount:	$HT_0 = 100$
Average per capita cost for development staff:	$C_0 = \$.11M$
Ratio of maintenance/total headcount:	$M = .4$

Table 2. Assumed changes during forecast period

Annual Average Headcount Growth:	$R = 5\%$
Annual Average Inflation Rate:	$I = 4\%$
Annual Average Change in Maintenance/Total Headcount:	$S = +2\%$

appreciation of the overall benefits and conditions that could prevail within their software development environment if software engineering tools were introduced into the organization.

STATUS OF THE CASE

The appendix in this chapter details the characteristics, parameters, and equations that were defined and utilized in the implementation of the Cost Benefit Forecasting Model. Based on these, the CBFM was implemented and used (1) to analyze the software development environment for Company X, and (2) to justify the company's strategic decision to adopt software engineering products and methodologies.

Table 1 summarizes the current status (Known Values) of Company X. Table 2 lists the assumed values necessary to use the CBFM to model Company X.

"No Change" Forecast

Using the "no change" formulas shown in Figure 2, the effects of the data shown in Tables 1 and 2 were forecast. Table 3 contains that forecast and illustrates that after five years Company X will have spent more than $66 million on development personnel. By 1999 there will be 22 additional software engineers of which 18 will be assigned to software maintenance! This increase can be expected to result from the assignment of more programmers to support new, poorly designed code developed from 1995-1999.

"If Change" Forecast

To compare and contrast the effects of introducing new methodologies into Company X, it was necessary to presume some of the likely benefits that the company would realize. For the sake of this analysis it was assumed that as a result of the new

Table 3. Five year "no change" forecast

YEAR	NON MAINT H.C. (HD_n)	MAINT H.C. (HMn)	TOTAL H.C. (HT_n)	LABOR COST (M$) (C_n)	TOTAL LABOR (M$) (CT_n)
1995	60	40	100	0.11	11.00
1996	61	44	105	0.11	12.01
1997	62	49	110	0.12	13.12
1998	63	53	116	0.12	14.32
1999	63	58	122	0.13	15.64
			GRAND TOTAL:		$66.09

Table 4. Efficiency improvement rates

YEAR	NEW CODE EFFICIENCY(%) (QNn)	MAINTENANCE EFFICIENCY(%) (QMn)
1995	0	0
1996	10	8
1997	20	16
1998	25	25
1999	30	30

practices, the proportion of programmers working on maintenance projects would decrease at a rate of 2% over five years (S = -2%).

Over the five-year period Company X must continue to develop new products in order to remain competitive or even to avoid being shut down! To do this, the company would have to hire new programmers, all of whom will use the new tools and methodologies. As their proficiency with the new tools grows, so will their coding efficiency. This efficiency would begin to level off in the third or fourth year.

As newly developed products are released into the marketplace, those new products would constitute a growing percentage of products requiring maintenance. However, because of the improved quality of those products, programmers responsible for their maintenance should be able to perform with increasing efficiency. Table 4 illustrates one set of hypothetical efficiency trends that might be expected.

As described, improvements in efficiency come about as a consequence of additional expenditures in hardware, software, and a continuing training program. Table 5 contains a set of values that would be associated with these items.

Table 5. Associated cost factors

1. New Hardware And Software	
New Hardware Cost per capita:	CH = $7,500
New Software Cost per capita:	CS = $4,000
Initial Equipment Replacement Rate:	$E_0 = E_1 = 50\%$
Annual Equipment Upgrade Rate:	$E_n = 15\%$ (n = 2,3,4)
Depreciation Schedule:	$D_0 = 12.5$; D1 = 25
	$D_2 = 25$; $D_3 = 25$; $D_4 = 12.5\%$
2. Training	
Weekly Training Cost:	WTC = $2,200
Initial Training Period:	NT = 2.5 weeks
Steady State Training Period:	ST = 1.5 weeks
Headcount after Attrition:	B = 90%

Table 6. Costs of new methodologies

YEAR	EQUIPMENT COST (M$)	TRAINING COST (M$)	TOTAL COST (M$)
1995	0.42	0.55	0.97
1996	0.50	0.34	0.85
1997	0.26	0.32	0.59
1998	0.32	0.34	0.66
1999	0.30	0.35	0.64
TOTAL	$1.80	$1.90	$3.71

Based on the efficiency improvement factors and the additional costs inherent in introducing new software engineering practices, CBFM recalculated the additional costs to be realized over the forecast period. Table 6 indicates that after five years Company X will have spent an additional $3.71 million to implement new software engineering products and methodologies.

However, the efficiencies of the new practices provide significant benefits in terms of headcount cost savings that will result from the efficiencies assumed in Table 4. Table 7 illustrates the effects of those efficiencies. A comparison of this table with the "no change" forecast shown in Table 3 reveals that there is approximately $12 million saving in personnel reduction and costs due to the new efficiencies ($66.09 million versus $54.27 million). Furthermore, as Table 8 illustrates, in the fifth year a larger percentage of all developers will be assigned to non-maintenance programming tasks.

Table 9 illustrates the total budgetary comparison of the "no change" versus "if change" scenarios. An examination of the data reveals that over the five-year period

Table 7. Five year "if change" forecast

YEAR	NON MAINT H.C. (HDn)	MAINT H.C (HMn)	TOTAL H.C. (HTn)	TOTAL (M$) (CTn)
1995	60	40	100	11.00
1996	59	37	96	10.98
1997	56	33	89	10.59
1998	57	30	87	10.76
1999	58	27	85	10.94
			TOTAL:	$54.27

Table 8. Headcount comparison

YEAR	HEADCOUNT REDUCTION	NO CHANGE MAINT RATIO	IF CHANGE MAINT RATIO
1995	0	.40	.40
1996	9	.42	.36
1997	21	.45	.37
1998	29	.46	.34
1999	37	.48	.32

Table 9. Overall budget comparison

YEAR	NO CHANGE HEADCOUNT TOTAL (M$)	IF CHANGE HEADCOUNT TOTAL (M$)	IF CHANGE OTHER TOTAL (M$)	ANNUAL TOTAL (M$)	TOTAL SAVINGS (M$)
1995	11.00	11.00	0.97	11.97	(0.97)
1996	12.01	10.98	0.85	11.83	0.18
1997	13.12	10.59	0.59	11.18	1.94
1998	14.32	10.76	0.66	11.42	2.90
1999	15.64	10.94	0.64	11.58	4.06
TOTAL:	66.09	54.27	3.71	57.98	8.11

Company X would realize a savings of more than $8 million dollars. However, it is likely that during the first two years, when large capital outlays are more obvious than gains in productivity, that senior management may view the introduction of the new practices at best, with suspicion, and at worst, with outright hostility. However, if the storm can be weathered for those two years, dramatic improvements start to be realized at the end of the third year.

SUCCESSES AND FAILURES

CBFM, which is essentially a decision support system (DSS) was designed to permit Company X analysts to quantify categorized, anecdotal benefits of software engineering practices and products, particularly CASE tools. The value of the model lay in the ease with which the effect of "programmer productivity" on the company's software engineering labor force could be examined. Users of the model did not require a substantial detailed database, but instead could intuitively incorporate their own knowledge-base into assumptions necessary to use the model. CBFM required the analysts to quantify such issues in terms of a straightforward, multi-year software developer productivity profile from which they could quickly determine the effects of their assumptions. Furthermore, the level at which the model analyzed data and reported its results was oriented specifically to the interests and concerns of senior corporate management in Company X.

One of the major limitations of CBFM was that it only focused on costs associated with specific activities within specific phases of the traditional system development life cycle that involved software engineers. The model did not address issues and the related costs that might be associated with integrating software engineering methodologies across all dimensions of the corporate structure.

In fact, until we are better able to articulate what those issues are, it will be extremely difficult to understand the cost implications of corporate-wide adoption of software engineering practices. Orlikowski (1993) conducted a study of two corporations in order to address this problem and developed a theoretical framework that reflects the corporate complexities involved in this problem. Huff (1992) developed a budget framework which emphasizes the entire life cycle costs for CASE adoption. However, even ignoring the fact that this is a work in development, using the framework requires the selection of particular environmental-specific cost drivers.

CBFM was really concerned with a subset of such cost drivers — those related to costs and benefits of acquiring and using software engineering products rather than fully integrated systems and practices. In any case, until and unless fully integrated software engineering products become standard, practical, and cost effective, there is probably no real value in attempting to develop a computerized cost benefit model that considers "soft" costs that may perhaps one day, in some utopian environment be relevant to software development.

In its current form, one could argue that CBFM lacks precision. For example, the model does not distinguish among different levels of programmer. If it did, it might be argued that the introduction of modern software engineering tools may effect the productivity of experienced, senior programmers differently than that of inexperienced, junior programmers. Modification of this model to include more than one level of programmer would not be difficult. However, requiring an analyst to provide such a model with a different multi-year productivity profile for each programmer level might prove to be a terribly onerous chore with questionable, if not diminishing returns for the effort.

Another dimension of precision not inherent in the design of this CBFM is consideration of analyses on a project-by-project basis. For example, as described in this chapter, the CBFM for Company X was used to compare overall effects of development resources under the "if change" and "no change" scenarios. More precise results might have been be forthcoming if, for example, equivalent analyses had been performed on a project-by-project basis, then consolidated into a corporate aggregate.

A similar argument might be made regarding the fact the CBFM used the same assumptions across all Company X development centers. It did not consider that there might be individual differences from center to center based on such factors as geographical, cultural, local economic, or even historical differences.

Finally, it might be argued that the model's lack of facilities to enable analysts to explicitly quantify those intuitive issues discussed earlier in this chapter were a drawback to its utility. For example, the model did not explicitly address the difficult learning curves associated with the introduction of new software development tools. It did not specifically address realities of vendor products that include poor product support, inadequate documentation, buggy products, ever-changing standards, etc. Certainly, if the model had contained such quantifiable measures within a useful analytical paradigm, it would have been able to assist analysts in focusing on issues that lie at the heart of many of the controversies that have surrounded software engineering.

Despite the absence of these or other related factors, from a practical standpoint it remains to be seen whether or not the inclusion of such items would have produced substantially different results from the current model. And herein lies another drawback to this CBFM, either in its current form or as it might be extended. Any model requires validation through the use of real data from real environments. At the present time, considering the lack of uniformity of software engineering practices within the software development community, the kinds of data required to validate this model are simply unavailable. Despite the widespread accessibility of products such as CASE tools, many companies are unwilling to divulge the proprietary competitive advantage that their software development methodologies have provided them. Could it be that they are unwilling to confess to the difficulties and failures they have experienced?

EPILOGUE AND LESSONS LEARNED

The cost benefit forecasting methods described in this chapter were developed to investigate possible long-term effects of improved productivity that might be realized from the use of modern software engineering tools. A computer-based forecasting model was designed based upon the development environment of Company X, a multinational corporation that manufactures embedded processor-based control system products. The primary purpose of the model was to generate comparative data to answer "what-if" questions posed by senior corporate management attempting to understand possible overall effects of introducing the new methodologies. The strengths of this particular analytical approach are reflected in the fact that the model provided comparative data regarding programmer labor costs, probably this company's most visible yet least understood line item in their monthly status reports. For Company X, the assumptions that were used to develop the CBFM were tailored to senior management's own priorities, hence the model produced comparative summaries that allowed Company X to make the decision to begin implementing new strategies for product development.

Improving productivity through automation is an excellent method of cost control. Automation, a proven, cost-effective approach in many production and manufacturing environments can be extended to support software production (Levy, 1987). Until recently, system developers have been so caught up in automating users' systems that they have completely overlooked their own (Case, 1986). In addition, some practitioners have perceived systems development as an art that cannot be crafted by machines. However, as economic factors begin to impose on the development process, there becomes an increasing need to automate labor-intensive activities and processes.

New products and methodologies extend software development through automation. Like most automated systems, these systems provide tremendous potential to reduce the costs of labor by increasing development productivity while at the same time enhancing product reliability. The fundamental concept is to support the various phases of the system development life cycle with an integrated set of labor-saving tools.

But do these new techniques really work? It is too early to come to a firm conclusion one way or the other. The adoption of the new systems does not always assure success. Some testimonies that report a positive impact of CASE products have been quite encouraging (McClure, 1989) while other reports have been mixed (Taft, 1989). There are similar debates regarding Rapid Application Development (Gordon & Bieman, 1995) as well as other software-development related matters (Andriole, 1995). One of the reasons for this may be attributed to the current nature of the products. But as the technology evolves, product shortcomings that include inadequate functionality or integration will be resolved.

Another and perhaps more serious problem may be management-related. In some instances there may be a misunderstanding about what certain software engineering products can or cannot accomplish. For example, corporate management may view CASE tools as the ultimate software development solution while neglecting to correct deficiencies in project management practices. Reasons for failure may also be due to a lack of training or to poor enforcement of existing development methodologies and standards.

Product such as CASE tools should not be promoted on the basis of the merits of their technology. Instead, they should be considered as an investment that has the potential to address the problem of the rising costs of software production and maintenance. Toward this end, the implementation of new software engineering methodologies must offer a company economic justification. In the absence of mechanisms to project the costs and benefits of these technologies, it is difficult to demonstrate such justification. Until there is broad, cross-industry experience upon which deterministic decisions can be made, the cost benefit forecasting models such as those used for Company X are tools that can be used to narrow the options.

QUESTIONS FOR DISCUSSION

1.   What are some of the pros and cons a company can expect from adopting formal software engineering methods?

2.   If a software development organization does not adopt software engineering methods, what kinds of changes can it expect to see over the next few years?
3.   What, if anything, is unique about using cost-benefit analyses to determine whether or not to adopt software engineering methods?
4.   Are there cost-benefit methodologies that can be mechanistically applied to any organization to ascertain accurately the costs and resultant benefits of adopting software engineering methods?
5.   In this case, what were the main cost drivers chosen by Company X, and why?

REFERENCES

Abdel-Hamid, T. K. (1990). Investigating the cost/schedule trade-off in software development. *IEEE Software, 7*(1), 97-105.

Andriole, S. J. (1995). Debatable development: What should we believe? *IEEE Software, 12*(40), 13-18.

Boehm, B. W. (1981). *Software engineering economics.* Englewood Cliffs, NJ: Prentice-Hall.

Case, A. F., Jr. (1986). *Information systems development: Principles of computer-aided software engineering.* Englewood Cliffs, NJ: Prentice-Hall.

Cash, J. I., Eccles, R. G., Nohria, N., & Nolan, R. L. (1994). *Building the information-age organization: Structure, control, and information technologies.* Boston: Richard D. Irwin.

Dreger, J. B. (1989). *Function point analysis.* Englewood Cliffs, NJ: Prentice-Hall.

Drucker, P. F. (1988, January/February). The coming of the new organization. *Harvard Business Review*, 45-53.

Ein-Dor, P., & Jones, C. R. (1985). *Information systems management: Analytical tools and techniques.* New York: Elsevier.

Gibson, L. G., & Huges, C. T. (1994). *Systems analysis and design: A comprehensive methodology with CASE.* Danver, MA: Boyd & Fraser.

Gordon, V. S., & Bieman, J. M. (1995). Rapid prototyping: lessons learned. *IEEE Software, 12*(1), 85-94.

Huff, C. C. (1992, April). Elements of a realistic CASE tool adoption budget. *Communications of the ACM,* 45-54.

Kemerer, C. F. (1987, May). An empirical validation of software cost estimation models. *Communications of the ACM,* 416-429.

Levy, L. S. (1987). *Taming the tiger: Software engineering and software economics.* New York: Springer-Verlag.

McClure, C. (1989). *CASE is software automation.* Englewood Cliffs, NJ: Prentice-Hall.

McGrath, M. E., Anthony, M. T., & Shapiro, A. R. (1992). *Product development.* Stoneham, MA: Butterworth-Heinemann.

McLeod, R. (1994). *Systems analysis an design: An organizational approach.* Orlando, FL: Dryden Press.

Orlikowski, W. J. (1993). CASE tools as organizational change: Investigating incremental and radical changes in systems development. *MIS Quarterly, 17*(3).

Putnam, L. H. (1980). *Tutorial, software cost estimating and life-cycle control: Getting the software numbers*. New York: IEEE Computer Society.

Smith, D. B., & Oman, P. W. (1990). Software in context. *IEEE Software, 7*(3), 15-19.

Statland, N. (1986). *Controlling software development*. New York: John Wiley & Sons.

Taft, D. K. (1989, May 29). Four agencies using case tools get mixed results. *Government Computer News*, 37-38.

Taylor, D. A. (1992). *Object-oriented information systems planning and implementation*. New York: John Wiley & Sons.

APPENDIX

The Case Cost Benefit Model

This section describes the characteristics and parameters that were developed to forecast and analyze alternative scenarios for Company X. The model and its description are relevant to any organization that is in need of making similar decisions, provided that the organization accepts the assumptions made by Company X.

The model is based upon a five-year period ($n = 0, 1, 2, 3, 4$) for which comparative headcount and related costs are forecasted. Known values for the initial baseline condition must be set. As in the case of Company X, these values can be determined from an operating or strategic plan. A second set of "assumed" values is required to forecast the effect of natural changes that might take place during the forecast period, whether or not new software engineering methodologies are introduced. Figure 1 lists the required parameters.

The model assumes that the initial and ongoing average per capita developer cost (C_0) is the fully burdened cost. That is, it includes salary, benefits, overhead, etc.

Organizations that investigate the use of new, software engineering tools are often concerned with the ever-increasing problem of maintaining older or poorly designed code. As these systems age, the need for maintenance developers to upgrade or fix bugs can become overwhelming. Setting a positive value for the variable S (in Figure 1) will reflect that situation.

Figure 1. Initial forecast values

1. Known Values	
Total Development Personnel Headcount:	HT_0
Average per Capita Cost for Development Staff ($):	C_0
Ratio of Maintenance/Total Headcount:	M
2. Assumed Values	
Annual Average Headcount Growth (%):	R
Annual Average Inflation Rate (%):	I
Annual Average Change in the Ratio of Maintenance/Total Headcount (%):	S

Figure 2. Resource forecasting relationships

Initial Maintenance Headcount:	$HM_0 = M \cdot HT_0$
Total Headcount in Year n:	$HT_n = HT_0 \cdot (1 + R)^n$
Maintenance Headcount in Year n:	$HM_n = HM_0 \cdot (1 + S)^n$
Non Maintenance Headcount in Year n:	$HD_n = HT_n - HM_n$
Per capita Labor Cost in Year n (\$):	$C_n = C_0 \cdot (1 + I)^n$
Total Labor Cost in Year n (\$):	$CT_n = HT_n \cdot C_n$
Maintenance Cost in Year n (\$):	$CM_n = HM_n \cdot C_n$
Non Maintenance Cost in Year n (\$):	$CD_n = CT_n - CM_n$

The "No Change" Condition

For the "no change" situation, the variables shown in Figure 1 can be used in the formulae shown in Figure 2 to calculate the total annual headcount and labor costs as well as the ratio of headcount allocated to maintenance versus the headcount allocated to the development of new code.

These relationships permit the following items to be forecast:

- Annual headcount and cost of maintenance developers.
- Annual headcount and cost of new code (non maintenance) developers.
- Total annual headcount and cost.
- Total headcount expenditure over the forecast period.

The "If Change" Condition

In order to examine the "if change" condition it is necessary to consider the benefits that software engineering methodologies can provide as well as the additional cost factors necessary to bring about those benefits.

The Benefits of Software Engineering

Productivity gains that should be realized from the application of new methodologies suggest that their use will alter the ratio of maintenance to new-code developer headcount. It should also improve the *productivity* of all developers. These effects can be shown in two ways.

First, there may be the opportunity to reduce the number of developers assigned to software maintenance. Within the model such a change can be explored by setting the variable S to a negative value. This variable is the annual average change of the ratio of maintenance headcount to total headcount; a positive value reflects growth and a negative value indicates shrinkage.

Secondly, there should be an improvement in the productivity of all developers whether they are developing new code or whether they are involved in maintaining newly developed and easier-to-maintain code that will eventually enter the maintenance phase of the system development life cycle.

The model reflects both of these possibilities through the definition of productivity improvement variables, QN_n and QM_n. These variables may be used to define the yearly efficiency improvement pattern or trend of both new code development and code maintenance. For each of the forecast years, the following may be set:

New Code Efficiency Improvement (%):	$QN_0, QN_1, ... QN_4$
Maintenance Efficiency Improvement (%):	$QM_0, QM_1, ... QM_4$

On an annual basis, the effect of these variables would be to reduce the total number of maintenance and new code developers, each by the percentage indicated by QN_n or QMn.

The Costs of the New Practices

There are two major costs that the implementation of new tools introduces. These are (1) the cost of new equipment (hardware and software), and (2) the cost to train and retrain all development personnel.

Hardware and Software Costs. The decision to implement new tools is likely to result in a number of additional expenses that would otherwise not need to be borne. One of the more significant costs is for the purchase of new workstation equipment that can effectively capitalize on the functions available in modern software engineering tools. There will also be the cost of the new software itself.

In a development environment it is simply not realistic to insist that all developers or all ongoing projects convert immediately to the new systems. So, while conversion must be an evolutionary process, the implementation of new systems also requires a considerable and immediate commitment by management. The forecasting model reflects the need for that commitment by assuming that there will be substantial hardware and software acquisition in the first two years. Thereafter, a proportionally smaller expense relating to such factors as hardware/software upgrade or maintenance will be required annually.

Equipment costs are determined by setting the following values:

New Hardware Cost per capita ($):	CH
New Software Cost per capita ($):	CS
Initial Equipment Replacement Rate (%):	E_0, E_1
Subsequent) Annual Equipment Upgrade Rate (%):	$E_n (n = 2, 3, 4)$
Annual Hardware Depreciation Schedule (%):	$D_0, D_1, ... D_4$

A logical constraint on the sum of the replacement rates (E_n) and the total depreciation (D_n) over the forecast period is that each sum must be less than or equal to 100%. Depreciation rates should be determined on the basis of the organization's financial policies.

Using these constants, CE_n, the annual expenditure for equipment (software, hardware, and maintenance) can be calculated for all development headcount (HT_n) as:

$$CE_n = HT_n \cdot E_n \cdot (CS + CH \cdot D_n)$$

The total equipment cost, TCE, over the forecast period is:

$$TCE = CE_0 + CE_1 + CE_2 + CE_3 + CE_4$$

Training Costs. Another cost that will be incurred when new software development tools and methodologies are introduced into an organization is for the retraining of existing personnel. The model assumes that initially, all currently employed developers will require a significant amount of new training (NT), the costs of which must be borne by the organization. CBFM assumes that in the first year there will be an extensive training program for all developers. Thereafter, all previously trained employees will continue to require annual training, but to a lesser degree (ST).

The model also considers that of the total pool of developers there will be an annual attrition rate. Given the realities of the employment marketplace for experienced software engineers it must be assumed that there will be a steady annual influx of less experienced developers. These replacement personnel, as well as other newly hired developers will require a greater amount of training in their first year than they will in subsequent years.

In order to calculate the costs associated with these training requirements, the following values are required:

Present Value of Weekly Training Cost ($):	WTC
Initial Training Period (weeks):	NT
Steady State Training Period (weeks):	ST
Remaining Headcount after Attrition (%):	B

Annual training costs can be calculated as the cost to train all new employees plus the cost to train the replacement employees; additional costs must be borne to train, but to a lesser degree, all previously hired employees. If HT_n is the headcount total for year n, then the annual training cost (ATCn) can be shown to be:

In the first year:

$$ATC_0 = WTC \bullet NT \bullet HT_0$$

In subsequent years (i.e., n > 0):

$$ATC_n = WTC \bullet [NT \bullet (HT_n - HT_{n-1}) + ST \bullet B \bullet HT_{n-1}]$$

Total Costs

The consequence of undertaking the use of new software engineering tools will be an ongoing training program as well as the continued acquisition of new hardware and software. The annual total cost (TAC) associated with these new and additional expenditures is:

$$TAC_n = ATC_n + CE_n$$

Because software engineering methodologies are intended to provide long-term benefits to an organization, it is unrealistic to expect immediate cost savings in the early years of the program. In fact, because of the extra expenditure required, it is more likely that in the early years the "if change" scenario will cost more than the "no change" scenario. Because of this it is more useful to examine and compare the total expenditure over the five-year period.

Sorel Reisman is a professor of management information systems in the Department of Management Science/Information Systems, Cal State Fullerton, and university extended education director of information systems and services. In addition to his MIS educational responsibilities, Dr. Reisman is actively involved in the development, implementation and management of client/server software and systems, multimedia systems and applications, the Internet, intranets, and World Wide Web. Dr. Reisman is a member of the editorial boards of the Journal of End User Computing *and the* Journal of Global Information Management, *as well as* IEEE Software *and* IEEE Multimedia, *in which he serves as a referee, product reviewer, and writes regular editorial columns concerning contemporary issues in computing. Dr. Reisman's book* Multimedia Computing, Preparing for the 21st Century *was published in 1994 and reprinted in 1996. Dr. Reisman confers with a variety of U.S. and foreign multinational corporations in the fields of management consulting, information systems management, and multimedia computing.*

This case was previously published in J. Liebowitz and M. Khosrow-Pour (Eds.), *Cases on Information Technology Management in Modern Organizations*, pp. 57-71, © 1997.

Chapter XV

Systems Design Issues in Planning and Implementation:
Lessons Learned and Strategies for Management

Mahesh S. Raisinghani, Texas Woman's University, USA

EXECUTIVE SUMMARY

Telecommunications Company (TC) [company identity is concealed] produced a sales management application through internal and contract resources. This application, schedule graph (SG) system, was designed to automate the sales schedule process that had previously been a paper and pencil process. The system was designed and implemented in a matter of months to reduce cost and deliver an application that was long overdue. The project had been proposed for years, but funding issues had routinely delayed initiation. The sales development organization worked on the design and development for this application for approximately six months. The application was released with numerous software, hardware and network problems. The effects on the customer community, the information systems department and other stakeholders were sharp and far reaching. This case study investigates the lessons learned with this application and the implications for theory and practice. It can be instrumental to information systems managers, academicians and students to learn from the success and pitfalls of other organizations related to information systems development and management.

BACKGROUND

TC is a Fortune 100 sales company in the telecommunications industry with 50 regional sales offices across the country. TC employs thousands of people with both domestic and international operations, however, the user base for the SG system is approximately 150 employees nationwide. TC wanted to automate a sales scheduling process. Previously, directory sales representatives had been scheduled in markets or canvasses by a paper and pencil process. This process was very time consuming and led to tremendous frustration among sales managers. The burdensome manual process took valuable time away from sales coaching and selling activities that produced revenue dollars. The sales calendar stems 12 months and is typically updated on a weekly basis as personnel and markets can change rapidly in their business.

SETTING THE STAGE

The sales managers had been requesting an automated solution for years in an effort to end what had rapidly become an administrative job rather than a sales job. The sales organization conducted business and processed sales online, a "Paper-less Automated Sales System." While the system itself was paper-less, the output as a hard copy report was paper-intensive in spite of online capabilities. While the automation was highly desirable and the efficiencies could not be argued, there were financial considerations and constraints that continued to push the project to the back burner for several years.

The sales organization, in a time of declining sales, realized it was time to redirect its focus back into the development of its employees in an effort to strengthen its sales position. This meant that the manual processes needed to be removed by investing in process automation. An application needed to be developed to automate the sales schedule process. Time and money were significant factors in the development of the SG system.

CASE DESCRIPTION

In-house application developers as well as contract resources were involved in the design and development of the scheduling system. Rapid application development (RAD) was used to produce the SG system. The systems development life cycle was significantly reduced to save time and money for a project that had been long awaited in the customer community. Many applications that are produced from the RAD framework are developed in isolation, since that contributes to its speed to market. The deliverables and outcomes of RAD are the same as for the traditional structured development life cycle (SDLC) — a systems development plan, which includes the application being developed, a description of the user and business requirements for the application, logical and physical designs for the application, and the application's construction and implementation, with a plan for its continued maintenance and support. However, the traditional SDLC is indifferent as to the specific tools and techniques used to create and modify each of the deliverables listed above; RAD puts a heavy emphasis on the use of computer-based tools to support as much of the development process as possible in order to enhance the speed of development.

In this case study, the focus is on the use of RAD instead of SDLC. However the SG System suffered from not involving other organizational business units. This is noted as a significant drawback with RAD because traditional development stages are able to have greater overall business understanding, as speed is not the primary concern (Hoffer et al., 1998). In fact, David E. Y. Sarna (Eskow, 1997), who is well known for his work with RAD, argues that network planning and monitoring are important issues that can often be overlooked when development takes place in isolation. Specifically for optimal application and system performance, server(s) must have adequate memory, processing capacity, and redundant array of individual disks (RAID) are recommended for back-up and data security.

After the SG system was released to production, significant time was spent rewriting the application, deploying upgraded hardware, integrating network technology and developing a support staff to maintain the application. Additionally, the project and development costs were significantly increased with the activities required to stabilize the system. Approximately $350,000 was spent to purchase additional hardware and maintenance, and almost $550,000 was invested in application coding changes for the system to run more efficiently. Although $900,000 was not a significant percentage of TC's overall budget, proactive systems planning could have minimized or eliminated this expense and caused less frustration for the approximately 150 SG system users nation-wide. It is also important to note that the additional maintenance cost is not uncommon in applications developed using the RAD methodology due to lack of attention to internal controls (e.g., inconsistent internal designs within and across systems), lack of scalability designed into the system and lack of attention to later systems administration built into the system (e.g., not integrated into overall enterprise data model and missing system recovery features). In contrast, Office Depot jeopardized short-term profitability in 1996 when the company decided to delay most of its new application development for almost a year.

The decision was made in an effort to stabilize an application portfolio before it was released. While the company suffered in the absence of short-term gains, the additional time spent on development has had a long-term positive effect on their bottom line due to better quality results from the applications produced (Hoffman, 1998). This is a sharp contrast to TC's results with their release of SG. As noted in the SG discussion above, speed was the primary motivator in the development process. In the long run, neither time nor money was saved as a result of the rewrites and additional hardware acquisitions, that were expended in an effort to rescue a system that was the product of poor planning and design.

The current system architecture of the SG system is illustrated in Figure 1 and described in the following section.

Current System Overview

Purpose:	Online interactive tool for sales force market scheduling
Software:	In-house vendor product
Tools:	Oracle Developer 2000: Oracle, Procedure Builder, SQL Plus & SQL Net
Client:	Sales Managers
Database:	Oracle version 8i

Figure 1. Current SG system architecture

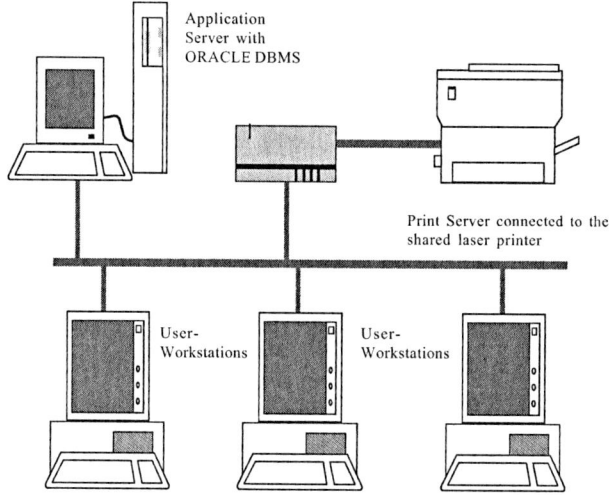

Application
Server with
ORACLE DBMS

Print Server connected to the
shared laser printer

User-
Workstations

User-
Workstations

System Architecture

DEC Server: Oracle database resides in a DEC cluster - VMS version 6.3.
Oracle Database: Users are connected to the Oracle database via TCP/IP and SQLNET.
User PC Configuration:

*   Processor: Pentium
*   Memory: 64 MB
*   Network: Token Ring or Ethernet
*   User Interface: Windows 95/98/2000
*   Communications: OnNet 1.1 FTP TCP/IP & SQL Net 2.1.4C

Application Server: Application is distributed on local Banyan Servers to the desktop.
Network Protocol: Application generates two significant types of network traffic.

Banyan traffic is generated from the PC to the local Banyan Server. Additionally, SQL Net traffic is generated across the wide area network (WAN) for database connectivity.

There are a number of considerations that must be reviewed when new software is being developed. Additionally, there are a number of technical and customer groups that must be involved in this process.

*   *Network:* The network will likely play a part in any new application or system that is developed today. In the case of SG, the network implications were not investigated before the system was released to production. The system was not written

to efficiently run across the network. When the application was released, it was determined that the network response was too slow because the application creates too much traffic across the wide and local area networks. The application required extensive bandwidth for customers to access as well as update the schedule graphs. Additionally, the client hardware was not powerful enough to mask the network response issues. To correct the current situation, the application was rewritten and eventually loaded and launched from the client's PC. The network services group created a special distribution utility that would load any software updates from the server to the desktop. The client checks the server at log-in and if there is a new version of software, it downloads the changes to the desktop. This process should have been in place before the application was released rather than after as damage control.

- *Hardware Considerations:* Applications for the most part are more sophisticated today with GUI interface and Web-enabled technology that dictates greater desktop processing power, memory, hard drive space, etc. With this in mind, it is important to understand what equipment your client community will be using to access the application you are designing. If an organization's customers have not yet upgraded to Pentium processors, you will need to address this in the design or possibility in the project budget as recommendation for new equipment may be appropriate. In the case of SG, the customer community had been accustomed to working on "dumb" terminals (DEC: VT420 & VXT2000) and only recently been given recycled personal computers that ranged in models from Intel 80386 to low-level Pentium personal computers. When SG was released, customers with 486 machines could use the application, but with very poor response, and those with 386 machines could not access the application because they did not have the minimum configuration. Needless to say, this caused significant customer issues and dissatisfaction from the client community. To correct this situation, the computer equipment group replaced the customer's equipment with Pentium II and III model computers. Again, a situation that could have been avoided with proper planning and requirements investigation.

- *Customer and Supplier Involvement:* Customer involvement is very important in the design of any new system. This is especially true with rapid application development as the development time is shorter. The need for customer input and involvement is significantly higher because there is no time for misunderstandings or multiple reworks of the requirements or the code design. Additionally, a strong partnership with customers will likely yield important feedback and help the information technology (IT) organization to produce quality software by better understanding customer needs.

Increasingly, organizations collaborate to complement their core competencies. New product development for TC, for example, is often a collaborative process, with customers and suppliers contributing complementary knowledge and skills. Information technology facilitates interorganizational learning since organizations collaborate closely through virtual integration. The role of information technology in lower and higher levels of interorganizational learning, cognitive and affective trust, and virtual and humanistic interorganizational collaboration should be leveraged (Scott, 2000).

- *Planning — Application Testing and Design*: The notion of "write once, run anywhere" may or may not be realistic yet in the software industry when many software vendors, of which Java is a good example, do not offer the same templates, training and features (Kobielus, 1998). However, planning is an important means to get there. Without proper planning before, during and after the designing and testing phases, there is little chance for implementation success. Quality in the form of customer needs is what must be planned in the design of any system software. The software must be functional and easy to use based on their frame of reference. Additional concerns include the speed at which the software performs based on customer equipment and interfaces. Anticipating future user needs and allowing flexibility in the design models are further examples of situations that should be addressed in the planning process to help ensure quality results. Thorough testing both from a technical perspective and a customer perspective are also vital events that require planning to achieve a successful implementation. Thorough testing of the technology including data integrity and product integration are key quality metrics. Additionally, customers must be involved in the testing process to verify functionality and utility of the system.
- *Maintenance and Support:* Before a system is released to production, there should be a clearly defined support organization to sustain the application and resolve any technical or training issues. In the case of SG, a support organization was not established or trained nor was any formal documentation in place with information about the system design. This was a difficult obstacle to overcome. It was almost six months before a technical review document was distributed and system overview sessions were held for key support groups including: operations, production control, network operations and help desk personnel. These activities should be part of a project plan for the release of any new software or system. These activities should be performed long before the system is released, as it will dramatically increase the success of the project and the satisfaction of the customer community.

The field of information technology moves very quickly, and it can be difficult to determine what new technology will be embraced by corporate America based on stability and business needs. Ravichandran and Rai (2000) identify top management leadership, a sophisticated management infrastructure, process management efficacy and stakeholder participation as important elements of a quality-oriented organizational system for software development. Businesses typically adopt a standard, as it is easier to implement, control and maintain. Over the next five years, "adaptive architectures" will be a primary design point for enterprise-wide technical architectures (Meta Group, Inc., 1998). The adaptive nature will allow more rapid change to be made to business applications and processes with reduced impact. It is predicted that by the end of next year, half of global 2000 businesses will cross-train technologists and business strategists (Meta Group, Inc., 1998). The merger should produce better information technology products and services which is a strong profit source. Global markets have increased significantly with Internet commerce and opened many more business opportunities. The Internet is and will be a driving force for years to come (Stevenson, 1995). Microsoft went through a reengineering process two years ago as part of a business strategy to dominate the Web as it has the desktop (Porter, 1997). Finally, as a result of business process reengineering and corporate transformation due to the digital economy, improving process has become an important area. From a systems development perspective, it is important to keep in mind that the issue is not process; it is what programmers are asking process to do, and where to apply it. Armour (2001) points out three laws of software process that are relevant in the context of this case study:

1.   Process only allows us to do things we already know how to do.
2.   People can only define software processes at two levels: too vague and too confining.
3.   The last type of knowledge to consider as a candidate for implementation into an executable software form will be the knowledge of how to implement knowledge into an executable form.

Current Challenges/Problems Facing the Organization

In the last decade, there has been a great deal of attention and discussion on Business Process Redesign that is largely tied to the Total Quality Movement (Lee, 1995). Work processes in many large corporations have been under executive scrutiny to improve quality and customer satisfaction. Continuous improvement in all business processes is the corporate goal. Business processes have two important characteristics: (1) internal and external customers, and (2) cross functional boundaries (Davenport, 1993). Developing new software is a good example of a high impact business process. If we relate this to the design of SG, there should have been internal and external customers and cross-functional involvement in the development process. Controlling business processes is important to their success and also a key to instituting incremental quality improvements. Information technology organizations have started to take on a customer advocate role as a means of controlling business and customer processes. "IT Capabilities should support business processes, and business processes should be in terms of the capabilities IT can provide" (Davenport & Short, 1994). Adherence to business processes maximizes efficiency and positive results.

Competition in the information technology field is fierce, specifically with respect to software development. This has had a strong impact on information technology organizations. Many system vendors are offering additional services to corporate clients including multi-platform support, application development and integration services as a means of increasing revenue beyond hardware sales. This is positive for users because it will likely drive prices down and create more product choices for customers. The supplier, however, will only gain competitive advantage if the product is innovative, quality rich, service friendly and priced reasonably. Hardware suppliers are recognizing the importance of service and emphasizing this with customers (Vijayan, 1997). There are several focal points that will determine the survival of specialized vendors or inter-company resources and they can be narrowed down to the following: (1) quality (features, ease of use, tools, etc.); (2) service (performance, ability to enhance, etc.); (3) price (Anderson et al., 1997). Competition can be tempered with current strategy and careful planning. Technology is a key ingredient in any corporate growth strategy (Way, 1998). The business units can be aligned properly to supply quality software on demand with aggressive planning including tools, requirements, design, testing, etc. Among the key networking O.S. companies, Microsoft has a long-term plan to fully integrate its development environments. Microsoft believes the result will reduce software development time and reduce product prices as a result of the efficiencies gained (i.e., training time and money, etc.) (Gaudin, 1997).

There is an inherent difficulty that surrounds software development. The nature of software is complex and unstable. Software, in spite of successful implementations, requires change over time whether it stems from customer request or vendor recommendation. "I believe the hard part of building software to be the specification, design and testing of this conceptual construct, not the labor of representing it and testing the fidelity of the representation" (Brooks, 1985). Realizing that complexity is an inherent and unavoidable obstacle, it becomes more important to focus on planning and designing the appropriate technology to avoid unnecessary updates. A well-designed system will lead to greater stability and reduce maintenance costs.

The release of SG was a challenging experience for the technical team as well as the customer community. How can future software releases be improved based on this experience:

- *Focus on Quality* — Quality is an overriding business strategy for most global companies. Customer satisfaction is a key success factor for application groups developing software in-house. If customers are not satisfied with the quality of the software produced, they can very easily find a contract vendor to develop any type of needed software. As previously mentioned, competition is fierce in the IT environment and customers have many choice which makes quality a distinguishing factor in success.

- *Importance of Planning* — While there is no easy cookbook for software development, planning is an essential ingredient to a successful software implementation. Proper planning will help minimize rework and contribute to quality and customer satisfaction. Planning will also provide more opportunity for reuse of models, coding, and testing.

- *Understand your Customer* — Extensive customer involvement is very important in the design and development of any new system, but particularly so with those developed under RAD. In the case of SG, customers should have been more involved in the requirements, design and implementation phases.

- *Initiate Cross-Functional Involvement* — Successful software development requires participation from multiple information technology groups as well as customer groups. A business process should be in place to coordinate the involvement of all appropriate groups.

It is suggested that within the next few years, "Architectural Webification" will replace "legacy-to-client/server migration" as the dominant design structure. In the next few years, it is predicted that enterprise-wide technical architectures will have quality measures for "time-to-implementation" and total-cost-of-ownership to benchmark: (1) logical horizon (linking any user to any node); (2) object heterogeneity (ability to share any information, service or process with objects from multiple platforms); and (3) systemic utility (scalability, portability, interoperability, availability, manageability, design transparency, and extensibility). It is expected that many Global 2000 companies will use a "software factory" model to implement new application systems (META Group, Inc., 1998). This approach will require software developers to focus on assembly and reuse as opposed to a craftsman approach.

With the strong influence of the Internet and electronic commerce on information technology, and more specifically its impact or potential impact on software development, systems designers and developers need to critically evaluate whether this medium would eliminate "middleware" issues and reduce development costs. Due to a constantly changing environment, a poor understanding of the user's needs and preferences, as well as a lack of willingness to modify existing organizational structures and decision models, the full economic potential of Web information systems (WIS) has not been realized. Reference models in the case of WIS, have to integrate a conceptual data and navigational model and — by choosing a system-specific optimal level of abstraction — should equally be applicable for structured as well as unstructured information. For that reason Martin (1996) introduces the term "intelligent evolution." With special emphasis on corporate business behavior, he compares three types of evolution with the classic Darwinian evolution based on the survival of the fittest:

a.   *Internal (r)evolution during the pre-deployment phase:*
     First order evolution, modifying a product or service (WIS) within a predesigned process and corporate structure.
     Second order evolution, modifying the process, methodology, or fundamental design of work (WIS methodology).
b.   *External Evolution:*
     Third order evolution, considering external factors outside the corporation (e.g., relationships with customers, other companies, governmental institutions, standardization committees, etc.).

One of the problems regarding modeling and developing hypermedia applications is the strong interdependency between presentation (user interface) and representation (explicit structuring) of published information. Many meta models and design methods for traditional client-server architectures lack the necessary object types for modeling this interdependency and are better suited for highly structured segments. Therefore, it is pertinent to design WIS with a close eye on the architecture of currently deployed systems and to compare them with each other.

REFERENCES

Anderson, J., Gallagher, S., Levitt, J., & Jurvis, J. (1997, January). Technologies worth watching-groupware, suites, security, data marts, and the ever-popular Web will be hot this year. *Information, 612,* 29-37.

Armour, P. G. (2001). The laws of software process. *Communications of the ACM.*

Brooks, F., Jr. (1987). No silver bullet: Essence and accidents of software engineering. *IEEE Computer, 20*(4), 10-19.

Eskow, D. (1997). Dealing with the aftershock of a new SAP implementation. *Datamation, 43*(4), 105-108.

Gaudin, S. (1997). Visual studio takes big step. *Computerworld, 31*(10), 2.

Hoffer, J., George, J., & Valacich, J. (1998). *Modern systems analysis & design.*

Hoffman, T. (n.d.). Office Depot endures app dev delays to ensure tech future. *Computerworld, 32*(11), 1, 97.

Kobielus, J. (1998). Write once, run anywhere: An impractical ideal. *Network World, 15*(9), 45.

Lee, T. (1995). Workflow tackles the productivity paradox. *Datamation.*

Martin, J. (1996). *Cybercorp: The new business revolution.* New York: Amacom.

META Group, Inc. (1998). *Enterprise architecture strategies* (White Paper). Retrieved from http://www.metagroup.com/newwhos.nsf/InterNotes/Link+Pages/eas+-+trends

Porter, P. (1997).Microsoft gets with the Web. *Software Magazine, 17*(7), 101.

Rai, A., & Ravichandran, T. (2000). Quality management in systems development: An organizational system perspective. *MISQ, 24*(3).

Scott, J. E. (2000). Facilitating interorganizational learning with information technology. *Journal of Management Information Systems, 17*(2), 81-114.

Stevenson, D. (1995, June). Positioning enterprise architecture. *ISWorldNet.*

Vijayan, J. (1997). Hardware vendors profit from integration push. *Computerworld, 31*(30), 28-29.

Way, P. (1998). The direct route to America. *Insurance & Technology, 23*(1), 35-36.

FURTHER READING

Bull, K. (1998). Making magic. *InfoWorld, 20*(25), 33.

DeCori, B. (1997). Bridging the gap. *AS/400 Systems Management, 25*(7), 47-50.

Hibbard, J. (1998). Time crunch. *Information Week, 691,* 42-52.

McGee, M. K. (1997). E-commerce applications in three months. *Information Week, 643,* 121-122.

Wilde, C. (1997). Do-it-yourself service. *Information Week, 659,* 87-90.

Mahesh S. Raisinghani was with the Graduate School of Management, University of Dallas, where he taught MBA courses in information systems and e-commerce. As a global thought leader on e-business and global information systems, he serves as the local chair of the World Conference on Global Information Technology Management and the track chair for e-commerce technologies management as well as a world representative for the Information Resources Management Association. He has published in numerous leading scholarly and practitioner journals, presented at leading world-level scholarly conferences and has published a book, E-Commerce: Opportunities and Challenges. *Dr. Raisinghani was also selected by the National Science Foundation after a nationwide search to serve as one the panelists on the information technology research panel and electronic commerce research panel. He serves on the editorial review board for leading information systems publications and is included in the millennium edition of* Who's Who in the World, Who's Who Among America's Teachers *and* Who's Who in Information Technology.

This case was previously published in the *Annals of Cases on Information Technology,* Volume 4/ 2002, pp. 526-534, © 2002.

Chapter XVI

Montclair Mutual Insurance Company

William H. Money, The George Washington University, USA

EXECUTIVE SUMMARY

Alan Rowne must plan and implement a number of information system (IS) upgrades at Montclair Mutual Insurance Company. This is a complex task given the evolving nature of IS developmental techniques, variety of vendor supplied tools and software, and industry organizational imperatives to modify the operations of firms to improve efficiency. He is concentrating on a decision to recommend either upgrading his present system or acquiring a new environment with new development tools. A new system development environment would offer Montclair Mutual Insurance Company the opportunity to develop information systems with strong system integration and interfacing capabilities that promise a high return on investment. This case presents data concerning the choices among information system development strategies, tools, systems which could be selected for upgrade or development, and implementation decisions for an insurance company facing a dynamic business environment.

BACKGROUND

Montclair Mutual was founded by community members of the Maryland farming area around Silver Hills, Maryland. The firm was originally formed (140+ years ago) to provide insurance (fire) for farms and buildings in the developing Maryland countryside. The company seeks to provide a high level of security and comfort to its policy holders in its commercial, residential, and farm insurance businesses. The annual report presents the firm's single guiding principle: to provide affordable reliable insurance for all policyholders; and to carefully balance assets against liabilities; strictly control administrative expenses; maintain a consistently high level of policyholder service; build customer confidence; and business growth in the years to come.

The company offers highly competitive insurance products in the seven major areas listed in Table 1. Table 2 presents the premiums and direct losses by state. Table 3 shows the business results for the previous five-year period.

The MIS Environment

Alan Rowne is the vice president of information systems at Montclair Mutual Insurance Company. He's facing a changing MIS environment and corporate pressure for performance improvements. He must decide what to recommend in order to address a number of systems development goals. His believes his broad options are to either apply the Systems Development Life Cycle (SDLC) methodology to upgrade the accepted mainframe systems used by the company for many years; or select and apply a new set of CASE tools, prototyping methodology, and database models to implement a new client server system.

The attractive new system components found in the client server environment are physically smaller machines that do not require specialized water cooled and air cooled facilities. When compared to a mainframe, the systems may house equivalent or greater amounts of CPU processing capability, disk space, network connectivity and memory at significantly reduced costs. The CASE and database tools in the systems marketplace are advertised to have broad functionality covering input required during the design of information systems, diagramming techniques, design specification components which can produce code when fed into code generators, and testing and debugging tools to speed the acceptance and testing of software.

Alan believes the cost of running a mainframe to support Montclair Mutual's system requirements is becoming unacceptable. He is well aware of the need to perform a feasibility assessment of all the costs and benefits of any new applications of technology since he has a technical undergraduate economics degree, work background as a financial analyst, and a masters degree in information systems from a large nearby university. A simple example of the apparent cost differences between the two options can be seen by comparing the cost of disk drives for a mainframe and a server. Disk drives that originally cost over $100,000 can be purchased for as little as several hundred dollars on a PC. It also appears that direct support costs such as power, cooling, specially prepared floor space, operating system licenses, mainframe systems support (from the manufacturer), and maintenance charges are combining to make the mainframe uneconomical for Montclair Mutual. However, Alan is not convinced that the client server environment will produce savings in indirect expenses since various trade studies have

Table 1. Direct premiums by type

Policy Type	% of Total Premiums	% of Losses Paid
Automobiles	27	32
Homeowners	21	26
Commercial Multiple - Peril	11	9
Workers Compensation	16	14
Farm Owners	8	12
Fire and Allied - Inland Marine	7	5
General Liability & Products Liability	9	2

Table 2. Direct premiums and losses by state

State	% of Direct Premiums	% of Losses Paid
Maryland	51	52
Virginia	22	17
North Carolina	17	21
District of Columbia	4	3
Delaware	4	5
Pennsylvania & West Virginia	2	2

Table 3. Five year growth (millions)

	1	2	3	4	5
Admitted Assets	67.4	75.3	80.7	84	87.8
Direct Premiums Written	46.7	49.6	57.6	64.6	65.7
Surplus	25.9	29.9	29.7	29.5	24.3
Direct Losses Paid	21.9	23.5	24.5	32.2	34.5

argued that there are significant hidden personnel, software, training, and networking costs involved in supporting a client server system. An additional disadvantage attributed to the mainframe systems used by Montclair Mutual is that the mainframe monthly maintenance costs have typically increased linearly as premium volumes increase. Modifying the system to add a new company, business unit, or line of insurance that is not provided by a wholly owned subsidiary (if a merger or corporate purchase were to occur) would require significantly higher new licensing fees and higher software support payments.

The decisions to be made are complex and have long term implications. The recommendations must all be presented to the Chairman and President Mr. J. David Adams and approved by the executive committee of the firm (consisting of the chairman, senior vice president, chief financial officer, vice president information systems, vice president marketing, vice president underwriting, and vice president claims).

The flow of information in the company is primarily hierarchical and sequential. The information flow supports a mainframe system geared for production activities, and is not oriented toward the production of management information. The actions on a policy application and potential acceptance are initiated in the mail room where the mail is opened, sorted, and routed to appropriate locations for action. A policy application is first directed to an underwriting team for review of completeness, resubmitted to an agency if incomplete, prepared for entry into the system (with completed or corrected data), and then entered into CICS screens when all data are complete. Simple editing is then initiated, error lists are reviewed and corrected, and the policy revolves through in-baskets until it is ready for scanning into the system as digitized paper. The 10-day-long process makes it difficult to track new business (the life blood of the company according to the president). Policies frequently have many errors, rejections, and corrections before they are approved. Error lists are continually maintained and updated to try to improve policy acceptance rates.

The systems used to implement the flow of information are best exemplified by a policy administration system (PAS) which supports policy/claims on an IBM 4381 mainframe, and a direct billing system (DBS). PAS is poorly structured; requires many manual codes, contains no data models, and has limited capability for rapid business or rule changes that are required when new and enhanced products are introduced by the firm. It does contain a stable database which has required very few changes or modifications for some time. However, Montclair is also being driven to consider alternative solutions because PAS will not operate when the year 2000 is reached. Date related programs and routines used for calculations and processing that will not work after the year 2000 are imbedded in PAS. The DBS is a full featured mainframe billing system that requires more system resources and support than the entire policy system (with a large monthly maintenance charge of $3,500). It is very risky to change the DBS, so modifications are only undertaken after completing extensive development and testing.

The systems must be coordinated by the "administrative staff" within Montclair Mutual through the imposition of rules, procedures, and behavioral standards that govern the client, policy, and charges for specific features in a policy, etc. Additional issues such as payment structures, discounts available to customers, advertising, sales techniques, and agent relationships are determined by state legislation and the "market environment" of the industry, but implemented by the business units of the firm.

Alan has identified several approaches to the problems presented by these old systems. Some of the approaches include: restructure the old systems into a tool based application (without rebuilding the systems); rebuild some or all of the applications to conform to new data definitions and business models based upon enterprise data modeling; do not convert any system to new client server technology, and build a bridge to the old systems using new technologies and data models; scrap everything and start over with an entirely new system; and purchase currently available new systems and/

or replacement products that can be integrated into the organization's current environment. Alan believes the analysis of which approach to take cannot be a simplistic assessment such as: does a system or subsystem work (leave it alone if it does), and does it incur high maintenance cost (restructure or rebuild if it does). Alan knows that the old systems can't be converted easily or supported forever; but he also knows its safe to assume that some old systems (if not all) will continue to be critical to the firm because even with their quirks and problems they must meet the ongoing business needs for the foreseeable future (or until completely effective replacements can be provided).

Role of Information Systems Department (ISD)

The role of the ISD at Montclair is to implement the strategic systems plan, and to collect, store, and provide access to the organization's data when it is required. Company databases and shared applications are funded through common overhead mechanisms, and corporate development projects must be supported by users who are members of cross functional business teams.

Alan is authorized to maintain a staff of 19 in the ISD. At present, this number includes an assistant manager, five programmer-analysts, four operators, one personal-microcomputer specialist, a project coordinator, two senior programmer-analysts, one systems programmer-database administrator, and two administrative employees. The staff is carefully supplemented (as needed) by responsive vendors who can provide additional reliable technical support.

Alan's management philosophy is to provide a work environment that is mentally stimulating to all ISD employees. He attempts to involve all of the staff in designing, building, and implementing successful products. The ISD staff is viewed as a special resource within the company. This group has a intra-company reputation of being very successful in bringing up new mainframe systems and modifications while maintaining existing ones. They also have an in-depth knowledge of the company and its products, and are of exceptional value at Montclair.

SETTING THE STAGE

Alan has identified four business limitations of the current system. First, there is significant information loss in the method used to store data. Many data elements such as children's names, ages, alternative phone numbers, previous addresses, and other predictors that may be used to track and identify credit or actuarial risks by experienced risk analysts and underwriters simply stay in the older systems' files. (As an example, agents know that individuals who may be poor risks are likely to change many things on their applications to a insurance firm; however, work phone numbers, home phone numbers, and old addresses will frequently remain the same on an application. These identifying variables could enable the firm's underwriters to link the high risk applicants to other policies or claims that have been made against the company).

A major file, the policy record, has grown extremely large over a long period of time with address fields with differently defined field lengths because it contains duplicate data that is stored in other records (Agent, Customer, Mailing Address and Policy Mailing address records). The data has to be changed in several locations when an address change for a customer is received. Separate customer address change requests are required to trigger changes to automobile, umbrella, boat, or jewelry policies.

Secondly, poor storage techniques require that large fields be maintained to store data (and remain blank if data doesn't exist). The attributes of the data stored in these field are relatively inaccessible because they can not be searched for specific values or conditions.

Thirdly, data terms and rules for action are translated into numbers and letters (codes) used to implement processing in the mainframe information system. This prevents the firm from updating profiles (and possibly selling add-on policies with increased premiums) of preferred customers who may now own homes with fireplaces and higher risk wooden (shingle) roofs, or pursuing new marketing strategies as the customer base demographics changes. From a marketing perspective, this also prevents searches which could identify inconsistencies in the insurance of a customer. This would mean that one individual could have extremely high insurance coverage on a very valuable home, and limited coverage on an automobile. However, the company is unable to approach the agent for the insured to determine if the coverage on the autos should be increased, and an umbrella policy offered to the customer to cover very large risks above the limitations on the combined house and auto policies.

Fourthly, the current system offers only a limited ability to manage the flow of information, and improve the service levels and performance of the individual departments. As an example, if an insured requests an increase in insurance coverage, the company first needs to locate a folder. The request for the policy/folder must be sent to the vault, and frequently the policy is "somewhere else," believed to be in transit, or missing (removed by someone from another department). The policy may have been sent to accounting, claims, or to someone's desk where it will remain in a drawer, stack of papers, or filing cabinet until their work is finished. Minimally effective tracking systems require manual entry of data from all the user locations. The constant updates required by the tracking systems are never fully complied with by employees. This confusion and complexity results in interdepartmental conflicts, customer frustration with the service delays, and possible errors because only one person can have the folder with the policy history at one time (to prevent loss of critical information or errors in the underwriting decisions related to coverage, claims, or the addition of insurance riders).

PROJECT DESCRIPTION

Potential Goals of New Systems

Alan believes he has to use this opportunity to construct a corporate data model to demonstrate the highly interdependent characteristics of Montclair's data and business relationships to the firm's various business units. Many complex information exchanges exist among the different departments (underwriting, claims, record storage (vault), and data entry/MIS) which have been assigned specific portions of the policy sales and management task. A new database model could use the policy entity as a central reference point for managing policy work, policy benefits, and all customer data. Coverage would be tracked to the basic data obtained from the policy. This major change could reduce the number of policy attributes because the stored data would be more concise. The new data model could also contain one customer number that may have multiple associated policies or policy numbers. For example, the customer number would associate the multiple customer addresses with dates to link changes in addresses with the appropriate policy data.

Alan is unsure of the rest of the modeling effort. However, he believes that the logical relationship could maintain an association between a policy and a customer number. The customer number would subsequently maintain an associated mailing address for that specific time. It even appears that similar processing logic could be used in managing the relationships for the coverages associated with a given policy. The business problem is that policies can be endorsed or changed many times during the policy term (and this has enormous implications for corporate liability). It is critical for Montclair business units to know what coverage was in force at a given time. Entities must be established to store the effective and expiration dates for each specific coverage and endorsement. In addition, the most current coverage with the coverage's effective and expiration dates identified for the last change in the policy must be maintained at some location. Singular relationships such as policy-address for one customer are easily understood by employees. However, if a customer purchases a second policy the same customer number is used again, and the address already stored is utilized by the system. If one assumes that the third, fourth, and other policies are obtained by the same customer more significant advantages are obtained. Even if a new policy is mailed to a different address (business, home, bank trust department, etc.), the entity policy address would be established. In that case, it would be used by the system logic as the default address unless it is not present. These detailed data dependent relationships are particularly important to the organization in the multiple policy business discount area. The current system has made it difficult to track multiple policies owned by one insured. Policies are purchased at different times or with variations in the name of the applicant (use of initials or middle names, etc.). The data are not always available for the agent (and different agents may have sold the policy to the owner). The present impact of this problem is twofold. The customer is overcharged; and, the state insurance department fines the company for discrimination and failing to follow its filed underwriting guidelines (a direct loss of profit for the firm).

Alan hopes there will be far fewer policy changes and missing address errors with this basic type of processing logic. He also hopes to address coverage concerns, readily compute correct discounts, and reduce incorrect charges. However, using all of this new logic would possibly introduce significant business and procedural changes for the business units, and many employees could find it difficult to accept that their unit is no longer exclusively in control of modifications to its data. In a survey of insurance systems to assess the options, Alan has reviewed the wide variety of available computer systems and software products. Lists of Property & Liability Insurance Systems available in the industry contain over 20 different systems available from more than a dozen different companies. The systems in the insurance area are written in many different languages with COBOL (for mainframe systems) being very common. Alan is concerned that the available commercial systems could use different data definitions, languages, databases and processes to implement the work of the organization. The risk is that the differences in the system databases used to store the data may ultimately require that the systems maintain contrasting data models, define the semantic meaning of the data differently, use conflicting values for specific information or cases, or implement incompatible data storage formats.

CURRENT STATUS: UPGRADE ASSESSMENT

There are a number of system and subsystem upgrades that could be included in any new system. The potential upgrades can be grouped into four major areas: data capture; operations and flow management; decision making (expert systems); and data storage and output.

Data Capture

Alan is considering recommending the acquisition of a screen development tool to use in building a new "front end" for all data entry. He has seen a tool called PowerBuilder from Powersoft. This package sits on top of a relational database (Oracle is used at Montclair) and could interface to a CASE tool. However, he has no personal experience with a GUI development environment, and wonders about its interfacing capabilities. He has been told that valid English term choices can be displayed for use during data entry. This would be an enormous change from the currently used codes associated with categories, policy types, and customer data.

Operations and Flow Management

He is also assessing tools to design and maintain an effective flow of information. The GUI front end screens could possibly be event driven, and enable one to enter data, present one with options based on this data, and perform consistency checks. In this way, the flow of data could be channeled at the time of entry by establishing selections available to the input staff based upon previous data value entries.

A system that controls the work would have to be very complex unless some tool or program can be found to support the development of this part of the system. It must establish mailboxes, define processing routines, set timing and triggers for the execution of routines, maintain comments, and construct forms to be used in systems processing, and effectively route images and files throughout the organization. Optional features include assigning a priority to the work, and placing the new business policy which requires review or approval/rating on the top of the list of items that must be handled in the underwriter's mailbox. The underwriter would have the ability and responsibility to act on the policy. It could be accepted, have comments placed on it in the file, or have an extract of the data sent to the CICS system where the data could be held in a staging area. Processing may also be impacted if an Optical Disk Storage System used for many years could be expanded using a window product with customized GUI screens. The optical image's goals would be to overcome the limits on the accessibility of the policy data. Currently, one department is able to locate and maintain control of a customer's file. What's needed is for the accounting department to have an image of the policy folder and all associated data on the screen while a claims representative is analyzing a claim. Multiple departments should be able to obtain a printed copy of the documents in the folder. Some newer optical systems may also maintain a "sticky note" or margin-comment capability that permits extensive comments or references to be associated with the information in the folder. This would permit the staff to document actions, note missing information, alert other department to problems, and even make comments on folders that are protected by special security levels. (Notes about jewelry or possessions of priority customers are only available to specific departments or underwriters at Montclair).

The obvious benefits from integrating and even expanding the optical storage of information include speeding up the customer service activities of the organization, and reducing the telephone "tag" that develops between the company, customer, and sales agent regarding a policy. With an optical storage subsystem there would be no repetitive editing of the data entered into the system that would force one to call back to the agent or customer to obtain more information.

Decision Making (Expert Systems)

Another option for upgrading the system is to address the underwriting decision-making area with an expert system. Alan recently viewed a vendor product written in C. It might be customized with the company's underwriting rules. (It can be purchased with some five hundred pre-developed business rules for one specific insurance line of business). This product operates in a OS/2 environment and uses DB2/2 as its relational database. Documentation for the system includes the rating information, underwriting manual, business rules for the company, filed ratings guidelines, and procedural manuals.

An expert system could replace the previous edit reports with system edits of inputs against the rules that are filed (with the state insurance commission) in the organizations underwriting guidelines. The rules would have to be incorporated into the screens used to collect the policy data. The expert system would attempt to rate a policy, and if it is not rated and accepted by the company, automatically refer the policy to an underwriter for further review and analysis. The analysis could possibly follow an exception rule, and only if the policy cannot be accepted would it be referred to the electronic mailbox of an underwriter (if the underwriters permit the introduction of this new process).

Data Storage and Output

Alan's view is that a client information file is potentially the primary source of management information. It could control all policy processing and still be a central file for other applications. Using this file, the company could develop increased managerial "what-if" capabilities that could be derived from the database and environmental information available to the business. As an example, "what-if" questions might include: what would be the impact of an increase in insurance deductibles for hurricane storm paths in particular regions of southeastern states? Similar questions might be developed for coverage limitations. The company would then be able to calculate the lost premiums from the increased deductible, and balance them against the eventual reduced payment and processing costs for not handling the smaller claims for less significant damages.

ORGANIZATIONAL IMPACTS

The planning of the firm must account for many organizational and business unit impacts associated with the complex changes contemplated.

Changes to Decision Rules and Process Improvement

The data corporate decision rules are now captured and recorded only by storing agreed to definitions or terms, and making these definitions available (via hard-copy memos) to all individuals in the company. This may be a problem for the company because the underwriters appear to disagree when presented with similar rating data. In effect, anything that creates standard data would create a more level playing field for the customer. One would then be able to obtain concurrence on a specific policy risk, or rating problem. Definitions would be documented for the policy questions and associated with the data that has been gathered from the customer. However, it is unclear if the underwriters will accept this new alternative. They may actually prefer the more ambiguous alternative and higher levels of discretion permitted when clear rules do not exist.

Training

A new system could radically change the training requirements of the data entry processors by eliminating the uses of arbitrary codes to represent field values, and by eliminating the need for data entry processors having to learn which codes to use under what circumstances. This training requirement has long been a problem during periods of clerical turnover. It is clear that experienced data entry personnel using the current procedures rapidly complete the entry of policy data. However, personnel with these requisite skills are difficult to acquire, train, and retain in the organization (since there is no career path for this highly specialized task). The system for training is now three-four months of training in codes, error messages, and data handling procedures. This could be significantly changed (hopefully reduced). Clerks would not have to memorize codes and information that would be presented to them on their screen.

This type of overall organization change is not without risk. The organization must be cautious of this change in operations and personnel. A future problem might develop if the highly trained underwriters are not present and future environmental shifts occur. New or less prepared staff underwriters might not have the experience to adjust the rules in an expert system to modify the criteria used to evaluate an application. This could be a cause for concern if the organization was required to change rules that are based on specific underwriting criteria rather than rules that are based on business or economics criteria.

Implementation Concerns

The question of "how" to implement any new system is also facing Alan. Decisions to implement changes have to be both strategic and incremental. The timing of the change must also be appropriate to the company's business success. Large investments cannot be expected in or after years where the company experiences significant losses. In addition, a high level of business and technical cooperation must be present between IS staff and the operating units. The tools, including the CASE, GUI front end, database, expert system, business rules, and optical scanning and storage capability all appear to be equally important. Finally, price is also a critical issue. The high cost of mainframe tools could make the adoption of a mainframe based CASE system far less advantageous than a less powerful client server CASE tool. However, the personnel and organizational costs of a new system or subsystem must be factored into any assessment and recommendation.

EPILOGUE

Alan firmly believes the company must remain flexible and able to make business decisions to respond to internal and external market demands and opportunities as quickly and economically as possible. This has been operationalized to mean that the firm's underwriters must be efficient and effective (not that clerical staff must be reduced as much as possible). The Montclair Mutual information system should not be a limiting concern in implementing this broad business goal. This goal implies that the company must be able to integrate multiple systems without specific system platform or contractual restraints. The client server platform appears to be the best platform for overall system integration, maximum flexibility, responsiveness, and cost savings.

The company has initiated the development of a full enterprise data model. The premise of this model is that all data for the company is defined and stored only once. Hopefully, this will eventually eliminate the various problems with the same information being in multiple places in various forms. A strong foundation for this model is relational database technology that provides the firm with better and faster access to information while enabling new applications and changes to be implemented at reduced costs.

Alan narrowed Montclair Mutual's options, and recommended rebuilding some (eventually all) of the applications to conform to new data definitions and business models based upon enterprise data modeling; and to building a bridge to some of the old systems using new technologies. The ISD has now developed the Maryland private passenger auto program primarily on a client server platform with the final feed of surcharge information to the mainframe policy system. The system is designed to use an Oracle relational database; PowerBuilder, for input screens; a workflow package; PC DocuMerge; and rating diskettes to produce a product that meets the specifications developed by a personal insurance lines task force. Alan believes the employees will like the new PCs, large color screens, and easy data entry expected with this new system. The design and development work has gone more slowly than expected. No implementation date has been set.

QUESTIONS FOR DISCUSSION

1. Identify criteria to be used to assess the recommendation.
2. What complex decision-making approach is required?
3. Describe the information flow and problems in the policy approval process.
4. Describe the firm's system procedures and coordination methods.
5. What are the limitations of the current system's data and information flows?

RECOMMENDED STUDENT CASE ASSIGNMENTS

1. Develop a Proposal for the Executive Committee and the President of Montclair Mutual.
2. Design the ISD organization structure which will best support the proposal you have developed for Montclair Mutual. Include job descriptions of individuals

assigned to any project development work, and reporting relationships of any new hires or consultants required by the proposal. In addition, identify any special training required and the costs of the training.

3.   Prepare a Development Schedule, Work Plan and Cost Estimate for all proposed development.

4.   Prepare a data model with appropriate entities and attributes for the customers of the company.

5.   Conduct a survey of potential commercial system development tools and options. Develop criteria appropriate to Montclair Mutual, and select a suite of tools appropriate for use on the development projects proposed.

William H. Money, PhD, is an associate professor of information systems with the School of Business and Public Management. He joined The George Washington University faculty in September 1992. His present publications and recent research interests focus on information system development tools and methodologies, workflow and expert systems, and the impacts of group support systems (GSSs) on organization memory, individual learning and project planning. Prior to his appointment to the GWU faculty, he supported the government acquisition of the Reserve Component Automation System (RCAS); served as program manager, and deputy program manager for the Baxter Healthcare Systems' Composite Healthcare System Program; and served as a product manager for Technicon Data Systems.

This case was previously published in J. Liebowitz and M. Khosrow-Pour (Eds.), *Cases on Information Technology Management in Modern Organizations*, pp. 9-18, © 1997.

Chapter XVII

Implementing Information Technology to Effectively Utilize Enterprise Information Resources

Yousif Mustafa, Central Missouri State University, USA

Clara Maingi, Central Missouri State University, USA

EXECUTIVE SUMMARY

This is a typical case of implementing information technology in order to assist an enterprise to effectively utilize its production information resources. The enterprise, a world-class leader in the pharmaceutical industry, currently keeps a large number of technical research reports on shared network media. These reports contain scientific specifications extremely essential to the enterprise's final products. In order to utilize these reports, a researcher has to navigate and literally read through each report to identify whether it is relevant to what he/she is currently working on. Often times, researchers find it more feasible to create their own reports rather than wasting time and energy on the searching process. Our solution to the problem is to create an information system which will keep track of these reports, provide a concise synopsis of each report, enable the researchers to search using keywords, and give a direct link to locate that report via a friendly Web-based user-interface.

BACKGROUND

The subject company is a world leader in life sciences focused primarily on two core business areas: pharmaceuticals and agriculture. Its dedication to improving life has been through the discovery and development of innovative products in the areas of prescription drugs, vaccines, therapeutic proteins, crop production and protection, animal health and nutrition. The company is also involved in the research, development, production, marketing and sales of organic and inorganic intermediate chemicals, specialty fibers, polymers, pharmaceuticals and agricultural chemicals. The company employs over 95,000 professional employees in more than 120 countries around the globe. Financial data are shown in Appendix A of this case.

SETTING THE STAGE

The company uses SAP Enterprise Integrated Software. SAP integrates and automates business processes starting with the procurement of raw materials, human resources, manufacturing and ending with the sale of the finished products. In order to manage the organization, the Decision Support Department frequently requires its employees, report developers, to generate various reports to respond to numerous types of queries. These reports are the major source of information for the organization to make decisions at any level of management. However, these report developers are not permitted to directly access the SAP database because of the following reasons:

1. Direct access of the SAP database would greatly slow down the SAP system performance.
2. The generic format and contents of the reports generated by SAP do not have specific use for most users.
3. Reconfiguring SAP to generate specific reports is very expensive since it is huge and written in ABAP (which is a German programming language), which makes it even more expensive to hire a programmer who knows ABAP.
4. Reconfiguring SAP would make it more difficult for the organization to easily upgrade to newer versions of SAP.

Therefore, the organization decided to set up a process in which data from the SAP tables are automatically copied to DB2 tables. The DB2 tables are immediately updated whenever the SAP data is changed. The SAP database is stored on Oracle tables on UNIX servers while the DB2 database are kept into IBM-DB2 servers. The company also decided to utilize a user-friendly report generator called Impromptu as their primary choice to access the DB2 database tables.

These reports cannot be generated by running a simple query on the DB2 tables because these reports often include computations which convert different sets of data into more complex information, such as calculating cycle-time for a product from the moment the raw materials are acquired in the warehouse to the moment the finished products are completed. This part of report generation takes the longest time because the formulas created must be tested for their accuracy. Due to the nature of the company, we are not at liberty to show samples of their actual reports. However, we have attached some general purpose sample reports that can be derived from the Impromptu report generator (see Appendix B).

Impromptu report developers individually generate their reports and store them on a shared network location. Currently, there are more than 5000 reports and close to 60-70 are created daily. However, storing these reports on a shared network location is of little or no use to the Impromptu report developers. Each time a report is needed, developers often start making an Impromptu report from scratch even though a closely similar report may have already been available on the network. Searching through the 5,000 plus reports is both time consuming and frustrating. A developer has to retrieve each report and read through it to determine whether or not it is relevant to his/her current needs. Almost all developers prefer to start from scratch rather than try to search the network. A single Impromptu report could be very costly since each may take anywhere from 15 minutes to 12 months to generate, depending on the complexity of the report. The cost of generating a report can be broken down into:

•   Searching the database tables for the required fields.
•   Analyzing and deciding on the logical combination of these fields, then generating the correct mathematical and statistical functions required for the report.
•   Testing the accuracy of the formulas on the report.
•   Fully documenting the report.

Each Impromptu report is saved in two formats, .pdf and .imr format. The .pdf format is a snapshot of the report that can only be viewed using Acrobat Reader. The .imr format, on the other hand, represents the executable version of the report which can be "run." The .pdf format is necessary as an Impromptu report developer can quickly glance to decide if it is the report that he/she needs. This is important because "running" an Impromptu report is a CPU time-consuming operation.

CASE DESCRIPTION

The first step in our problem-solving approach is to explicitly and clearly identify users' requirements. We used the personal interviews technique (described in Dennis & Wixom, 2000; Hoffer, George, & Valacich, 1999; Osborne & Nakamura, 2000; Whitten, Bentley, & Dittman, 2001), with the system users to identify the two following requirements:

1.   Providing developers with the capability of documenting and saving their reports in a searchable manner.
2.   Enabling the developers to search quickly and easily, via a user-interface, for a target report using different search items (which will be discussed in greater detail later).

Our solution to the problem is to develop an information system that would provide a rapid and easy tool to document, store, and search reports. Our system will keep a repository of searchable data of each report developed (in addition to the existing ones), including a link to the storage media where the report was saved. The system would enable developers to quickly search for any report using date report created, developer name, developing department, or a combination of key words.

In order for us to describe the processes of the new system, which we named the ImpromptuReport Dictionary, along with the data flowing between them, we used the DFD Gane and Sarson notations (described in Dennis & Wixom, 2000; Hoffer, George, & Valacich, 1999; Jordan & Machesky, 1990; Osborne & Nakamura, 2000; Whitten, Bentley, & Dittman, 2001). A DFD (Data Flow Diagram) is a graphical modeling tool used to depict the processes that a system will perform along with the data that flows in and out of each process. Figure 1 shows the context diagram of the system, which is the highest level of abstraction. Usually, the context DFD shows one process representing the whole system, the data which flows in and out that process, the origin of data (source), and the final destination of the data (sink).

Figure 2 shows level-0 DFD, which is a decomposition of the context DFD, where the system performs two major processes: updating the ImpromptuReport, and searching ImpromptuReport Directory Database. The diagram also identifies three data stores:

1. D1, our proposed searchable repository.
2. D2 and D3 are the locations on the network where the Impromptu reports (both the pdf and imr versions respectively) reside after they are submitted.

D3 is the location on the Impromptu report developers' personal computer on which they saved the Impromptu report so that they can modify it at later time if the developer chooses to.

Figures 3 and 4 depict further DFD decomposition in order to identify more processes.

Next, we modeled the data which the system needs to function properly using the ERD Chen's notations (explained in Dennis & Wixom, 2000; Hoffer, George, & Valacich, 1999; Jordan & Machesky, 1990; Whitten, Bentley, & Dittman, 2001). The ERD (Entity Relationship Diagram) is a graphical modeling tool used to depict the data entities, their attributes, and relationships.

Both the DFD and the ERD are excellent graphical tools for modeling purposes; they are also beneficial communication tools to validate that the software development team has accurate understanding of the system and users' requirements. Eventually each

Figure 1. The context DFD

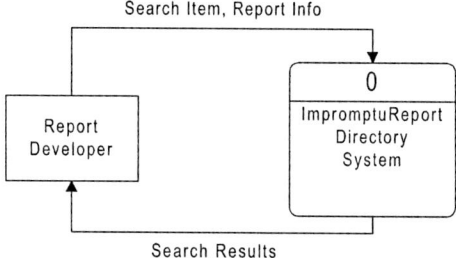

Figure 2. Level-0 DFD

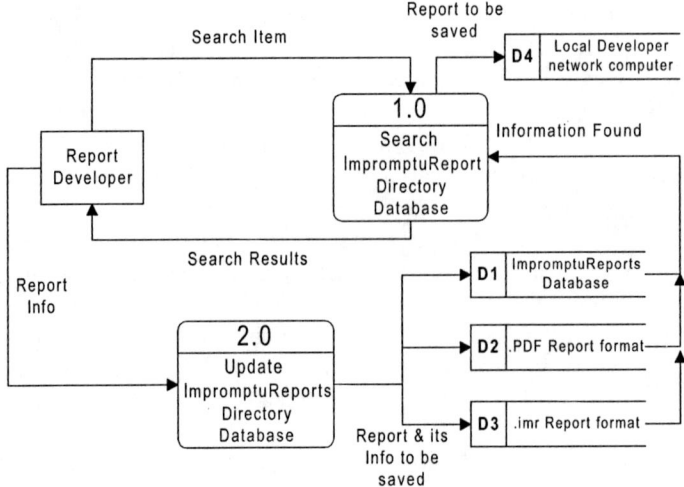

Figure 3. Level-1 DFD for Process 1.0

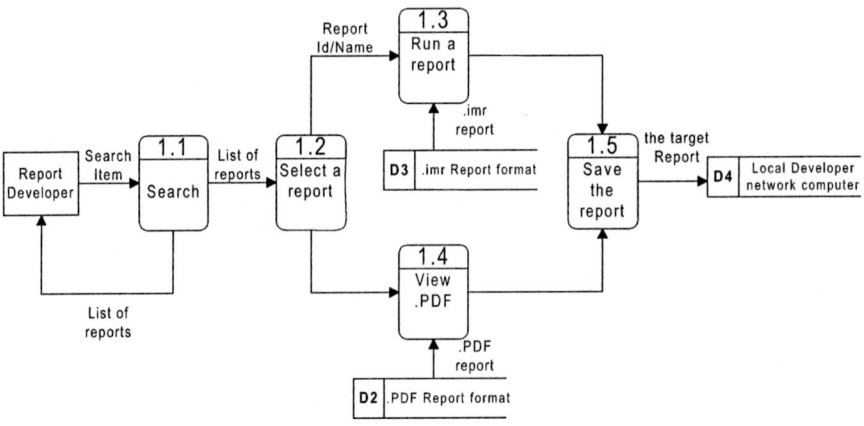

process on the DFD will be translated to a program, and almost every ERD may become a database table.

Figure 5 shows that the system contains three entities, which are relevant to the functions of our system, with only their partial attributes shown due to space limitations. However, to increase the efficiency and maintainability of our system, we made the decision to merge these three entities into one database table. This resulted in minimizing response time since querying one table is often quicker than navigating three.

We named the resulting table, ImpromptuReport, which has the following set of attributes:

Figure 4. Level-1 DFD for Process 2.0

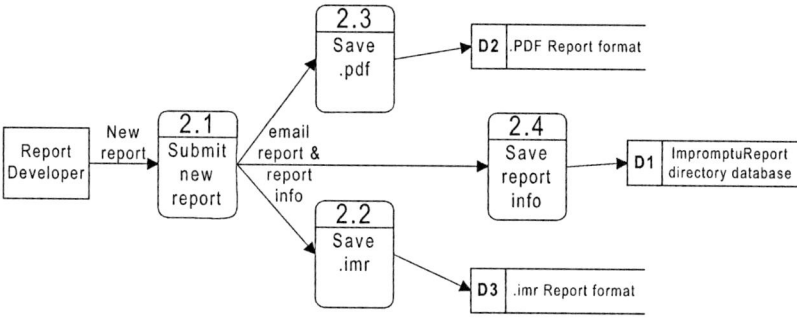

Figure 5. The ERD for the ImpromptuReport Dictionary system

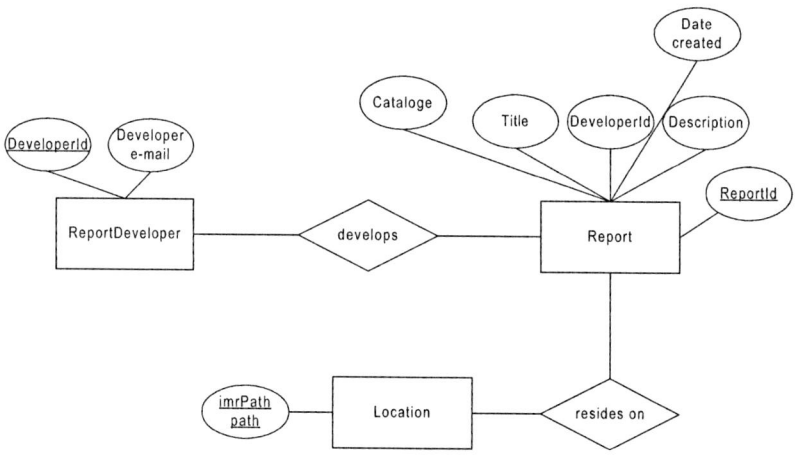

- **Report Id:** a unique numeric identification number which will automatically be generated whenever a report is archived. This is the primary key of the *ImpromptuReport* table.
- **Developer Id:** A string representing the developer's unique Id within the enterprise.
- **Business Function:** The department which the Report was made for (e.g., Inventory, Human Resources).
- **Catalog:** A string used to identify the various databases where certain reports are saved. Each business area within the company has its own database identified by a unique Id.
- **Report Title:** A string representing the title of the report as given by the developer.
- **Description:** A sting describing the functions and contents of the report.

- **imrPath:** A hyperlink to the .pdf version of the Impromptu report on the network.
- **pdfPath:** A hyperlink to the .imr version of the Impromptu report on the network.
- **HotFiles:** A list of the data files from the Oracle database needed in the Impromptu report. This data is not available from the SAP database.
- **Date Created:** The date when a report was created by the report developer.
- **Date Revised:** The date when the report developer revised a report.

In order to avoid any anomalous behavior (O'Neil, 1994) of this table, we had to make sure that the table is normalized in the third normal form (3NF) using the following tests (Ramakrishnan, 1997; Ricardo, 1990):

1. Since there are no multi-valued (repeating) fields, then the table is in the 1NF.
2. The table is in the 2NF if it is the 1NF and all the nonkey attributes are fully functionally dependent on the key. In other words, if the key is a single attribute, which is true in our table, then the table is in the 2NF automatically.
3. The table is in the 3NF if it is in the 2NF and no nonkey attribute is transitively dependent on the key. By examining our table, it is clear to us that the value of every nonkey attribute is only determined by the primary key of the table and not any other attribute.

Any further testing of a table which is in the 3NF is often unnecessary since many real-world databases in 3NF are also in BCNF (O'Neil, 1994).

A typical ImpromptuReport table would look like the one shown in Table 1.

Shown on the following pages are our Web-based graphical user-interfaces that users will use to provoke the various system functions. We followed the design

Table 1. An example of an Impromptutable

Attribute Name	Sample Value
Report Id	00990
Developer Id	Nm7435
Business Function	Inventory
Catalog	R3 Battg
Report Title	Manufacturing Goods Receipts
Description	Summarizes goods receipts from process orders for a product, product group, material type, month, year, and plant. Purpose is to provide data on manufacturing performance.
imrPath	\Reports\Planning
pdfPath	R3 Battrchkg.Pdf
HotFiles	Dbrport.MIS
Date Created	04/05/00
Date Revised	0/06/01

guidelines explained in Dennis and Wixom (2000); Hoffer, George, and Valacich (1999); Jordan and Machesky (1990); Navarro and Tabinda (1998); and Whitten, Bentley, and Dittman (2001).

Figure 6 shows the system Dialogue Diagram (as described in Dennis & Wixom, 2000; and Hoffer, George, & Valacich, 1999), where the system can be provoked via the company Web page. Developers will then be given the choice to search or submit a report for saving as shown in Screen 1. Upon selecting the search option, developers can use a number of search keys as shown in Screen 2. Once the system finds a match, the developer can then highlight the specific report and version to display.

Upon selecting the "submit a new report" option, Screen 4 will be displayed and all the information will be submitted to the system administrator. The system administrator, in turn, will use the same information given to save the report information to the ImpromptuReport database.

The final step in our case was to implement and operate our system. The following systematic steps were followed in order to materialize our design into a fully functional system which meets users' requirements stated in the beginning of the case.

1.   *Creating the database*
This includes creating the ImpromptuReport table using Oracle database, creating the form necessary to enter data into this database, and populating the table with data. All authenticated users (Impromptu report developers) will have read-only access to this database while a system administrator will have read-write access.

Figure 6. Dialogue Diagram for the ImpromptuReport Dictionary System

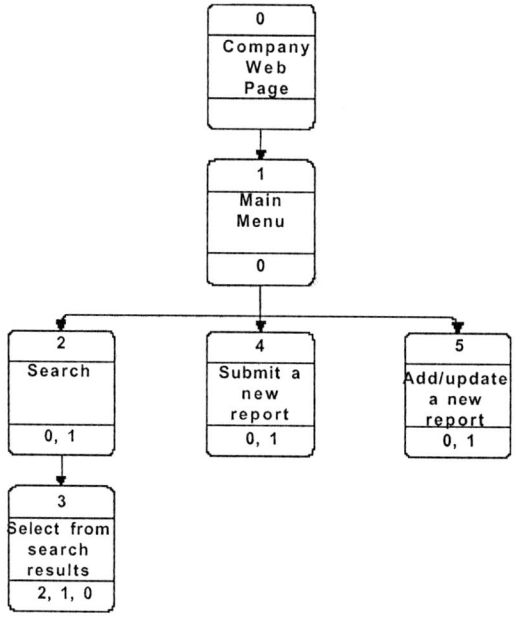

Diagram 1. Narrative view, Screen 1

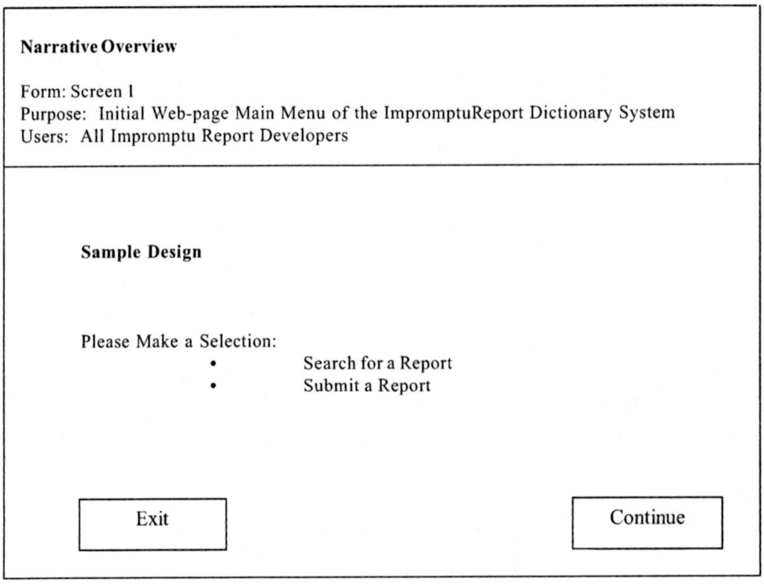

Narrative Overview

Form: Screen 1
Purpose: Initial Web-page Main Menu of the ImpromptuReport Dictionary System
Users: All Impromptu Report Developers

Sample Design

Please Make a Selection:
- Search for a Report
- Submit a Report

| Exit | Continue |

Diagram 2. Narrative overview, Screen 2

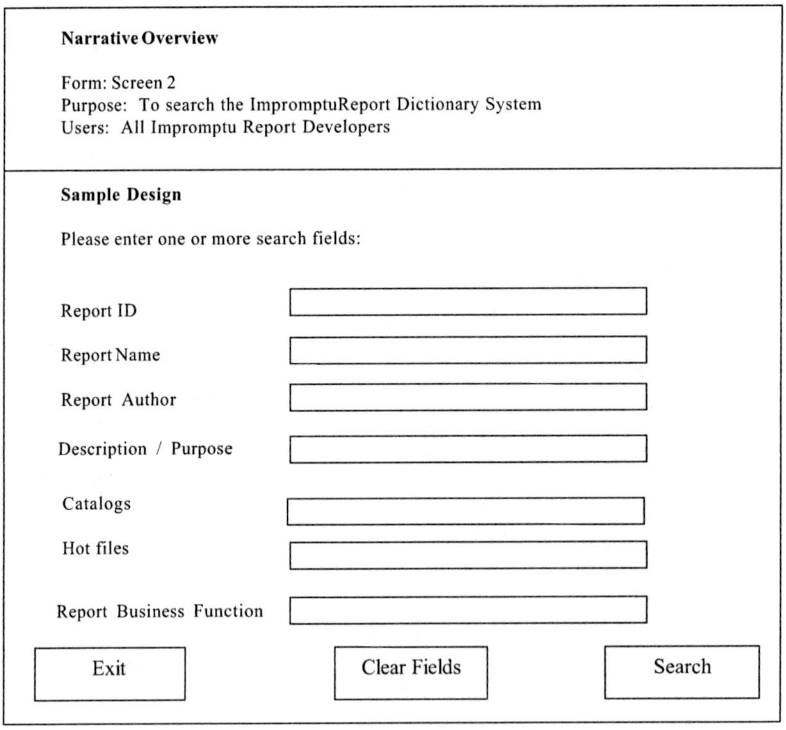

Narrative Overview

Form: Screen 2
Purpose: To search the ImpromptuReport Dictionary System
Users: All Impromptu Report Developers

Sample Design

Please enter one or more search fields:

Report ID

Report Name

Report Author

Description / Purpose

Catalogs

Hot files

Report Business Function

| Exit | Clear Fields | Search |

Diagram 3. Narrative overview, Screen 3

Narrative Overview

Form: Screen 3
Purpose: To display results obtained from searching the database
Users: Impromptu Report Developers

Sample Design

Click on the .pdf link to view the Impromptu pdf file in Acrobat Reader

Report ID	Report Title	Developer Id	Description	Date Created	Catalog	HotFiles	Business Function	pdf.Path	.imr Path

*If there are no matching reports this will be displayed by a text message.

Exit		Back

Diagram 4. Narrative overview, Screen 4

Narrative Overview

Form: Screen 4
Purpose: To submit a new Report to the ImpromptuReport Dictionary System
Users: All Impromptu Report Developers

Sample Design

Please provide the following information:

Report Title

Developer Name

DeveloperId

Description

Catalog

Business Function

HotFiles

Date Created
Date Revise (when applicable)

Exit	Attach .imr	Submit

Diagram 5. Narrative overview, Screen 5

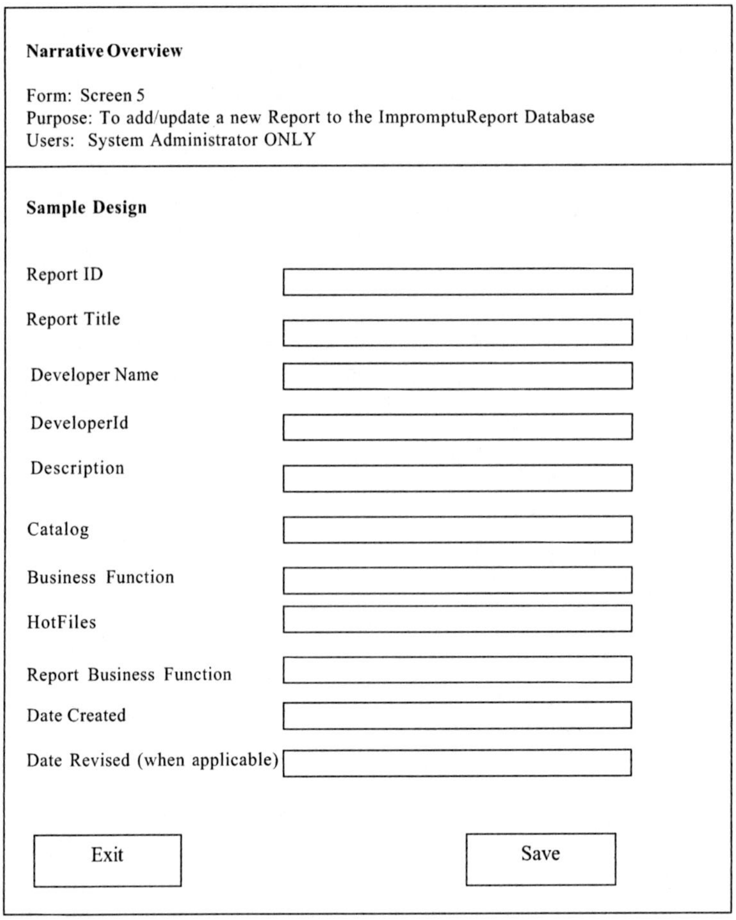

Narrative Overview

Form: Screen 5
Purpose: To add/update a new Report to the ImpromptuReport Database
Users: System Administrator ONLY

Sample Design

Report ID	
Report Title	
Developer Name	
DeveloperId	
Description	
Catalog	
Business Function	
HotFiles	
Report Business Function	
Date Created	
Date Revised (when applicable)	

Exit   Save

2.   *Creating the Web interface and the search mechanism*

A Web-based interface was created to be used to navigate through this system. Creating this Web interface includes creating an ASP form and developing all the codes which will be required to connect the Web interface to the database and enable the user to search the Oracle database by submitting a search on the HTML forms. Some valuable tips and procedures to execute this step were founded in Champeon and Fox (1999), Friedrichsen (2000), and Hentzen (1999).

3.   *Training and documenting*

All users will be trained to search for any report as well as submit their own reports for saving. A full-scale documentation of all aspects of the system, including operation and troubleshooting, was conducted as part of our project.

CURRENT CHALLENGES/PROBLEMS FACING THE ORGANIZATION

We believe that the company will face three types of challenges as a result of implementing our system:

1. **Cultural:** The system will enforce the concept of team work in which report developers have to adapt to reuse and build on top of other players' work. The system will also enforce the culture of personal accountability where each developer has the responsibility of fully and properly documenting his/her reports so that it can be utilized by other developers. Additionally, report developers will have to follow a standard procedure and format when developing and/or saving their reports.

2. **Operational:** The company must develop an operational procedure and allocate the required resources in order to maintain the system on a regular basis. Maintaining the database and the other files and providing developers with Ids are two examples on ongoing operational procedure.

3. **Technological:** Report developers have to face the challenge of learning and utilizing the advances of information technology in order to improve their performance. The company, on the other hand, will need to search fo the most efficient report development tool. SAP is about to release a new version that has more report development features, therefore the company will have to evaluate SAP development tool versus Impromptu.

REFERENCES

Champeon, S., & Fox, D. (1999). *Building dynamic HTML GUIs*. CA: M&T Books.

Dennis, A., & Wixom, B. (2000). *Systems analysis and design*. NY: John Wiley & Sons.

Friedrichsen, L. (2000). *Access 2000*. AZ: Cariolis Group, LLC.

Hentzen, W. (1999). *Access 2000 programming*. CA: Osborne/ McGraw-Hill.

Hoffer, J., George, J., & Valacich, J. (1999). *Modern systems analysis and design*. MA: Addison-Wesley.

Jordan, E., & Machesky, J. (1990). *Systems development*. MA: PWS-Kent Publishing.

Kowal, J. (1988). *Analyzing systems*. NJ: Prentice Hall.

Navarro, A., & Tabinda, K. (1998). *Effective Web design*. CA: Sybex Inc.

O'Neil, P. (1994). *Database, principles, programming, performance*. CA: Morgan Kaufman.

Osborne, L., & Nakamura, M. (2000). *Systems analysis for librarians and information professionals*. CO: Libraries Unlimited.

Ramakrishnan, R. (1998). *Database, management systems*. NY: WCB/McGraw-Hill.

Ricardo, C. (1990). *Database, principles, design, and implementation*. NY: Macmillan.

Whitten, J., Bently, L., & Dittman, K. (2001). *Systems analysis and design methods*. NY: McGraw Hill Irwin.

APPENDIX A

Financial Summary
For the six months ended on 06/2000, net sales rose 9% to EUR11.09 billion. Net income applicable to Common before U.S. GAAP rose 57% to EUR337M when compared to 1999 results. Results reflect increased life sciences sales.
Recent Earnings Announcement
For the 3 months ended 09/30/2000, revenues were 5,429; after tax earnings were 126. (Preliminary; reported in millions of Euro.)

Statistics at a Glance – NYSE:AVE		As of 5-Dec-2000
Price and Volume	**Per-Share Data**	**Management Effectiveness**
52-Week Low	Book Value (mrq*)	
on 8-Mar-2000		Return on Assets
	$11.60	
$45.50	Earnings	N/A
Recent Price	N/A	Return on Equity (ttm)
$79.25		1.23%
	Earnings (mrq)	
52-Week High	$0.14	**Financial Strength**
on 30-Nov-2000		
$79.938	Sales	Current Ratio (mrq*)
	N/A	1.06
Beta	Cash (mrq*)	Debt/Equity (mrq*)
0.46	$0.15	1.39
Daily Volume (3-month avg)	**Valuation Ratios**	Total Cash (mrq)
126.0K		$120.9M
Daily Volume (10-day avg)	Price/Book (mrq*)	**Short Interest**
166.0K	6.83	As of 8-Nov-2000
Stock Performance	Price/Earnings	Shares Short
AVE 5-Dec-2000 (C)Yahoo!	N/A	658.0K
big chart [1d | 5d | 3mo | 1yr | 2yr | 5yr]	Price/Sales	Percent of Float
	N/A	0.1%
	Income Statements	

(continued on following page)

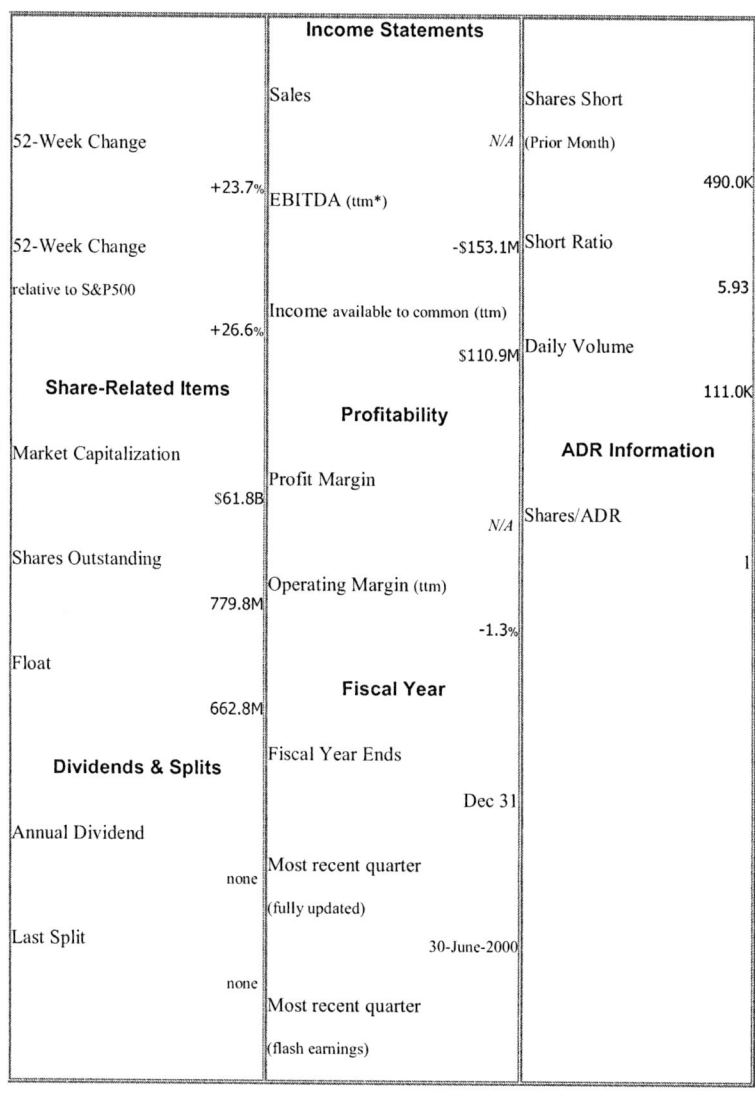

	Income Statements	
	Sales	Shares Short
52-Week Change	*N/A*	(Prior Month)
+23.7%	EBITDA (ttm*)	490.0K
52-Week Change	-$153.1M	Short Ratio
relative to S&P500		5.93
+26.6%	Income available to common (ttm)	
	$110.9M	Daily Volume
Share-Related Items		111.0K
	Profitability	
Market Capitalization		**ADR Information**
	Profit Margin	
$61.8B	*N/A*	Shares/ADR
Shares Outstanding		1
779.8M	Operating Margin (ttm)	
	-1.3%	
Float		
662.8M	**Fiscal Year**	
Dividends & Splits	Fiscal Year Ends	
	Dec 31	
Annual Dividend	Most recent quarter	
none	(fully updated)	
Last Split	30-June-2000	
none	Most recent quarter	
	(flash earnings)	

APPENDIX B

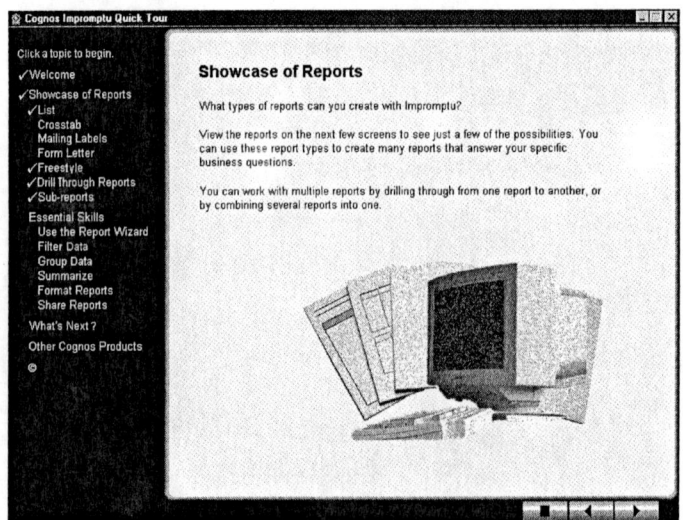

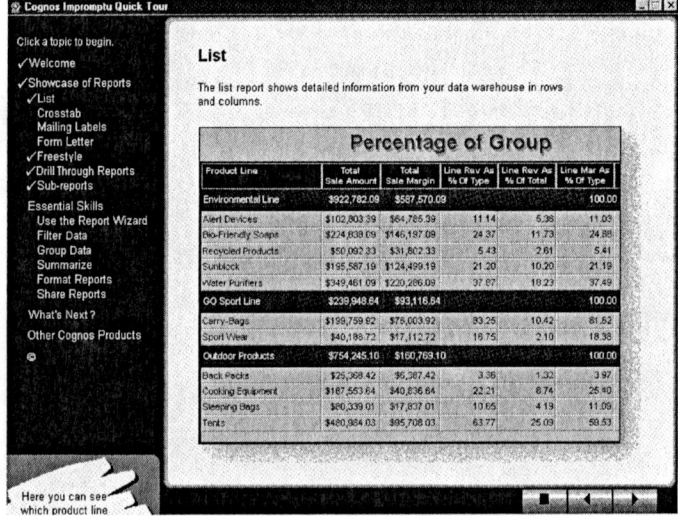

(continued on following pages)

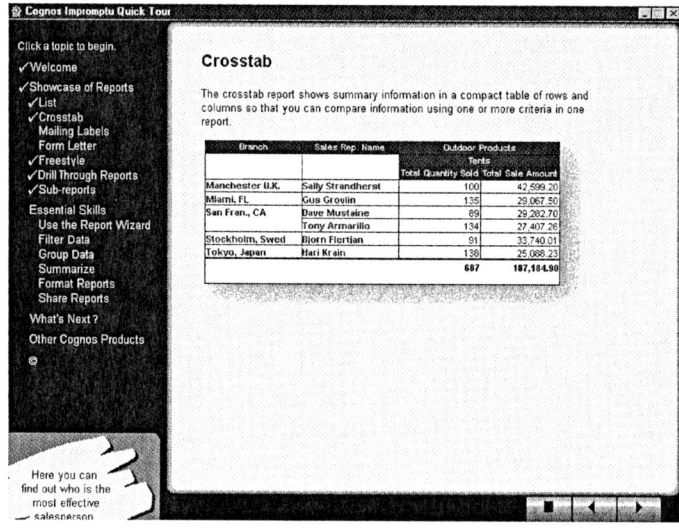

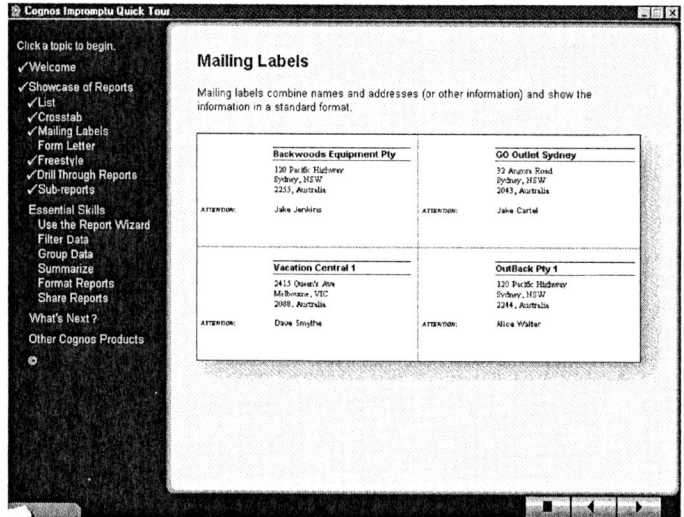

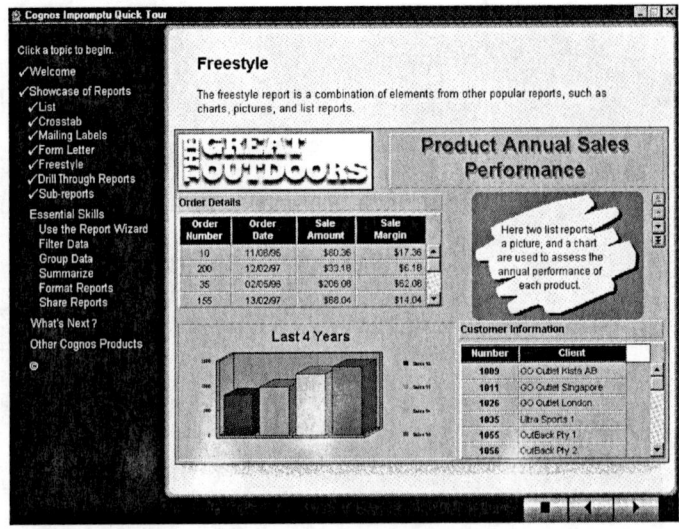

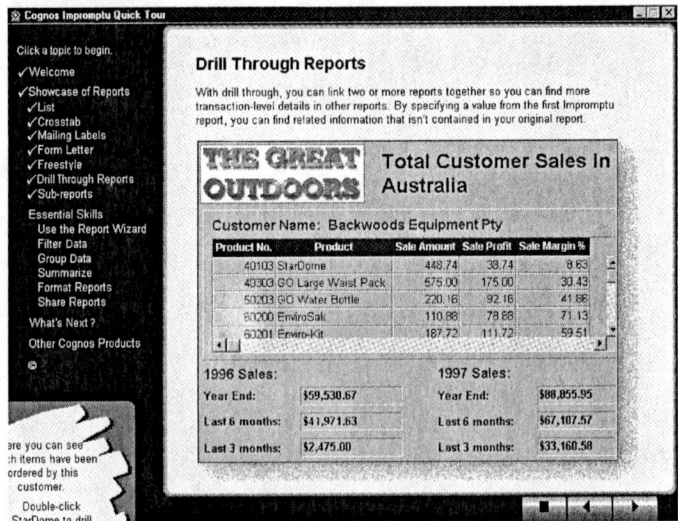

Yousif Mustafa earned a PhD in industrial and manufacturing engineering in 1998 and an MS in industrial and manufacturing engineering in 1993 from Wayne State University, Detroit. Dr. Mustafa is currently an assistant professor with the Computer Information Systems Department of Central Missouri State University, Warrensburg.

Clara Maingi is a senior with a double major in CIS and accounting at the College of Business of Central Missouri State University, Warrensburg, Missouri. Currently, Clara is doing her internship as an application developer at the Information Systems Department of Aventis Pharmaceuticals, Kansas City, Missouri.

This case was previously published in the *Annals of Cases on Information Technology*, Volume 4/ 2002, pp. 84-102, © 2002.

Chapter XVIII

Evolving Organizational Growth Through Information Technology

Ira Yermish, St. Joseph's University, USA

EXECUTIVE SUMMARY

The Service Employees International Union, Local 36 Benefits Office, provides service to over 3,500 union members in the Delaware Valley area. In addition to managing the collection of dues and other funds through employers, the Benefits Office administers several insurance programs and funds. From 1979 until 1996, this office has grown in sophistication and service efficiency primarily through the leadership efforts of Joseph M. Courtney, its only administrator during this period. Starting with an organization with no technical sophistication, Courtney identified critical areas where technology could make a difference in service levels to the Local membership. This case study describes the gradual evolution of the use of information technology, first to support basic transaction processing, and ultimately to support the strategic issues that such an operation faces. Through the careful use of a number of outside vendors and consultants, and through the slow growth of internal talent, Courtney was able to shepherd the operation from a purely manual system to one where every employee has a PC workstation connected to a network of internal servers and external services. Issues that will be raised in this case include questions of internal versus external development of applications, the relationships among various vendors and consultants, and the growth of internal expertise without significant information technology staff. As the case closes, a new administrator, Michael Ragan, looks at the operation and considers alternatives. He is very much concerned that their primary vendor, Benefit Systems, is no longer responsive to Local 36's needs.

BACKGROUND

As March 1996 comes to a close, Joseph M. Courtney looks back on his 20 years as the administrator of the Benefits Office of the Service Employees International Union (SEIU) Local 36. Retirement approaches and this former steam-fitter can look back with pride on his accomplishments and the services that been extended to the members of the Union. From a completely manual operation, the Benefit Funds office has been transformed into an efficient operation using multiple computer systems to perform the various operations on behalf of the membership.

THE LOCAL

The Local 36 Benefits Office provides a number of services to over 3,500 active union members and 2,000 pensioners in the Delaware Valley area. Of primary importance is the collection of dues through the member's employers. What makes this so much of a challenge is the nature of this union. The Union represents janitorial, window cleaning and maintenance workers, many of whom are on the bottom of the economic ladder and who frequently do not speak English. Exhibit 1 shows the relationship of the Benefit Office to the Local. In addition to being responsible to the Union, which is part of the International Union, two boards of trustees are responsible for oversight of the two major Funds managed by the Benefits Office: Building Operators Labor Relations, Inc. (BOLR) and the Building Maintenance Contractors Association (BMCA). These two funds

Exhibit 1. Organizational relationships

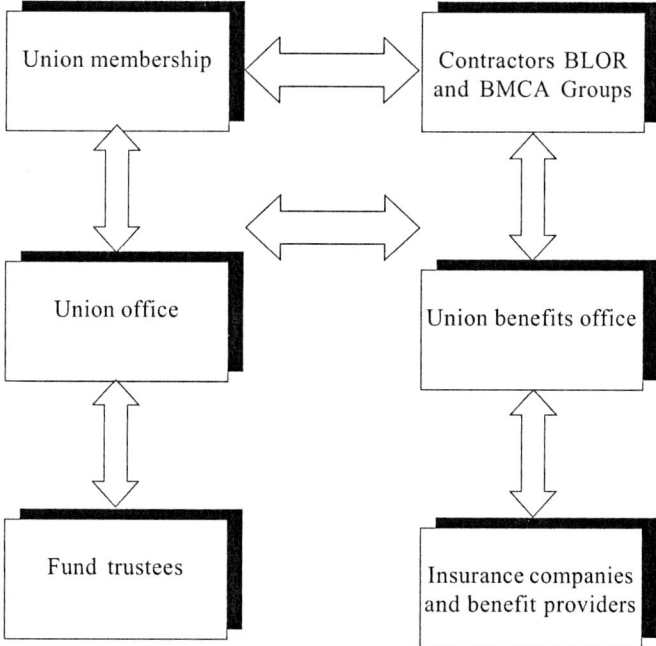

Exhibit 2. Organization charts

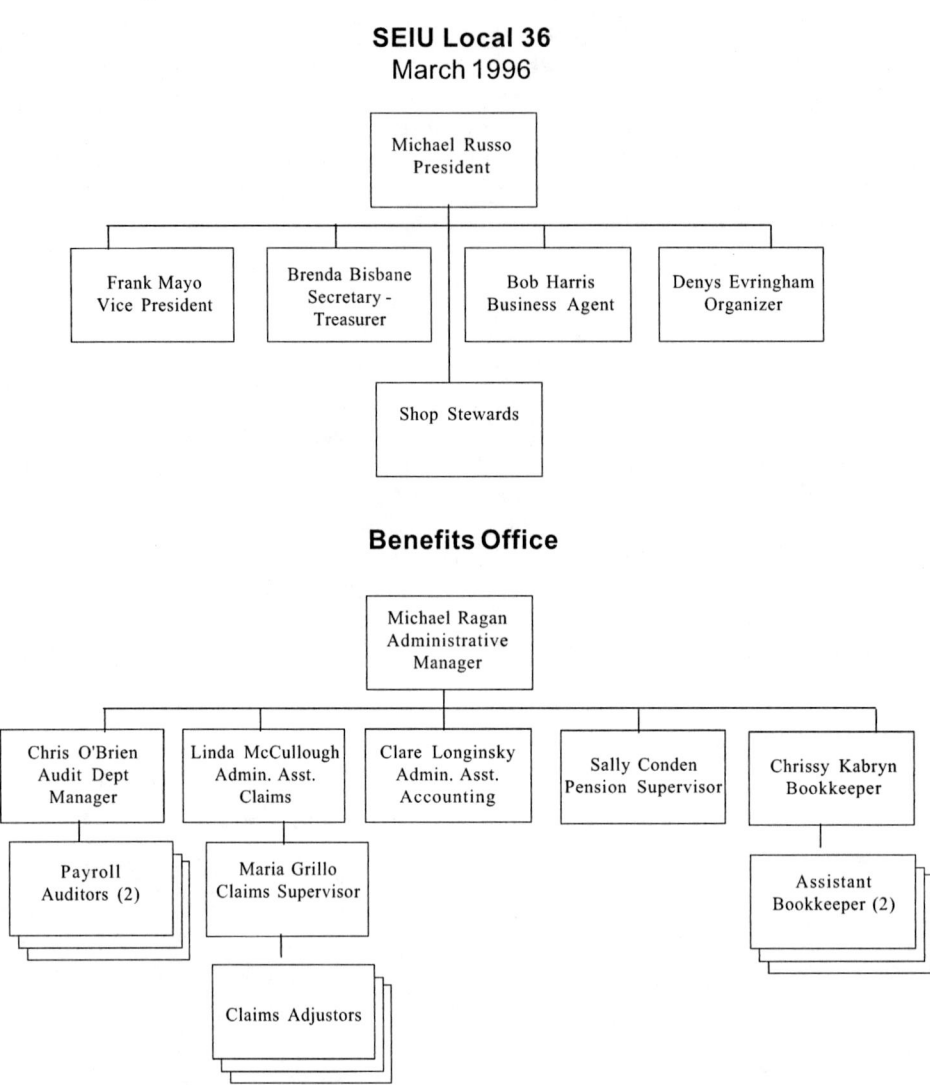

SEIU Local 36
March 1996

Michael Russo
President

Frank Mayo
Vice President

Brenda Bisbane
Secretary -
Treasurer

Bob Harris
Business Agent

Denys Evringham
Organizer

Shop Stewards

Benefits Office

Michael Ragan
Administrative
Manager

Chris O'Brien
Audit Dept
Manager

Linda McCullough
Admin. Asst.
Claims

Clare Longinsky
Admin. Asst.
Accounting

Sally Conden
Pension Supervisor

Chrissy Kabryn
Bookkeeper

Payroll
Auditors (2)

Maria Grillo
Claims Supervisor

Assistant
Bookkeeper (2)

Claims Adjustors

represent different contracts and groups of employers. The two contracts have different characteristics and benefits, the results of separate collective bargaining agreements. Exhibit 2 shows the organization structure of the Union and the Benefits Office. Exhibit 3 contains an excerpt and the contents of the BOLR collective bargaining agreement.

The Benefits Office must provide services to the members that have been negotiated for in the agreements. In addition, the Trustees of the Funds must be convinced of the fiscal soundness of the operations and their ability to supply the services to the members. Clearly, there are a number of pressures that the Benefits Office must face. First, they are

Exhibit 3. BOLR Collective Bargaining Agreement (excerpt)

Agreement

(Office Buildings)

This Multi–Employer Agreement, entered into the first day of November, 1993, by and between BUILDING OPERATORS LABOR RELATIONS, INC. (hereinafter called the "Corporation"), acting for and on behalf of such of its Member Buildings as are listed in Schedule "A" attached hereto (each of whom is hereafter referred to as "Employer"), on the on hand, and SERVICE EMPLOYEES INTERNATIONAL UNION, Local #36, AFL–CIO (hereinafter called the "Union"), on the other hand.

ARTICLE I   RECOGNITION

Section 1. The Employer recognizes the Union as the sole and exclusive agent for the purpose of collective bargaining and for those of its employees in the following classifications:

> Janitorial Employee, Class 1
> Janitorial Employee, Class 1 (Day Matron/Day Attendant)
> Janitorial Employee, Class 2
> Janitorial Employee, Class 2 (Lobby Attendant)
> Elevator Operator
> Elevator Starter
> Combination Elevator Operator and Job Class No. 2 Cleaner
> Foreperson
> Mechanics and Maintenance Workers
> Licenses Engineers and Operating Engineers

Such definition shall exclude supervisors, clerical employees, confidential employees and armed guards as defined in the National Labor Relations Act, and whose operating engineers and maintenance mechanics who are covered under a separate collective bargaining agreement. Whenever the word "employee" is used herein, it refers only to those employees for whom the Union is the recognized bargaining agent.

Section 2. The Employer Shall have the right to hire new employees from any source whatsoever. All new employees shall be on probation for ninety (90) days after employment and during such probationary period the Employer shall be the judge as to whether or not such new employee is qualified to continue in its employ and the Employer may discharge such employee for any reason at its discretion. Employees hired on or after November 1, 1993, shall not be entitled to holiday pay, paid funeral leave or jury duty benefits during their probationary period.

ARTICLES

II	UNION SECURITY AND CHECK–OFF	XX	SPLIT SHIFTS AND ASSIGNMENTS
III	RIGHTS OF MANAGEMENT	XXI	GRIEVANCE AND ARBITRATION
IV	NO DISCRIMINATION		PROCEDURE
V	WAGES AND OVERTIME	XXII	GOVERNMENT CONTRACTS
VI	HOLIDAYS	XXIII	SUBCONTRACTING AND
VII	VACATIONS		REDUCTION OF FORCE
VIII	CONVERSION AND SEVERANCE PAY	XXIV	NO STRIKES OR LOCKOUTS
IX	FUNERAL LEAVE	XXV	OTHER LEGAL ENTITIES
X	JURY DUTY PAY	XXVI	MOST FAVORED EMPLOYER
XI	UNIFORMS	XXVII	INSPECTION OF RECORDS
XII	TEMPERATURE WORKING CONDITIONS	XXIX	SAFETY
XIII	HEALTH AND WELFARE PLAN	XXX	SEPARABILITY
XIV	PENSION PLAN	XXXI	HOURS
XV	FAILURE TO REMIT DUES OR TRANSMIT	XXXII	PREPAID LEGAL SERVICES PLAN
	WELFARE AND PENSION CONTRIBUTIONS	XXXIII	BREAKS
XVI	INDUSTRY PROMOTION FUND	XXXIV	JOB POSTING
XVII	SENIORITY	XXXV	TOOLS
XVIII	DISCHARGE AND DISCIPLINE	XXXVI	TERMS OF AGREEMENT
XIX	UNION ACTIVITIES IN BUILDINGS		

responsible for collecting dues and fees from the Employers (contractors). Employers are responsible for identifying new employees, making dues and other payments. Given the nature of the industry, there are many opportunities for abuse. The Benefits Office is responsible for identifying these abuses and making sure that all members have been properly identified and that the moneys have been collected.

On the other side of the operation, the Benefits Office is responsible for administering the various insurance programs and pension plans. There are questions of eligibility and abuse by service providers and members. Contractual relationships between these providers, e.g., Blue Cross, is based upon utilization. The Benefits office must be sure that the services provided are appropriate and cost effective. The rapid growth in healthcare costs affects its members more severely than many other cohorts.

The Funds represent large sums of money maintained on behalf of the membership. Some of these funds provide for self-insurance coverage for some of the benefits. These funds must be tracked and analyzed to be sure that the investments are safe and meeting the needs of the membership.

Exhibit 4 outlines most of the major functions of the Benefits Office. Activities on behalf of the members and their direct employer relationship are handled by the Union Office. The Union Office is responsible for organization and contractor-member conflict resolution.

Looking back on his tenure as administrator, Courtney realizes how important information technology was to making the Benefits Office a model of responsible and fiscally sound operation, recognized by others as forward thinking and creative.

DEVELOPING IT IN THE BENEFITS OFFICE

The history of information technology in the Benefits Office tracks the use of technology in other organizations. First, basic transaction processing is automated. Next, tactical management issues are addressed and finally, strategic planning issues are attacked. A timeline of hardware and software developments are summarized in Exhibit 5. Exhibit 6 describes the current hardware configuration and Exhibit 7 outlines the distribution of applications on the various hardware platforms.

In 1979, a consulting company, MagnaSystems, Inc., was asked to review a contract for a new software system to be installed at the Benefits Office. Up until that point, the office had been completely manual, with file cabinets full of documents, slow processing and meager services. They were swamped and were looking forward to getting out from under the piles of paper. They had looked at a number of software systems and had identified a package appropriate for managing union benefits operations. The vendor, Benefit Systems, Inc., was located in Baltimore and had already installed their software at a number of offices, one of which was in the Philadelphia area. Given that the Benefits Office had no experience whatsoever with information technology, it was important that they install a system that did not require in-house expertise. The staff did not include any college graduates, but given the leadership of the administrator, they were loyal and willing to move forward. After some negotiations and plans for some modifications to meet the specific needs of Local 36, the contract for a minicomputer-based system was

Exhibit 4. Benefit office operations summary

Dues and Contribution Collection

Each month the bookkeeping department sends a remittance form to each of the contractors, showing all of the known members and the required dues and fund contributions. The contractors are responsible for remitting the appropriate funds which they are responsible for and the funds that are supposed to come out of the employee's pay. They are also responsible to update the list with terminations and new hires. The remittance form is then processed into the transaction processing system to maintain a complete record of all fund contributions which is used for benefits eligibility.

Insurance and Benefits Claims Processing

The major function of the Benefits Office is to coordinate the member insurance and benefits services. Some of the insurance services are provided by third party insurance carriers such as Blue Cross and Blue Shield. Other services are provided through self-insurance programs that are funded by the members' contributions and administered by the Benefits Office. The Claims processing operation is responsible for determining eligibility and appropriateness of benefits. Each month the office sends out checks to providers and members for benefits not covered through the third party insurance providers.

Coordination of Benefits Processing (COBRA)

Occasionally, a member may have other insurance outside of the Union plans. Furthermore, the insurance must be provided for a period of time after a member leaves the employ of a contractor in the system.

Pension and Disability Processing

At this time, the Benefits Office keeps track of those members on pension and disability but contracts out to a bank for the processing and distribution of the monthly checks. Pension benefits must be determined for members and surviving spouses.

Investment Tracking

Millions of dollars of member's funds are invested to provide the long-term viability of the welfare and pension funds. The funds are also managed by outside investment firms, but it is the primary responsibility of the Administrator to see that these funds are invested wisely and in a way that will make it possible to improve the benefit packages to the members.

Bookkeeping and Fund Accounting

This activity coordinates each of the above to keep track of all of the transactions and fund balances.

Contractor Audits

It is important to make sure that each of the contractors is making the appropriate payments to the Benefits Office on behalf of the membership. There are many opportunities for fraud and for honest mistakes. This activity attempts to match up the actual payroll records at the contractor sites with the records processed by the Dues and Contributions activity described above.

Provider Audits

Third party insurance carriers are responsible for checking the validity of claims submitted by hospitals, physicians and other service providers. However, there are many opportunities for abuse. One of the Bookkeeping Department's functions is to audit the activity in these areas to determine if duplicate payments have been made or if inappropriate services are provided. Since the costs if insurance are directly determined from actual usage, it is important to keep the costs as low as possible by the constant audit of actual service. Once offending institutions and providers are identified, it is possible to automatically monitor these for subsequent problems.

Planning and Negotiation

As each collective bargaining agreement nears its end, the process of renegotiating on behalf of the membership begins. The Union constantly strives to enhance the pay and benefits to for its members. The Contractors try to keep their costs down through pay and benefits reductions. One of the planning functions of the Benefits Office is to determine the impact of possible benefit changes on fund viability. Given the complex relationship that exists between the Union, the Contractors, the Fund Trustees and the Benefits Office, these negotiations can be quite complex. The Benefits Office has developed the tools to test the effects of changes in benefits over time to determine the changes in employee or employer contributions needed.

Exhibit 5. System implementation timeline

1979   Initial Hardware/Software Installation

A Microdata minicomputer was purchased to run the standard union office package supplied by Benefit Systems., Inc. (now BASYS). This system handled the processing of dues collection, fund tracking, benefit eligibility and claims processing. The software was written in a business dialect of BASIC and included some report writing capabilities.

1983   COMPAQ Portable and Contractor Audit Software

A transportable computer was purchased to be used by the first payroll auditor hired by the Benefits Office. Benefits Systems provided some customized software to permit the downloading of contribution data to the portable computer through a direct serial port connection. A consultant developed a series of data translation programs and LOTUS spreadsheets to assist in the preparation of the audits.

1987   Minicomputer Upgrade

As additional services were provided to the membership, the original system became inadequate to meet the needs of the Benefits Office. More terminals were required and additional storage was required to maintain the growing historical database of fund contributions. A McDonnell-Douglas system based on the PICK operating system was purchased and the existing software and databases were ported to the new environment with little problem. The vendor, Benefit Systems, Inc., provided the necessary support for this transition.

1988   Planning Model Development

The Administrator with the help of the consultant, developed a planning model for the larger fund (BOLR). The model was implemented in LOTUS 1-2-3 and installed on a 386. There was extensive testing of the model before its acceptance by the Union as part of the negotiation process. A model of the smaller, but more complicated fund, BMCA quickly followed.

1989   Insurance Subrogation Tracking System

Another standalone PC system was purchased and database software (using Foxpro) was developed to handle the Insurance Subrogation Tracking function of the Claims Department.

1990   Provider Audit Software

Another standalone PC system was purchased and software was developed to provide more extensive analysis of the claims handled by the minicomputer system. A procedure to transfer the Blue Cross claims information from the minicomputer to the PC was implemented. Subsequent software was developed to analyze the Prescription claims and the Blue Shield claims.

1993   PC Network

As the number of standalone PCs grew, it made sense to connect them with a Novell network. The installation of the network coincided with a move to new, expanded quarters. The planning model, subrogation system, provider audit software, contractor audit software and standardized word processing and spreadsheet software were installed on the network.

1994   Sun SparcServer Upgrade

Again, to meet the growing needs of the base contribution and claims processing system, the centralized minicomputer was upgraded to a Sun SparcServer. The software was directly ported to the new environment. Unlike the prior systems, this system used the UNIX operating system which promised more flexibility.

1995   Network Integration

Using the TCP/IP options of the Sun SparcServer, the Novell network was integrated with the BASYS applications software. Dedicated terminals were eliminated in favor of PC workstations connected to both systems for transaction processing, word processing and other functions. Using Windows 3.1, clerks could process data on the transaction processing system, use the network word processor, or perform other complex reports using the data warehouse functions available on the Novell network.

Exhibit 6. Current network configuration

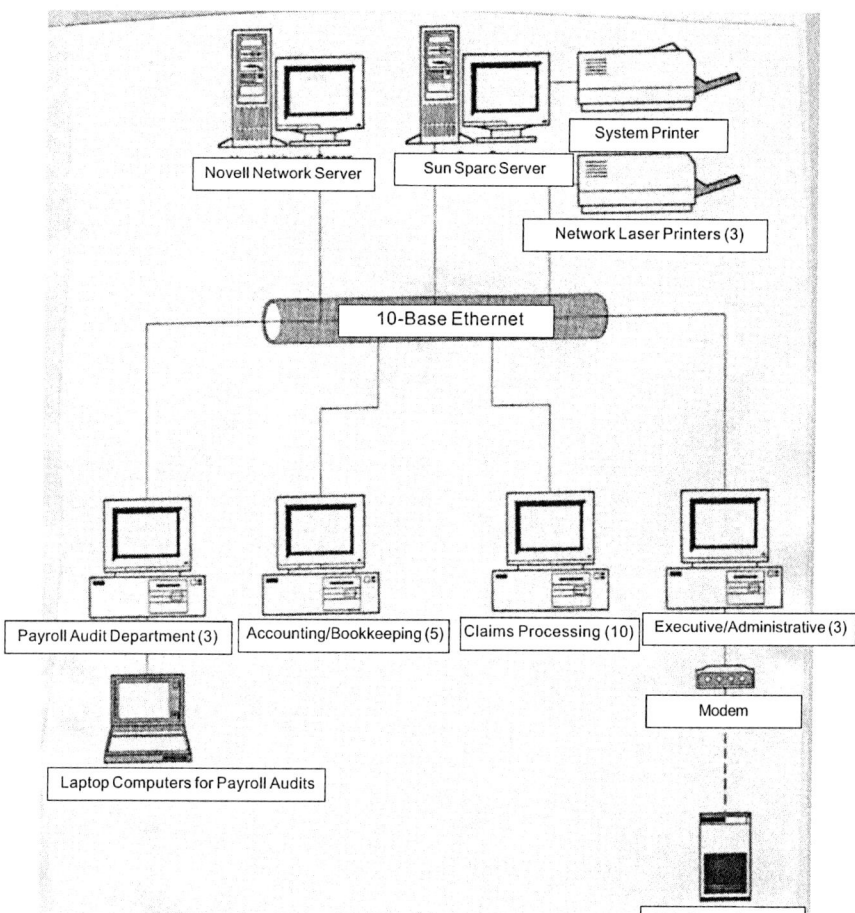

awarded and work started on building the database of membership, contractor and contributions. Within a very short period of time it became evident that they had made the right decision. The tedium of tracking thousands of contribution records was handled well by the system.

Courtney had gambled that the technology would work and he was right. Contributionswere processed more efficiently, benefits were distributed more rapidly. After a year or so of operation, getting the initial bugs out of the system, and making it meet the Local's specific needs, he identified another area where the technology could be of some help. It was well known that the contractors could exploit the system because of the nature of the workforce and the kinds of work involved. It was not unusual for contractors to hide workers in their payrolls and thus avoid some of the employer based contributions. It was also possible for employers to pocket contributions from members and not forward them in a timely fashion to the Union. The system was based upon the

Exhibit 7. Current software configuration

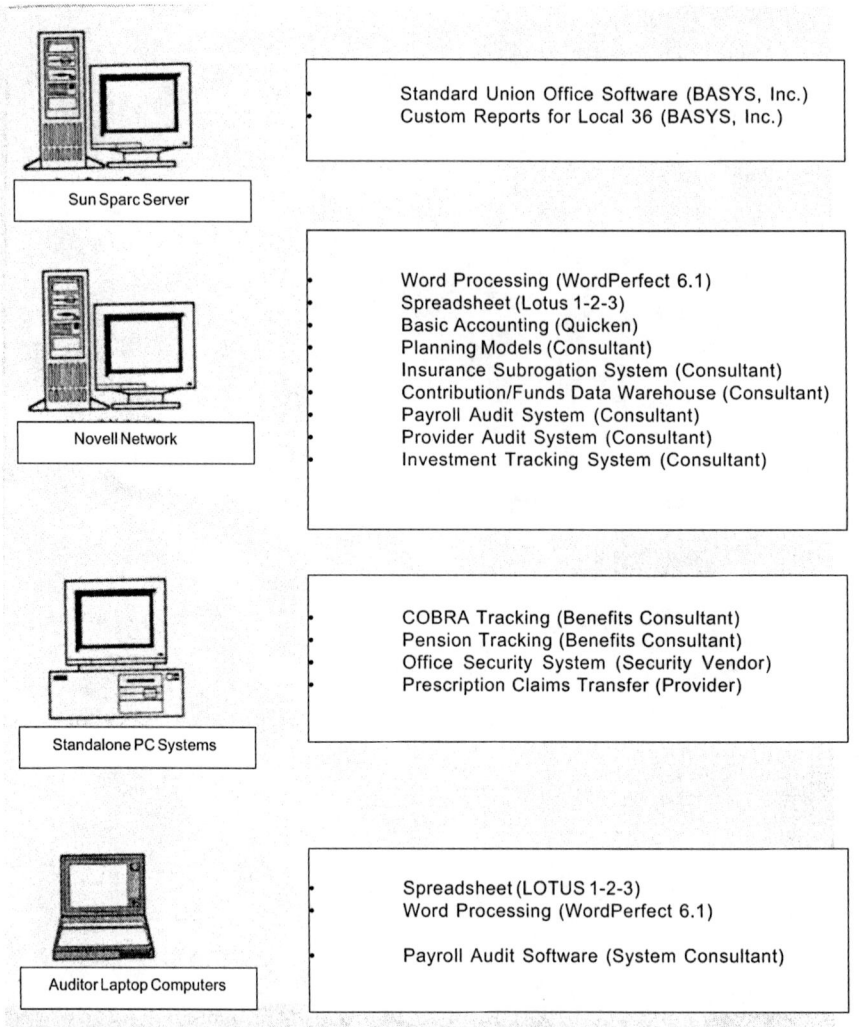

good faith of the employers to provide the information and contributions in a timely and accurate fashion; a dubious assumption if there ever was one. Courtney saw the volumes of data on contributions that were accumulating and wondered whether this data could also be used to improve the collection process. In other words, could the data be matched against actual employer records to confirm that all contributions were accurate? He was faced, however, with a problem. Who could do these *audits* and with what tools?

It was 1983 and the IBM PC and its most interesting competitor, the COMPAQ portable, were just starting to make their presence felt in industry. The consultant was asked to examine the feasibility of using these tools in conjunction with the minicomputer to provide an audit capability. Certainly, it was technically feasible, but who would do the work? No one in the organization had any experience with personal computers or the new software technology: *spreadsheets.* For the first time, the Benefits Office hired a college graduate with accounting and computer background to perform the audits. A special program was supplied by Benefit Systems to download contractor data to a COMPAQ portable, and the consultant wrote a Lotus 1-2-3 macro to transform the downloaded contribution data into a standardized spreadsheet format. With these software tools and the "portable" computer, the auditor started visiting contractors, causing quite a stir. It wasn't long before the Benefits Office was able to recoup its investment with adjustments and settlements.

In 1995, for example, there were 36 audits resolved, with a total of nearly $200,000 recovered. One audit alone was resolved for a total of $87,000. Recent years were even more dramatic:

Year	Audit Resolution
1990	$182,832
1991	$113,650
1992	$435,863
1993	$310,410
1994	$359,887
1995	$199,008

Today, the audit professional staff includes a manager and two auditors who, using notebook computers continue to guarantee that the contractors are meeting their obligations. All of the contractors are visited on a regular basis. Young college graduates with MIS or accounting degrees have filled these positions admirably over the years, some moving on to other areas where their acquired expertise has been valued. However, it has been clear to Courtney that these young people have provided a significant path of professional growth for the organization and its relationship with the contractors and service providers. The initial hardware and software investment has paid for itself many times over.

The next several years saw a continued dependence on the technology. After about seven years with the initial system, a major CPU upgrade was required. The platform on which the software was running (McDonnell-Douglass minicomputer with the Pick Operating System) was well suited to meet the needs of the basic centralized processing environment. Using the built-in report writer, ENGLISH, it was possible for at least one of the Local 36 staff to generate specialized management reports. The hardware and software were not inexpensive, but could be justified on the basis of their cost-effectiveness. There were concerns, however about the proprietary nature of this environment. At that time, given the software involved and the lack of real alternatives, the upgrade was made but with the concern that we would monitor the environment carefully for changes.

Adding to this concern a problem arose that frequently plagues users of vendor supplied application software. The vendor wants to keep each of its customers happy, but there is also the pressure to maintain a standardized, maintainable package. Each of the vendor's customers may see its case as being unique even though, in this case, they are all Taft-Hartley union operations supporting benefit operations. The vendor is responsible to maintain the software in light of changing government regulations and passing these changes to the customers in a timely fashion. But how do these changes get installed when many of the users have requested and gotten special modifications to the software? As a vendor grows, each of the customers becomes a smaller share of its business and apparently less critical for the vendor's survival. This may translate to a perception of indifference and alienation. On the other hand, the vendor is critical to the success of the customer, having become strategically dependent on that vendor for its basic operations.

At that time (1988), there didn't seem to be any other reasonable choice. It often took a long time to get specialized reports programmed. The programming charge for these reports and updates required hard negotiations. The "captive audience" expressed concern but could do nothing but maintain vigilance.

Around this time, Courtney also began to address other strategic issues. One of his primary responsibilities at contract negotiation time was to determine the feasibility of new benefits and contribution structures. These were often complex negotiations where fund viability could be jeopardized by an overaggressive negotiator. The fund trustees were often faced with emotional decisions and could not see the financial implications. Courtney came to the consultant with the idea for a planning tool that could be used directly in the negotiation process. He wanted to be able to play "what if" games to determine the effects and contribution requirements of the decisions. A 386 PC was purchased and a planning model was developed for the BOLR fund. This model forecasted the results of decisions based upon the actuarial forecasts of expense changes and contribution commitments. Fund balances were displayed and presented. The model was very effective in negotiations and planning. Over the year the model was updated on a quarterly basis with actual results. This helped spot possible expense trouble spots before they could become a real problem. After the BOLR model was tested and used successfully for a year, a similar model was built for the smaller, but more complicated BMCA fund. This planning model operates as a central repository for the operations staff to record quarterly summaries of operations. Exhibit 8 contains a summary of the Benefit Offices welfare operations and forecast as extracted from the planning models.

EVOLVING INFORMATION TECHNOLOGY

In 1988 Courtney looked carefully at his operations, trying to identify where information technology could further enhance his operations. The terminals attached to the minicomputer were doing the routine transaction processing, posting collections, identifying eligibility, processing insurance claims. One of the first additional applications that he saw involved the tracking of insurance subrogation cases. In these cases where a member may have been involved in an accident, claims against other parties must be tracked. Some of these cases involved very large insurance claims that could severely impact the rates paid to Blue Cross and Blue Shield. Of primary concern was the exposure for not being able to collect the funds from other insurance carriers.

The manual system previously maintained by Linda McColllough, the Administrative Assistant for the Claims Operations, was unable to generate timely status reports. Several choices for this system were identified. The primary vendor, Benefit Systems,

Exhibit 8. Summary of benefits operations (welfare only)

	Actual					Estimated				
	1991	1992	1993	1994	1995	1996	1997	1998	1999	2000
BOLR										
Number of Participants	3,192	3,115	3,084	2,949	2,820	2,820	2,820	2,820	2,820	2,820
Contributions	11,168,395	11,550,627	11,380,577	11,684,595	12,839,586	13,163,760	13,163,760	13,163,760	13,163,760	13,163,760
Interest Income	209,812	161,905	50,935	40,619	139,989	645,930	765,904	844,901	876,301	852,633
Total Income	11,378,207	11,712,532	11,431,512	11,725,214	12,979,575	13,809,690	13,929,664	14,008,661	14,040,061	14,016,393
Blue Cross Expenses	5,351,621	5,354,700	5,701,999	4,143,611	3,998,150	4,308,632	4,647,997	5,018,670	5,424,353	5,869,308
Blue Shield Expenses	745,797	1,020,545	1,449,933	1,373,234	1,458,993	1,619,482	1,797,625	1,995,364	2,214,854	2,458,488
Prescription Plan (PCS) Expenses	1,238,934	1,343,472	1,287,407	1,264,839	1,313,455	1,455,714	1,613,510	1,788,547	1,982,716	2,198,114
Self-Insured Plan Expenses	1,301,745	1,466,646	1,240,701	1,236,984	1,158,197	1,262,009	1,377,126	1,505,002	1,647,296	1,805,913
Medical Center	333,423	344,493	348,516	337,068	302,661	338,400	338,400	338,400	338,400	338,400
Loss of Time Disbursements	646,816	652,047	659,273	657,276	648,412	687,317	728,556	772,269	818,605	867,721
Administrative Expenses	629,194	755,585	852,757	851,629	1,229,115	1,290,571	1,355,099	1,422,854	1,493,997	1,568,697
Total Disbursements	10,247,530	10,937,488	11,540,586	9,864,641	10,108,983	10,962,125	11,858,313	12,841,106	13,920,221	15,106,641
Net Change Gain/(Loss)	1,130,677	775,044	(109,074)	1,860,573	2,870,592	2,847,565	2,071,351	1,167,555	119,840	(1,090,248)
Adjustments	470,732									
Year-End Fund Balance	5,414,362	6,189,406	6,714,203	8,995,196	11,817,782	14,665,347	16,736,697	17,904,252	18,024,093	16,933,844
Months of Reserve	6.3	6.8	7.0	10.9	14.0	16.1	16.9	16.7	15.5	13.5
BMCA										
Number of Participants	654	596	563	491	420	424	424	424	424	424
Contributions	439,921	706,697	747,471	644,158	649,260	582,072	607,512	607,512	607,512	607,512
Interest Income	46,351	16,459	11,385	16,977	21,654	28,050	21,241	11,356	(2,650)	(21,345)
Total Income	486,272	723,156	758,856	661,135	670,914	610,122	628,753	618,868	604,862	586,167
Benefit Expenses	542,762	614,522	475,684	481,116	516,606	576,716	632,973	693,077	758,419	831,871
Administrative Expenses	102,307	136,992	139,804	137,214	138,305	154,174	163,424	173,230	183,623	194,641
Total Disbursements	645,069	751,514	615,488	618,330	654,911	730,890	796,397	866,307	942,042	1,026,512
Net Change Gain/(Loss)	(158,797)	(28,358)	143,368	42,805	16,003	(120,768)	(167,644)	(247,439)	(337,180)	(440,345)
Adjustments	(45,547)									
Year-End Fund Balance	401,674	317,857	549,625	620,223	639,269	437,546	269,902	22,463	(314,717)	(755,063)
Months of Reserve	7.5	5.1	10.7	12.0	11.7	7.2	4.1	0.3	-4.0	-8.8

proposed a solution to be integrated with the minicomputer system. However, the consultant pointed out that there was really little connection between the data tracked in that system and the requirements for the application. Instead, the consultant developed a stand-alone PC-based application (using the Foxpro database application language) to track these cases. A PC was purchased for McCollough and the database application was developed and operational within a few months.

In this time frame, the actuarial consultants and auditors, the Wyatt Company, provided two other stand-alone applications to support the Benefit Office. The first of these helped Sally Condon, the Pension Supervisor, track the status of pensions. This is critical for testing the viability of the pension funds. Identifying the ages and status of the pensioners determines the funding requirements for the pension funds.

Chrissy Kobrin, Bookkeeper, was a heavy user of the Benefit Systems transaction processing system. Her department was responsible for the processing of dues and fund contribution collection. She developed the skills to generate special reports from the minicomputer system using the ENGLISH report writing language. In addition, she was given a PC and some basic accounting software (Quicken) to track some of the funds. Furthermore, her department's responsibilities included the tracking of the coordination of benefits (COBRA). The Wyatt Company provided a stand-alone PC application for the tracking of these cases.

Another major responsibility that Kobryn's department has is the identification of insurance claims problems. In essence, they were also responsible for the auditing of the third party carriers and the service providers themselves. For example, hospital billing departments would submit the same claim more than once which might not be picked up by Blue Cross. A physician might submit a claim based upon a diagnosis that was unacceptable to the plan which might not be identified by Blue Shield. At first, Benefit Systems provided a number of reports to assist in this process, but they lacked the flexibility needed to meet their needs. The ENGLISH report writer was too difficult to use for the current staff. Benefit Systems was reluctant to provide changes to their basic system. Given the proprietary nature of their software and the operating system, Courtney expressed concerns about the future of the operations, particularly with respect to the flexibility needed.

In 1990, Courtney met with the consultant to explore the possibilities. The information industry pressure was away from proprietary minicomputer platforms and towards industry standard platforms based upon UNIX. Unfortunately, Benefit Systems could make no commitments that they would "port" their system to this platform. Furthermore, the sharing of applications and data via PC local area networks had become routine. The Audit Department manager at that time, John Matekovic, had an undergraduate degree in management information systems. Perhaps his background could be used to support additional internal development to meet the long-range needs of Local 36's operations.

It was agreed that it was too early develop a complete operational system in-house. The costs would be too great and the expertise was not there. However, to meet the needs for the third-party audit operations, a large amount of the data captured on the minicomputer system needed to be processed. The most critical area was in identifying problems with Blue Cross payments which were being loaded monthly onto the minicomputer system from tapes supplied by Blue Cross. Benefit Systems was asked to provide software to extract this data from their system and supply it in a form that could be loaded directly onto a PC. This became the first step in a possible transition to a completely home-grown operations system. At this time the design was to include the capabilities for multi-user processing. Matekovic, Kobryn and her assistants, helped prepare a list of the kinds

of functions and reports neededto improve their operations. Given the years of experience they had gained with the minicomputer system and their familiarity with more
current, mouse driven, PC-based software, they were quite efficient and doing this design
work. The consultant developed a standalone PC application (in Foxpro again) to provide
for the much more powerful and user-friendly analysis programs needed for the third-
party audits. This system would be flexible enough to provide for the audit of other
insurance providers such as Blue Shield and Prescription Claim Services (PCS).

THE THIRD GENERATION OF
INFORMATION TECHNOLOGY GROWTH

In 1993, the Benefits Office was contemplating a move to new, larger quarters. As
part of this move, the technology issues were identified. Though they had not done so
yet, it was clear that additional networking capabilities would be desirable. However, they
were faced with a problem, the need to provide the cabling for the minicomputer system
as well as a possible PC network. The consultant recommended a parallel path. Both serial
(RS-232) cables would be run to meet the needs of the minicomputer system and twisted-
pair (10BaseT Ethernet) cabling would be run for the PC network. Most of the workstations were "dumb" terminals for claims processing and bookkeeping. Other stations
would be PCs with both a direct serial connection to the minicomputer and a LAN
connection to the PC network. At this point, a user could, within the Windows 3.1
environment, have a terminal connection to Benefit Systems software and connections
to the PC LAN software.

Finally, recognizing the industry trends, Benefit Systems implemented their software on a Sun SparcServer using Solaris, the Sun version of UNIX. This opened up a
number of interesting opportunities. First of all, it was possible to eliminate the serial
connections and provide all of the terminal TCP/IP connections from the PCs to the Sun
and the PC network. This notably improved the terminal response time. It also simplified
the connections to the systems. Finally, it improved dramatically, the connection abilities
for transferring data between the Benefit Systems software and the PC network applications.

Through 1994 and 1995, additional applications were installed on the network and
others contemplated. At his retirement banquet at the end of 1995, Joe Courtney was able
to look back with pride on his accomplishments and the growth his organization. Much
of the success he attributed to information technology. Only one member of his senior
management was not with him at the beginning of this growth back in 1979. Each of his
staff was routinely using the technology for reporting and analysis purposes as well as
standard word processing. New applications for pension processing and for investment
tracking were on the boards. Fund balances were well under control and the relationships
with other organizations were strong.

FUTURE GROWTH AND DEVELOPMENT

As Michael Ragan, the new Administrator, sits in his office, his PC connected both
internally to the computer networks and externally to data information services, he could

foresee many ways to improve the operation and gain further control. He expresses concern over the unresponsive character of their primary information technology vendor and is looking at alternatives. His strong educational background and experience as administrator of the operating engineers gives him some confidence to take additional steps. In his former position he led the way for the internal development of operating software over which he had complete control. He looks forward to the possibility of developing more efficient claims processing software, of building stronger relationships with vendors via information links, and perhaps, developing more internal information systems expertise. Looking down Chestnut Street there are many avenues open and potholes to contend with.

EPILOGUE AND LESSONS LEARNED

The Union Benefits operation continues to operate efficiently using the technologies that it has acquired over the period described in the case. There is special emphasis in improving the access to the claims data supplied by the insurance providers (Blue Cross, Blue Shield, and PCS). There is also a serious consideration for switching primary software vendors based upon the high cost for custom software modifications and the lack of responsiveness. This step, however, must be taken very carefully because of the tremendous economic and operational impact it would involve.

The staff has grown comfortable with the Windows 3.1/Netware/Unix environment, switching easily between applications running on the Netware fileserver and the Unix server. CD-ROM databases are being used for research and external access to the service providers is improving. Throughout the case, the consultant and management were careful to implement software and hardware tools gradually. Each incremental step has met with little resistance because the normal flow of work was not changed significantly, though over time, the changes have been dramatic. Within the staff, the skills needed to use the new tools were acquired as they were needed to assure productivity. For example, when a switch from Word Perfect for DOS to Windows was made, staff were given the opportunity to take courses on the software during work hours.

The technology provided the basis for organizational improvement, but it was the management style that encouraged the growth of the staff's use of that technology that has been instrumental in the success of this operation.

QUESTIONS FOR DISCUSSION

1.   What are the similarities and differences that organizations such as the Benefits Office face compared to typical for-profit operations? What forms of executive motivation are appropriate for an organization like the Local 36 Benefits Office?
2.   In this case, how has productivity been affected by the growing information technologies? How should this productivity be measured?
3.   Up to the end of the case the Benefits Office has not had a formal internal information systems staff, though most of the Audit Department has some MIS education. From both the organization and the individual perspective what would

be the advantages and disadvantages of such a staff? When, if ever, should such a staff be developed?

4. Consider the dual role that the consultant took, both as an expert to help identify and evaluate alternatives and as a supplier of some of the software and hardware solutions. What are the operational and ethical considerations of such a relationship?

5. What are the pressures on the primary software vendor who develops "vertical market" applications? How does the relationship between vendor and customer change over time? Consider the relative growth rates of the vendor and its clients.

6. What direction should the new administrator take to assure continued organizational growth? To what extent should he institute changes to demonstrate his own talents and experience?

REFERENCES

Fitzsimmons, J. A., & Fitzsimmons, M. J. (1994). *Service management for competitive advantage.* New York: McGraw-Hill.

Inmon, W. H., & Hackathorn, R. D. (1994). *Using the data warehouse.* New York: John Wiley & Sons.

Keen, P. G. W. (1991). *Shaping the future: Business design through information technology.* Cambridge, MA: Harvard Business School Press.

Dr. Ira Yermish is an assistant professor of MIS at St. Joseph's University in Philadelphia. His teaching and research areas include systems analysis and design, data base management, data communications, information resource management, decision support and expert systems, and business policy and strategic management. In addition to designing the undergraduate and graduate curricula in MIS, he has been active in the executive MBA program and the executive programs in food marketing and pharmaceutical marketing. He was the designer of the College of Business microcomputer network and has provided continuing technical support for microcomputer applications in the college in wholesale-distribution and manufacturing systems.

This case was previously published in J. Liebowitz and M. Khosrow-Pour (Eds.), *Cases on Information Technology Management in Modern Organizations*, pp. 194-208, © 1997.

Chapter XIX

IS Strategy at NZmilk

Paul Cragg, University of Canterbury, New Zealand

Bob McQueen, University of Waikato, New Zealand

EXECUTIVE SUMMARY

NZmilk is a small, fresh milk supplier that is contemplating using IS to a greater extent to become more competitive. Following deregulation of the industry in 1997, supermarkets and home delivery contractors could purchase milk from wherever they chose, rather than a required local manufacturer. This had opened up both competition and expansion opportunities within the industry. NZmilk recognised that they would have to fight hard to retain and increase their share of the market. They had already lost some of their local market to competitors coming in from outside their region, but had also gained a contract to supply Woolworths supermarkets outside their traditional market area Improvements to production facilities and distribution systems were in place, but NZmilk knew that a fresh look at how they interacted with their customers would be needed. Their general manager was convinced that information systems had a greater role to play at NZmilk beyond just the accounting and order processing that was presently undertaken. A new direction in using information systems to support NZmilk's rapid growth and new strategy was needed, but he was unsure of which way to proceed.

BACKGROUND

Whangarei Milk Company was formed as a private company in 1946 to supply home delivery milk in the growing town of Whangarei. In 1990, the company changed its name to NZmilk, and became a fully owned subsidiary of Northland Milk Products, an established, progressive dairy cooperative operating in the Northland region of New Zealand. This relationship with Northland Milk had brought benefits in terms of a guaranteed supply of whole milk. Previously, a number of dairy farms were directly contracted to supply NZmilk 365 days of the year, so NZmilk had to make use of all the milk provided each day from these suppliers. Now NZmilk could request the volume of milk it required by obtaining a milk tanker delivery from Northland's major processing factory (during most of the year) on relatively short notice.

Another advantage of the association with Northland Milk Products had been the ability to call on their resources when needed, particularly in the managerial, technical and financial areas. The parent company required NZmilk to submit monthly reports on their operations, and any major initiatives required approval from the Directors of Northland Milk Products.

By 2000, NZmilk had become the fourth largest supplier of fresh white milk in New Zealand, with annual sales of $25 million. Milk had always been the heart of their business, but they had recently increased their product range to include fruit drinks and fruit juices, and were considering developing other food products to add to their product range.

NZmilk occupied a modern plant on the outskirts of Whangarei, in one of the fastest growing regions of New Zealand. It employed 80 people, plus a distribution system involving an additional 36 vendors. These vendors were self-employed contractors who delivered on a daily basis to supermarkets, convenience outlets, and homes.

SETTING THE STAGE

Up until 1997, the home delivery of milk had been tightly regulated. Licensed local processors set the retail price of milk but were compelled to provide a home delivery service regardless of economics. Each home delivery milk processor had sole rights to a district. For NZmilk, this effectively meant no competitor could supply milk into the Whangarei region. Any firm could compete outside their restricted territories with other products like flavoured milk and yogurt. However, fresh milk was still the major product sold.

Although the fresh white milk industry in New Zealand was worth about $400 million per year, milk consumption was slowly falling and losing market share to other beverages. Sales of flavoured milk were helping to slow the decline. New Zealand's largest dairy company, New Zealand Dairy Group (NZDG), with revenues of $2 billion, was mainly focussed on the export of powdered milk, butter, cheese and other manufactured milk products, but also had a dominant market share of the pre-deregulation fresh white milk market in New Zealand's North Island, where about 80% of the New Zealand population of 3.5 million lived. NZDG's stated strategy was to become the low-cost leader in both the NZ domestic market as well as for the export products.

Deregulation was forced on the industry by the government, rather than the industry choosing deregulation. Many milk companies initially resisted deregulation, but some, like NZmilk, saw deregulation as a business opportunity and a way for the company to grow. After deregulation, milk companies began to supply their products into competitor's previously protected regions. Supermarkets were one target for additional sales outlets, but convenience outlets and home delivery drops in remote regions were less attractive, as more complex warehousing and distribution systems were required.

The move into markets outside their region had been anticipated when NZmilk changed its name, and was further reflected with the introduction of "NZ Fresh" as the Company's major brand.

Pricing was another area that had changed. Prior to deregulation, pricing was controlled to the extent that prices in supermarkets had to be within three cents of the price of home-delivered milk from vendors. Deregulation removed such controls. At times, competitors had cut prices of milk, particularly during the spring and summer when milk was plentiful. This meant that pricing policies had to be flexible and able to respond to competitive pressures in the marketplace, particularly in supermarkets.

Home delivery vendors had seen further erosion of their sales to homes. Prior to deregulation, supermarkets supplied less than 10% of fresh white milk, with the balance through home delivery and convenience outlets. The supermarkets' share had risen considerably since deregulation. Various initiatives were taken to protect home delivery systems. For example in Nelson, most home delivery vendors had purchased hand-held computers so they could respond easily to changes in price and demand. Elsewhere, NZ Dairy Group had begun rationalising its distribution system by reducing the number of route contractors and amalgamating various supply companies.

While supermarkets now had the advantage of a number of suppliers eager to sell them fresh white milk at competitive prices, they were not solely interested in stocking the lowest priced product. Reliability of supply, product quality (freshness), ease of ordering and obtaining product quantities matched to daily store demand, delivery frequency and ability to minimize the paperwork required for head office payments were all part of the equation.

CASE DESCRIPTION

NZmilk had grown from a small home milk supply company providing a single product in glass bottles for a local market, into a progressive, highly sophisticated, multifaceted organization which even manufactured its own plastic milk containers. Every day, tankers brought about 85,000 litres of fresh milk to the NZmilk plant. Within only a few hours of arriving, most had been processed into full cream, homogenised or low-fat milk varieties, packaged in plastic bottles or cardboard cartons, and trucked for delivery to retail outlets and homes around the region. The product range included fresh cream, yoghurt and flavoured milks.

Developing high standards of product quality was an important priority. NZmilk had established a quality assurance section which changed the emphasis from a random sampling "quality control" philosophy into an ISO 9002-accredited total quality management (TQM) program for ensuring top quality products. The emphasis on quality had helped NZmilk win a major contract to manufacture milk shake and soft-serve ice cream mix for McDonald's restaurants throughout the North Island of New Zealand. Another international food company, Cadbury Schweppes, has its famous Roses Lime Juice bottled on contract by NZmilk.

Innovation was another important characteristic of the company. NZmilk manufactured its plastic bottles for its own use, but had sufficient excess blowmoulding capacity to be able to sell containers to outside firms. It had pioneered a scheme for the collection of plastic milk bottles which were then recycled by plastics manufacturers into other plastic products, which had been successfully copied by other milk processors throughout the country. Their in-house product research and development programme had produced a New Zealand first in honey-sweetened natural yoghurt.

NZmilk planned to grow by competing on quality and service to extend their sales through supermarkets while defending their current local vendor network. Sales of fresh white milk had been falling since the mid-1970s, but it was still a profitable market. The home delivery market might fall by another 50% in the next 10 to 30 years, but there were large barriers to new entrants as good delivery systems were essential. The market for home delivery and convenience outlets seemed to be relatively price insensitive.

NZmilk had seen a decline in volume sold through independent home delivery vendors, with a corresponding reduction in the number of vendors from 45 in 1991, to 36 in 1995, to 20 by 2000. Sales in 2000 averaged 300,000 litres per vendor. The selling price of products to vendors varied by the type of end customer. Prepayments were made to NZmilk by vendors weekly, and at month-end vendors reported the exact number and type of product sold in each category. A credit or debit was then issued with payment due on the 20th of the following month. The sales through their main channels for 2001 were expected to be about 25% to home delivery via vendors, 25% to local shops and small supermarkets delivered by vendors, and the rest to large supermarkets through bulk delivery and direct invoicing. Total sales for 2001 were expected to be 21 million litres of milk, and 4 million litres of other products.

However, with growing interest in Internet shopping, and in particular the trials of grocery shopping over the Internet which highlighted the requirement for effective home delivery logistics, NZmilk saw a potential opportunity to supplement the home delivery of milk with the delivery of other grocery products. Thus, maintaining a healthy and profitable existing vendor network was important to both current operations and possible future strategic initiatives, and could not be discarded lightly.

CURRENT CHALLENGES

Doug Allen was NZmilk's sales and marketing manager, and William Edwards the sales manager. The customer relationships they had to manage were with the home delivery vendors, the owners of local shops and small supermarkets, and the buyers, product managers and local store deli managers of the large supermarket chains. There were a host of daily problems and complaints that had to be handled promptly and professionally which soaked up most of their time each day. In addition, they had to try and look to the future expansion of sales volumes in the context of a steady annual decrease in per capita consumption of milk products, and increased potential for competition in their own local area.

Liaison with the 20 vendors was one of William's responsibilities. It was important not to take these vendors for granted, as they were the point of contact with NZmilk's customers and consumers. Most vendors were small operators. The owners usually drove the delivery trucks themselves, hired some students to assist with the evening deliveries and did the paperwork when they returned home at night. Some of these vendors had computers to assist with their accounting, but there was a wide spectrum of computer capability, and computer-to-computer links were only possible with some vendors.

Doug and William were aware that reducing administrative burdens of invoice checking and reconciliation was a high priority for supermarket managers. At Woolworths, shipping dockets had to be initialled by department managers, and then reconciled in store with invoices submitted by suppliers. The invoices were batched, and submitted to the head office in Auckland, where the invoices were collected together with others from that same supplier, batched together and then approved for payment. While it seemed at odds with the need to reduce administration, the supermarket head office was indicating that daily invoicing of goods received at each store was the way of the future. Daily invoicing would allow better control over in-store stocks, and avoid some of the batching and reconciliation steps presently required.

Some snack food vendors had instituted off-truck invoicing, where an invoice was printed by the company truck driver from a computer in the truck as the goods were delivered. This allowed for flexibility in meeting the needs of the store at the exact time the delivery was made. However, this system was somewhat less attractive for perishable products like milk which had short duration "best before" dates.

There had been some discussion of electronic data interchange of invoices between suppliers like NZmilk and the supermarket head offices. However, the transfer of the invoice was only a small part of the process of ordering, checking of incoming goods and reconciliation of invoices that was undertaken at each individual supermarket, and systems that supported the whole process electronically seemed still some time away.

Most stores now had checkout scanning systems that read the bar code label on each product, and kept track of how much of every product went through the checkouts every day. NZmilk had been assigned unique bar codes for each of its products and sizes, so it was theoretically possible to tap into these checkout systems, either to individual stores or to the head office host computers to determine volumes of product sold each day and how much inventory remained in the store. However, it was an area of rapid change for the major supermarket chains, and it was not known whether they would be keen to provide access to their computers to outside firms like NZmilk. A further complication in this area was that the "best by" date was not presently bar coded, so that tracking of shrinkage from in-store inventory could not be exactly matched to the daily deliveries made.

Manufacturing Facilities

NZmilk's plant contained four filling production lines where containers were filled, labelled, capped and crated. Two lines were dedicated to plastic bottles, one to cardboard containers and the fourth to fruit drinks. Other parts of the factory manufactured plastic bottles and food mixes.

Once packed in crates, the milk was transferred to a cooler from which orders were picked and assembled for their customers. The fridge pickers started at 6:30 am, and the first truck for the day would leave at about 8:30 am. Two other large trucks left at about 11 am for local vendors and the town of Kamo, and smaller trucks for local destinations left at 9:30 am, 10:30 am and 12:30 pm. Other trucks departed for other towns during the afternoon, some only on alternate days. During the evening, one large truck left at 10:30 pm for Auckland. All loads went as a mix of specific orders for vendors or supermarkets, with the exception of the Auckland load for Woolworths, which was a bulk order which was then broken down for specific supermarkets. Often what was delivered to individual Woolworths supermarkets was different from what was ordered the day before, following a last minute telephone call to the Auckland depot to change the order. Seven lock-up depots were located in the Whangarei region, but each vendor had their own secure area for collection at their convenience.

John Tobin was the production manager overseeing all of the manufacturing areas, as well as raw material and finished goods stock holding, and distribution. There were 34 production staff, plus nine in blowmoulding. John had implemented changes in the manufacturing process to reduce unit costs, but more efficiency could be gained if longer runs and fewer setups were possible. He was concerned that the production schedule seemed to be disrupted every two or three days in order to meet urgent delivery needs. The daily production schedule determined which products and sizes were to be run on which packaging line. The daily plan often had to be made before all orders for that day had been received, and what went out on a truck was a mix of product that had been in the fridge from the previous night's production, and what was just coming off the packaging lines as the order was assembled.

Vendors and other customers preferred to receive consignments where all product had the same "best before" date. Therefore, production was sometimes disrupted to change the date back one day to be consistent with stocks held overnight. Such "best before" dates were conservative estimates; if kept refrigerated, milk could last considerably longer. Because of this dating scheme, it was impossible to stockpile product for future use. What went out on a given day had to be packaged that day or the night before, to the requirements of orders just received that day.

Once all orders for the day had been received (sometimes as late as 11 am), it took time to work out if there was likely to be a shortage of any product or size. Therefore, the schedule was not finalised until well after the production run for the day had started. If a shortage was expected, then two changeovers were required; one to run the shortfall product, and another to return to the original plan. Although a shortage was never ignored, mixed "best before" dates were tolerated to avoid additional setup delays, despite concerns expressed by the marketing staff. These disruptions extended production times and led to total overtime hours of 170 to 200 hours per week. John felt that better production planning could reduce overtime to 50 to 70 hours per week.

Fridge capacity was limited, but workable. However, the layout presented difficulties to pickers, with only 90% of orders being picked correctly. This led to one or two complaints each day from customers, so NZmilk was investigating total automation and a move to just-in-time manufacturing. However, initial enquiries suggested that this would require an investment of at least $500,000 in additional packaging and computing equipment. As a result, it seemed that NZmilk was unlikely to move in this way within the next three years.

Order Entry

Vendors and major supermarkets were expected to fax, e-mail or phone their daily order to NZmilk, either the night before or up to 10 am on the day of delivery. Order data was collected in both the factory office and the front office. When Diane McCann arrived to work in the factory office at 5 am, she collected the order faxes and the messages left on the answering machine overnight. Some orders had to be ready by 6:30 am for loading the first truck. She analysed the orders just received to check the production plan that had been prepared for that day. Many orders for afternoon deliveries were taken during the morning of the same day.

Local vendors telephoned, faxed or e-mailed their orders to Diane directly, while Jane Roberts and one other in the front office typically spent two hours every morning telephoning the 30 Woolworths supermarkets to obtain that store's order details. They rarely got through to the right person (such as the deli manager) in the store the first time they called, and often had to wait on the phone while the person was located, the store needs worked out and the order finalised. If the deli manager was not available, they had to deal with one of the other junior employees in that area, and this sometimes caused problems with over or under ordering of required products. Jane spent a further hour each day collating the Woolworths order data.

An accurate picture of the orders required to be manufactured and shipped that day was often not available until late in the morning, which meant that the planned packaging run of some products in the plant had already been completed. This meant that additional short runs of products might have to be done late in the day, which entailed significant time to clean out and re-configure the packaging lines for these products.

Invoicing and Accounts Receivable Control

Joan Proudfoot was in charge of NZmilk's major computer system, which was called Milkflex. Because of the very different business processes and billing arrangements with vendors that had arisen from the previously regulated environment, off-the-shelf order and billing systems were not suitable for NZmilk's needs, and the development of a custom system had been required. Milkflex had been developed about five years previously by a local software services company in Dataflex, a PC database language. Milkflex had been designed to specifically incorporate NZmilk's existing business processes during the time of government regulation. The major function of Milkflex was to produce invoices and accounts receivable reports, after sales order data had been entered from the order sheets. Milkflex was originally written to invoice home delivery vendors, but was modified to include centralized weekly billing for Woolworths, and to incorporate different pricing policies. About 50 supermarkets were invoiced weekly, and about 65-70 other customers invoiced monthly. The unique system of discounts and credits for the home delivery vendors complicated this billing cycle. Furthermore, as supermarkets were billed weekly, but often supplied with product from vendors' trucks, these orders were initially entered into order forms by vendor, later entered into Milkflex under the vendor code, then later recoded by supermarket. Typically, vendors were charged for what they ordered, while supermarkets were charged for what they sold.

Other changes had been made to Milkflex over the years. The range of reports was extended to include daily stock reports and a production plan based on stock on hand and past sales. Monthly sales analysis reports were typically available by the middle of the following month. Further minor changes were still outstanding; for example, it was not easy to prepare monthly summaries by supermarket. Joan found that Milkflex worked well for her, and wondered why production and marketing were not keen on their new reports, after asking her for them for months. Angela and Joan assisted Les Brown, the financial controller, with the wages each week, using a Milkflex module, which worked well, despite limited reporting functions and very demanding formatting.

Financial Planning and Control

NZmilk had been using modules of the standalone PC-based Charter accounting package since the early 1980s for creditors, general ledger and fixed assets. The system had been upgraded in 1995 after Les Brown looked for a more user-friendly product, but decided to stay with Charter as the only obvious replacement (CBA) could not be supported locally. Only Les and Joan had access to Charter, and much of the data came from the Milkflex system printouts and was transferred manually. Data input times were typically less than 15 minutes each time. The Charter system met most accounting needs, although its reporting features were very limited. NZmilk had about 600 creditors accounts, and 150 payables cheques were sent out each month.

Les and Joan also used Excel spreadsheets to produce forecasts, budgets and plans, typically for use by the whole management team. Much of this data was extracted from other reports, either from Charter or Milkflex.

Production Planning

A new report from Milkflex had been created to assist with production planning. However, this report had not been accepted or used by production staff, as the computer data never matched the physical closing stock data. Instead, production staff used a manual approach to plan the day's production, and hoped that Joan and Les would sort out the problems with the new report. As Robert Kokay put it, "The system seems to be right for the front office, but not user friendly for us." Production planning was made even more difficult as there were high sales at weekends, but there were no deliveries on Sunday. Ideally they wanted to generate a production plan by packaging line, by product, by time of day, for every shift.

Brian English and Diane McCann drew up the production plan late in the afternoon for the following morning. This had to be done when none of the next day's final orders were known. Instead, data from the corresponding day two weeks earlier was used as the best estimate. A two-week period was used because some of the more distant vendors collect on alternate days, and some social security benefits are paid on a two-week cycle.

The preparation of the production plan started after a physical stock-take. The planning task was demanding as there were about 50 to 60 products to consider, with most having to be made on a daily basis. Much of the afternoon was spent preparing the next day's plan, and it took an hour just to determine the bulk milk needs for the following day.

During the day, the plan was checked frequently as incoming order data became available. Revisions to the plan were made during the day if needed. It was desirable to finish a day's production with sufficient finished goods in the refrigerator overnight to satisfy the first three truck loads the next day, but on 80% of days, this stock was inadequate to completely fill the orders for these trucks. Part of the Milkflex system was designed to assist with the control of finished goods. However, it was not user friendly, failed to help when stock data did not balance and had a poor data entry screen.

Quality Assurance

Tony Fineran, Quality Assurance Manager, had a staff of two. The team conducted tests in the laboratory on samples of both raw and finished product. The results were analysed and presented on typed sheets, but Tony was trying to make greater use of a PC spreadsheet. Quality reports were also available, but only McDonalds requested them for the products made for them. No data could be automatically transferred from the testing equipment to the computer. As a result, Tony was concerned that some patterns/trends might go unnoticed. Furthermore, Tony received about 200 complaints from customers each year, and a further 150 from staff at NZmilk, all of which had to be handled. His team dealt mainly with serious cases, and had time to investigate only some of the others. Tony suspected there was an opportunity to use computers to assist in the monitoring and handling of complaints, which was currently a paper-based system.

Present Use of Information Technology

Up until the present situation, NZmilk would not have been considered to have been an intensive or strategic user of information technology. The early systems installed were financially focused, and to a large extent were automation of existing manual billing and accounting systems. The firm relied on external local firms to supply software, hardware, programming and technical support when needed. Only a few employees were knowledgeable about the systems that were operated, and they generated ideas and extensions for the custom software, which was then contracted out to an external firm. The focus of these changes was more about tailoring the software to fit existing business processes and manual procedures, rather than thinking about the re-design of business processes (supported by IT) to support company strategy. Most changes were oriented to make life easier or provide additional reports for the internal employees interacting with the system, rather than provide information useful to external customers.

Major hardware and software additions occurred in early 1991 when NZmilk installed a network of Windows-based PCs, in early 1995 when Milkflex was expanded, and in 1998 when the Internet began to be used for e-mail and WWW browsing.

There were no employees of NZmilk dedicated to support applications or develop software, although the hiring of an IT coordinator was being considered. Some training of employees on IT packages had taken place. For example, Les Brown and Joan Proudfoot had spent some time with consultants from Baycom, looking at the ways to change Milkflex report formats. Otherwise, most have learned by being users of the packaged software, and a few were able to specify changes required for custom applications like Milkflex, but the company had to rely on outside people to provide anything beyond these basic functions.

Ed Doughty was the local agent for the Charter accounting package, which had about 60 sites installed in New Zealand's North Island. He had an accounting background, and believed that NZmilk's needs were unique because of their vendors and their billing requirements.

The Charter package used by NZmilk was written in QuickBasic, and could accept datafiles straight into the General Ledger. There was a Bill of Materials module which might be able to help with bulk milk forecasting, but there was no easy way within Charter to use order data to determine a production plan. The general ledger and creditors modules that NZmilk were using were the latest versions. Other modules exist for Charter which were not presently used by NZmilk, such as a financial report writer. There was also a Global Report Writer, but this was for application developers rather than end users. Modules averaged about $1,300 each.

An external person who had regular contact with NZmilk was Hugh Gurney, who used to work for Whangarei Milk, and developed the first version of Milkflex after forming his own software company called Baycom. Baycom provided support to NZmilk and a large number of other clients. Baycom used various tools for system development, including Dataflex, a 4GL relational database programming language designed for experienced software developers. Baycom also sold hardware and other software, and had installed the Novell Netware software at NZmilk.

Over the years, Baycom had built various systems for their clients using Dataflex, including modules for general ledger and creditors, but not fixed assets. Some of these could be adapted for use by NZmilk if required. Baycom also had products for legal practices, as well as EDI experience with a local bakery for order entry via modem.

Hugh Gurney's business partner Graham Jackson felt that NZmilk's needs were unique, so no existing product could be used to meet all of their needs. Graham regretted missing out on NZmilk's contract for the upgrade to their local area network, and attributed it to various factors, including spending too little time determining requirements, and not providing top service to NZmilk at times. He would like to have extended Milkflex beyond the upgrades, and saw this as a viable option rather than NZmilk trying to develop their own systems. He was keen that NZmilk should retain Dataflex as their base technology, and would be happy to offer training so that users could more effectively use the Milkflex report writer for simple inquiries. FlexQL would be needed for more complex relational queries, but this had a steeper learning curve.

Neil Dickie, a local, independent consultant won a recent contract to supply and install three new PC's for NZmilk, and a notebook for Maurice Lloyd, the general manager. When extending the network, Neil noticed that the network required better configuration. Security and access were not well set-up or managed, and all the Microsoft files were in one directory. He also wondered why some applications were set up to work on only one PC. Neil expressed interest in spending a day sorting out these problems, which likely resulted from the network being set up by numerous people at various times, with no plan in mind.

Potential Uses of Technology

Maurice Lloyd was convinced that information technology could play a key role in NZmilk's growth strategy. There were a number of exciting ideas for using computers which included:

- EDI with supermarkets, although it seemed that firms like Woolworths were not likely to force this, at least within the next few years.
- Internet-based ordering with vendors and supermarkets (some already placed orders via e-mail).
- Invoice at point of sale through in-truck computer systems.
- Support for production planning and forecasting decisions.
- A fully automated warehouse.
- Business process reengineering.

- A telemarketing system to contact stores, solicit orders and sell additional products.
- Addition of home delivery of groceries via telephone or computer ordering.

Pressure for Change

There had been approaches by several New Zealand representatives of manufacturing and distribution software packages (both MRP II and ERP), and the operations and sales people in NZmilk were clearly interested in looking at what might be done. Maurice Lloyd had been exposed to the use of IT for strategic advantage in the part-time MBA degree he had been undertaking, and also had been following reports in trade magazines about manufacturing and distribution software systems, the industry transforming impact of the Internet, and the rise of customer relationship marketing. Their parent company, Northland Milk, also had some experience with the purchase of both packaged and custom-developed software, primarily in the manufacturing and financial areas.

However, Maurice was unsure whether their focus should be on solving their manufacturing and distribution problems with a tried and proven off-the-shelf packaged system tailored and modified to their requirements, or whether they should take a step back, and try and understand the impact of the new ways they might be doing business in the future, and the new trust and information exchange relationships they might develop with both present and future customers and business partners. He could see that a significant investment in IT was looming, and probably critical for their survival and growth in an increasingly competitive market. However, the potential was also there for an expensive disaster if an appropriate and realistic path was not taken.

The key question he kept turning over in his mind was: "How should I get this process underway?". But he was not sure how to proceed, what process he should follow and who he should involve, including who should be project leader. If NZmilk was to grab the opportunities that were available, and avoid the pitfalls, he had to make the right decision.

ADDITIONAL READING

Bergeron, F., & Raymond, L. (1992). Planning of information systems to gain a competitive edge. *Journal of Small Business Management, 30*, 21-26.

Currie, W. (1995). *Management strategy for IT: An international perspective.* Pitman.

Earl, M. J. (1989). *Management strategies for IT.* Prentice-Hall.

Earl, M. J. (Ed.). (1996). *Information management: The organizational dimension.* Oxford UP.

Galliers, R. D. (1991). Strategic information systems planning: Myths, reality and guidelines for successful implementation. *European Journal of Information Systems, 1*(1), 55-64.

Horne, M., Lloyd, P., Pay, J., & Roe, P. (1992). Understanding the competitive process: A guide to effective intervention in the small firms sector. *European Journal of Operational Research, 56*, 54-66.

Luftman, J. (1996). *Competing in the information age.* Oxford Press.

Martin, E. W, Brown, C. V., DeHayes, D. W., Hoffer, J. A., & Perkins, W. C. (1999). *Managing information technology: What managers need to know* (3rd ed.). Prentice-Hall.

Papp, R. (2001). *Strategic information technology: Opportunities for competitive advantage.* Hershey, PA: Idea Group Publishing.

Robson, W. (1997). *Strategic management of information systems* (2nd ed.). Pitman.

Thong, J. Y. L., Yap, C. S., & Raman, K. S. (1996). Top management support. External expertise and information systems implementation in small businesses. *Information Systems Research, 7*(2), 248-267.

Ward, J., & Griffiths, P.(1996). *Strategic planning for information systems* (2nd ed.). Wiley.

WEB RESOURCES

Customer relationship management. Customer Relationship Management Research Centre. http://www.cio.com/forums/crm/other_content.html

Gartner Group Report on CRM. http://www3.gartner.com/1_researchanalysis/executive/premierecrmmgt.pdf

Manufacturing Resource Planning (MRP II). A guide to MRP and ERP. http://www.bpic.co.uk/erp.htm

List of over 500 manufacturing software vendors. http://www.softmatch.com/manufact.htm

Site of a prominent manufacturing software vendor. http://www.qad.com/

Strategic information systems Planning. Information Technology Management Web. http://www.itmweb.com/

Milk industry in New Zealand. New Zealand Dairy Foods, a manufacturer similar to NZmilk. http://www.nzdf.co.nz/

New Zealand. General information about New Zealand. http://www.nz.com/

APPENDIX: PRODUCTION CAPACITY

Typically the filling lines are working from 5 am to 2:30 pm. 55,000 litres of milk are held overnight. Daily supply averages 85,000 litres, typically from 24,000 litre milk tankers. Raw milk is tested for quality on delivery. Typically only three or four loads are rejected per year, and problems occur rarely during peak periods.

The blowmoulding plant works 24 hours per day, with one machine producing 1 litre bottles, the other 2 litre bottles. Bottles require four hours to cool and shrink, otherwise they expand and thus take more milk to fill.

Filling capacities are:

Line one:
60 units per minute (2 litre) or 112 units per minute (1 litre)
Line two:
38 units per minute (1 litre) or 75 units per minute (300 ml)
Line three:
25 units per minute (1 litre carton)

The production cycle moves from low fat milk first through to higher fat content products. During the last few years changeover techniques have improved, so that change over times are now a maximum of 5 minutes rather than the previous 15 minutes. The change to cream requires a flush of the system, taking about 20 minutes. On average, there are about eight product changes per line per day.

Paul Cragg is an associate professor in information systems in the Faculty of Commerce, University of Canterbury, New Zealand, where he teaches in the MBA program, as well as with the BCom, MCom, and PhD degrees. He previously was on the staff at the University of Waikato, New Zealand, and the Leicester Polytechnic, UK. Dr. Cragg's research centers on small firm computing. Current studies focus on IT alignment, benchmarking, IT sophistication, and adoption and use of the Internet. He has published in many international journals including MISQ, EJIS, Information & Management, *and* JSIS.

Bob McQueen has a BApSc in electrical engineering from the University of Waterloo, an MBA from Harvard Business School, and a PhD (computer science) from the University of Waikato. He is presently an associate professor in the Department of Management Systems, University of Waikato in Hamilton, New Zealand. He has been living in New Zealand since 1988. He has also worked with IBM in Toronto and Digital Equipment in Vancouver. Dr. McQueen is an enthusiastic proponent of case method teaching as an effective inductive learning approach. Thirteen cases have been developed in the IT policy area under his supervision, using the Harvard Business School approach, for teaching IT policy at fourth year undergraduate and graduate (MBA) levels.

This case was previously published in the *Annals of Cases on Information Technology*, Volume 4/ 2002, pp. 73-83, © 2002.

Chapter XX

Life After a Disastrous Electronic Medical Record Implementation:
One Clinic's Experience

Karen A. Wager, Medical University of South Carolina, USA

Frances Wickham Lee, Medical University of South Carolina, USA

Andrea W. White, Medical University of South Carolina, USA

EXECUTIVE SUMMARY

The majority of users of an electronic medical record (EMR) at a family medicine clinic located in a small city in the western United States are currently quite dissatisfied with the system. The practice experienced a disastrous implementation of the EMR in 1994 and has not recovered. Although the level of dissatisfaction varies among the practice employees, several influential physicians are pushing to "pull the plug" and start over with a brand new system. The authors of this case studied this practice during a more comprehensive qualitative study of the impact of an EMR system on primary care. The practice's negative experience was particularly noteworthy, because the other four practices in the larger study were satisfied with the EMR system. As with most system failures, there are multiple organizational and other factors that have contributed to the frustrations and dissatisfactions with the use of EMR within this practice.

BACKGROUND

In his textbook *Managing Information in Healthcare*, John Abbott Worthley (2000) discusses three "realities" that impact the use of information technology to manage information in healthcare and in other organizations. The first reality is "the general pervasiveness of information technology in managing information..." The second reality is that "computer-processed information continues to offer wonderful opportunities for significant improvements in organizational and social life, and in healthcare in particular." The final reality is that "the actual experience...with managing information technology in organizations and in society often has been disappointing and problem laden...In many cases the potential for using information technology in organizations has not been realized." Consider Burch's (1986) tongue-in-cheek life cycle of modern information systems. Stage one is wild euphoria; stage two is mild concern; stage three is broad disillusionment; and stage four is unmitigated disaster. Unfortunately, this life cycle is often more recognizable to veterans of an information system implementation than the more traditional life cycle of systems planning, analysis, design, implementation and evaluation.

Characteristics of Healthcare Information and Information Systems

What are the unique challenges to managing information in a healthcare environment? The healthcare industry has been notoriously slow to adopt information technology to manage patient information. Multiple reasons have been discussed for this lag behind other industries. Healthcare information is sensitive information. Many users are concerned about security and confidentiality issues. Patient information is documented in medical "language" and is narrative in nature. The scope of patient information has been all-inclusive; rarely are specific criteria outlined for what will or will not be documented. As a legal document the patient record contains a lot of details about patient care.

Healthcare organizations, particularly physicians' offices, also differ from other business enterprises. In a private primary care practice, the physician or physicians own the practice. Traditional medical education has not emphasized management practices. While many practices have business managers, often these people are responsible for billing and clerical functions. Patient documentation is seen as a function of patient care under the direct control of the individual physician.

Millions of dollars have been spent exploring the use of computers to manage patient information. Today's patient record systems, however, often look very much like they did two or three decades ago — they are maintained in paper form within manual filing systems. The data the physician and other care providers seek are often illegible, fragmented, incomplete or altogether irretrievable.

Current Status of EMR Systems in Healthcare

Following a comprehensive study of the myriad of problems associated with paper-based patient record systems, the Institute of Medicine recommended that the computer-based patient record (CPR) or EMR become the industry standard by the start of the new millennium (Dick & Steen, 1991). Although this has not occurred, there is an increasing interest among physician practices to adopt this technology (Anderson, 1992; Anderson, Aydin, & Kaplan, 1995; Balas et al., 1996; Edelson, 1995; Hammond, Hales, Lobach, & Straube, 1997;

Wager, Ornstein, & Jenkins, 1997; Yarnall, Michener, & Hammond, 1994). Rapid changes in healthcare and the influx of managed care plans have led to an increased need for accurate, timely patient care information, both at the patient-specific level and at an aggregate level (Dick, Steen, & Detmer, 1997).

The terms EMR and CPR have been used interchangeably to describe many different levels of computer systems used to maintain patient care information. One of the most widely accepted definitions comes from the Medical Record Institute (MRI), an organization dedicated to the promotion of the EMR (Waegemann, 1996). The MRI describes an attainable EMR as generally limited to a single enterprise and possessing at least the following functions:

1.   An enterprise-wide system of identifying all patient information,
2.   The ability to make all enterprise-wide patient information available to all caregivers,
3.   The implementation of common workstations to be used by all caregivers, and
4.   A security system to protect patient information.

The EMR is more than scanned images or online forms made available electronically. The structure of the EMR is established to allow for electronic information storage, exchange and retrieval.

The advantages of EMR systems include legibility, accessibility to data for display, reduction in time spent recording data, and improvement in data management (Rittenhouse & Lincoln, 1994; Safran et al., 1996; Spann, 1990; Yarnall et al., 1994). In addition, EMR systems are excellent tools for monitoring health maintenance data such as suggested screening time frames and normal clinical values by age and gender (Safran et al., 1993). They can also issue reminders to primary care providers and patients (Ornstein, Garr, Jenkins, Rust, & Arnon, 1991).

The PTX EMR System

The PTX system is one of the most established practice-oriented EMR systems on the market today. It was developed by a physician more than a decade ago and is used by hundreds of physicians and ancillary staff across the country. Data are not available for the actual satisfaction levels among all PTX customers, but the product enjoys a good reputation as an effective practice-oriented EMR system.

The PTX software includes all features typically found in a primary care medical record. Because it is in electronic format, however, the data elements are stored within and retrieved from a computer database. This database allows easy retrieval of patient information on both the individual level and in an aggregate form. Specific features of PTX include problem lists, progress notes, vital signs, medical history, social history, family history, medication lists, immunizations, health maintenance, laboratory results, and sections for reports of ancillary studies. Disease-specific progress note templates are available for facilitating direct progress note entry; however, direct data entry is not a requirement for using the system. Transcribed notes can also be incorporated. Templates for gender and age-specific health maintenance can be automatically applied to new patient records.

Two of PTX's features that complement the record itself are a prescription writer and an internal e-mail system. The prescription writer allows prescriptions to be entered into and tracked in the patient record while automatically checking for interactions with other medications and documented allergies. Limited drug cost information is also available to the user. Patient-specific e-mail messages, written and answered within the system's e-mail feature, can become a permanent part of the patient's record.

Table 1. Summary characteristics of participating practices

Description	A	B	C	D	E
Type of Practice	Solo Internal Medicine	Group Family Practice	Group Family Practice	Group Family Practice	Solo Family Practice
Total Staff Size	6	9	34	25	5
# physicians	1	1	7	4	1
# physician assistants	0	1	2	1	0
# nurses	1	2	8	5	2
# support staff	4	5	17	15	2
Patient Population	Primarily Medicare patients and self-pay. Small % with no insurance.	Primarily managed care, private insurance or self-pay. 20% Medicare.	Primarily Medicare, Medicaid and managed care. 10% self-pay	Primarily managed care and Medicare patients. Small % with no insurance.	Primarily managed care and Medicare patients.
Geographic location	Urban	Urban	Urban	Urban	Rural

EMR Implementation Critical Success Factors

What makes an EMR implementation successful? This question was addressed in a recent qualitative research study conducted by Wager et al. (2000). Wager and her colleagues designed a qualitative research study to study the impact of implementing the PTX EMR system in primary care physician practices. The study design included onsite visits and semi-structured interviews with practitioners and administrative staff in five primary care physician practices across the country. Table 1 outlines the demographic characteristics of the five practices studied.

Family Medicine Clinic (FMC) is the subject of this case study. It is identified in Table 1 as practice D, because it was the fourth practice visited during the study period. The visit to FMC was pivotal to the entire research project. Up to this point in the study, the vast majority of physicians and staff working in the other practices were satisfied with the EMR system. Within FMC, however, both physicians and nurses had very negative views toward the system and were seriously looking to replace it. Why was the same EMR system viewed so differently here than it was at the other sites? What organizational factors or characteristics led to EMR success or failure? This analysis continued following completion of the site visit to practice E.

Wager et al. (2000) discovered that the organizational context is a very important component in understanding the EMR's impact upon the practices. Within these five practices, the EMR system's perceived success or failure appeared to be related to a number of factors that are referred to as critical success factors (CSF). See Figure 1.

Figure 1. EMR success criteria

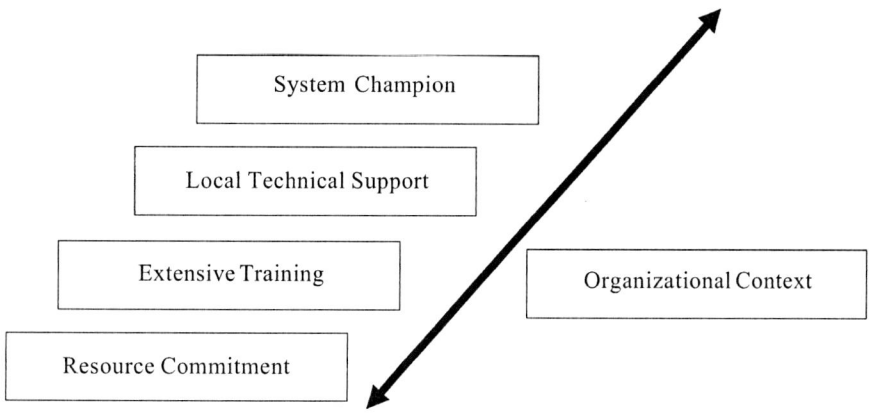

The practices that viewed the EMR system as an overall success (A, B, C, E) had a **system champion**. In each case, the champion was a physician who clearly served as a major advocate. He or she was instrumental in gaining "buy in" among the various user groups and in helping people overcome their fears and apprehensions. This person was also viewed as a leader in the practice — someone who was well-respected, knowledgeable, committed to the system's success, and powerful enough to make things happen. Generally, the system champion took the initiative to learn the intricacies of the EMR and offered assistance to others.

A second CSF was the availability of **local technical support** — someone available, preferably within the practice - who knew the intricacies of the software and who was able to handle hardware and network problems. Even those without a technical computer background were able to work with the vendor and others to find solutions to system problems.

A third CSF was **training**, both initial and ongoing. Staff who felt that the initial training was poor or less than adequate expressed many frustrations. Those who were pleased with the training they received had been given time to get comfortable using computers. In one practice, the staff was allowed to play games on the computer months before the system went live. They felt that it was helpful to begin by learning some basic functions and skills. Once they mastered these skills, they would be introduced to new concepts or functions. The staff was also provided with self-learning programs designed to teach typing skills. In another practice, the staff learned "a little at a time," but waited until everyone was at the same level before introducing new functions.

Besides being given an opportunity to get comfortable with computers and learn to type, many staff indicated that it was important to have intensive training just before the system went live so that concepts were fresh in their minds. They spoke of the need to have a trainer who can "talk to the level of the novice user." Even though the initial training provided by the vendor was felt to be a very important introduction for the staff, the ongoing training, once the vendor left, was equally important.

A final factor identified as key to an EMR's success was **adequate resource commitment**. The resources allocated or committed to the EMR included not only the up-front investment in hardware and software, but also the time and people needed to support it.

SETTING THE STAGE

FMC and Its Use of the PTX System

FMC is located in a small city (population 25,000) in the western United States. As a primary care practice, FMC serves a wide range of patients with multiple healthcare problems. A high percent of FMC patients are covered under managed care plans. The practice also serves Medicare recipients and a relatively small proportion of patients with no insurance. Patient volume is approximately 25,000 per year. On average, the five providers each see 20 to 30 patients per day.

FMC is an independent physician group practice. It is physically located next to the area's only hospital and all physician members have staff privileges, but there is no financial arrangement between the hospital and the practice. FMC currently employs 15 support staff, five nurses, four physicians, one physician assistant (PA), and an office manager. From 1993 until 1995 the practice nearly doubled in size, both in patient visits per year and in number of providers. In 1993, there were two physicians and the patient visits were 12,870. By 1995 the practice had added two physicians and a PA and patient visits had increased to 24,457.

The physical space occupied by FMC is spread across two floors of a building. The upper floor is where the physicians and nurses see patients. The professional offices, the office manager and the receptionist are located on this floor. The billing personnel are located in the basement area, as is the employee "lunch room." The atmosphere on the two floors is quite different. The upper floor is very hectic with a high level of activity. The billing offices downstairs are quieter and have much less traffic.

The DOS-based version of PTX was originally implemented by FMC in 1994. Since that time a Windows version has been released. FMC has installed the Windows version on 9 of its 30+ workstations. The providers and two nurses have the Windows version in their private offices or work areas. All other workstations are DOS-based. These workstations are located in the patient exam rooms, nurses' stations, reception and billing areas. There are no immediate plans to upgrade the remaining DOS-based workstations due to the cost. The PTX acquisition costs totaled $140,000, and the financial statements indicate that the practice has incurred an additional $260,000 in personnel, upgrades, equipment, and resources to maintain the system.

The practice uses the EMR as the primary form of patient data capture and retrieval; however, a paper medical record is still maintained and pulled for every patient visit. The paper medical record includes documents that are generated outside FMC, including lab results, X-rays and consultations. FMC has an operational lab interface, but the physicians did not like the format for viewing lab results in the EMR. Therefore, the original lab results are maintained in the patients' paper medical records. The practice has begun to scan some outside documents, but the process is still new and not yet working satisfactorily. The overall result is that half of any patient's information is in the EMR and half is in the paper medical record.

The physicians continue to dictate the majority of their visit notes, with templates used only 10-20% of the time for data entry. The PA is the only provider within the practice who consistently uses the templates for documentation.

Figure 2. Family medicine clinic organization

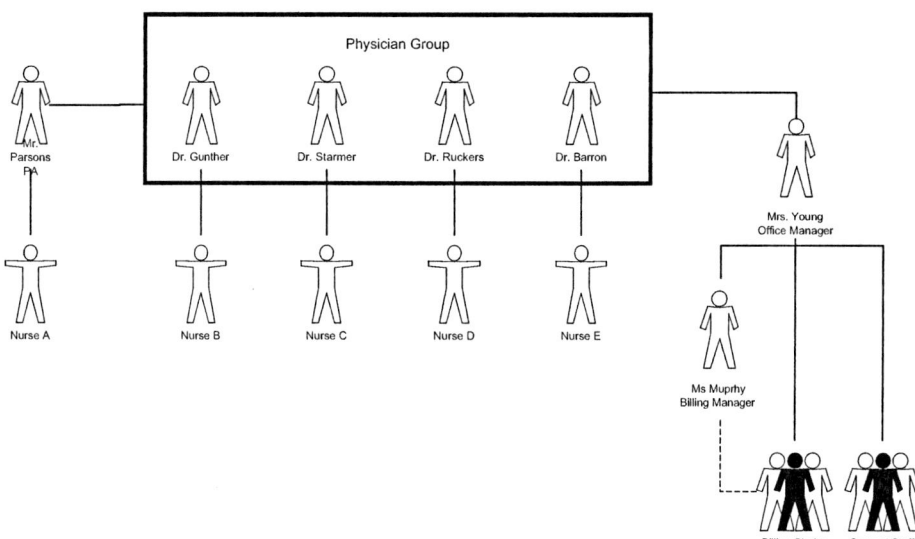

As a relatively large practice, the roles and job functions of each office staff member are fairly well defined. However, the organizational structure is not a typical hierarchical structure. Figure 2 is a graphical representation of the key players in the organization.

Providers

Dr. Barron is considered the lead physician because he started the practice. However, each physician operates fairly independently when seeing patients and the physician group makes decisions about the operation of the practice by consensus. Each provider (physician and PA) has one nurse assigned to work with him.

Dr. Barron is in his late 50s and the most experienced physician in the practice. He began as a solo practitioner nearly 25 years ago and since then has expanded the practice to include three other physicians and a PA. Other clinicians and staff members respect him for his knowledge, experience, and strong clinical skills. He is also well liked and respected by patients. His leadership style, however, is often described as "laissez-faire." He does not exert his opinion into discussion, but rather, in a quiet, gentle manner, Dr. Barron shares his opinions on key issues impacting the practice during periodically scheduled staff meetings. He is reluctant to push any issue, including the use of the EMR within the practice. Dr. Barron is also involved in professional association activities within the region's medical society. Through his work with this society, Dr. Barron hopes to convince other physicians in the region to implement the EMR within their organizations. He believes that his practice can only go paperless (and, thus realize the full benefits of the EMR) when other physicians and healthcare organizations in the community adopt this technology and are able to share patient information electronically.

Dr. Gunther joined the practice in early 1990. As a fairly young physician in his mid-40s, he admits to having the least computer experience among all the providers. His outgoing personality and love of conversation often leads to his being behind on his work at the end of the day. His patients are often seen 30-40 minutes late. Stacks of paper medical records are scattered throughout his private office waiting for dictation. He works long hours and is often weeks behind on dictating visit notes following patient encounters. He has formed a strong working relationship with the newest physician in the group, Dr. Starmer. The two of them often have lunch together and can be found conversing in their private offices at the end of the day.

Dr. Starmer joined the group in 1994, a month after the PTX implementation. He came to the practice from a large multi-specialty clinic with a high patient volume. Most of Dr. Starmer's computer experience was with Macintosh programs. In fact, he admits to being an "avid Mac user" and wonders why the "rest of the world never got it." Dr. Starmer is a young physician, in his early 30s, with a great deal of energy and drive. He is also very vocal. Although he feels that the EMR had no influence on his decision to join the practice, he admits he was intrigued by it. Having come from a large practice with "paper charts everywhere," he saw the potential of the EMR to make records more readily available to multiple providers at once.

Dr. Ruckers, the fourth physician in the clinic, joined the practice following the EMR implementation. He was not available for interview or observation during the data gathering of the case study; thus, his perspective is not presented.

Another key member of the clinical staff is Mr. Parsons, the Physician Assistant. Mr. Parsons is in his early 40s and, like Dr. Barron, is well liked and respected by the staff and his patients for his quiet, gentle manner and compassion for others. Mr. Parsons looks to the physicians for guidance and leadership in the practice. He consults with them frequently on patients and values their advice. Mr. Parsons worked in the practice at the time the PTX system was implemented. Prior to that time, he handwrote all his visit notes. As a proficient typist, Mr. Parsons welcomed the opportunity to directly enter visit notes at the point of care using the PTX system. He continues to directly enter visit notes using the templates and is able to keep current with documentation. Nearly all his visit notes are complete within a day or two following the patient's visit.

Nursing Staff

The five nurses also have a relatively long history with the practice. All but one of the nurses worked in the practice at the time of the PTX implementation. None of them had much prior computer experience. The nurses do not rotate from provider to provider; rather, each provider/nurse team develops its own routine for seeing patients. In general, the nurses are responsible for ascertaining the patient's reason for the visit, capturing information related to the patient's problems, responding to calls for refills on prescriptions, communicating patient questions to the provider, and documenting care. They are also responsible for handling referrals, which means that they need access to both insurance information and clinical information.

Office Manager and Support Staff

At the time the PTX system was installed, the current Office Manager Mrs. Young was working in the practice as a temporary staff person hired to help file reports and register patients. When one of the nurses discovered Mrs. Young had a degree in management and experience working in a specialist practice, she told Dr. Barron. It wasn't long afterwards that

Mrs. Young was offered the position as office manager. FMC had been looking for an office manager for quite some time — and Mrs. Young simply "was at the right place at the right time." She believes that her calm nature and strong work ethic landed her the job. Her appointment was apparently not favorably viewed by some of the support staff who had more experience and had worked in the practice longer. One receptionist left the practice because of Mrs. Young's appointment.

Mrs. Young rarely uses the EMR and spends most of her time with the financial component of the system. She prefers to use "sticky notes" rather than the internal e-mail system. Despite her limited use of the EMR, Mrs. Young serves as the point of contact within the practice when providers or staff members are having technology problems. Her primary role is to contact the technical support staff at the PTX headquarters to help resolve these problems. She also works with the billing manager, Ms. Carol Murphy, to troubleshoot system problems.

Ms. Murphy began working in the practice two years before the system was implemented, and she is responsible for conducting backups, monitoring system errors, and assisting staff having problems. As expected, Ms. Murphy is most familiar with the billing component of the system — and only uses the EMR when looking up patient diagnoses or problem lists to substantiate claims. Although interested in learning more about the EMR, Ms. Murphy feels that her time is "spread too thin" trying to manage billing and provide adequate computer support to staff. At the time FMC implemented the PTX system, Ms. Murphy was asked to assume additional responsibilities related to system management, yet none of her other responsibilities related to billing were taken away.

Mrs. Young and Ms. Murphy share responsibility for managing the support staff. All of the support staff (including transcriptionists, receptionists, data management and billing clerks) report directly to Mrs. Young on issues related to performance, schedules, and requests for annual/sick leave. The transcriptionists are primarily responsible for transcribing dictated reports directly into the EMR. The receptionists are responsible for answering the phones, making patient appointments using the PTX scheduling component, responding to patient inquiries, pulling paper medical records, and filing outside reports. Although much of their time is spent using the PTX system, most receptionists continue to spend between 10-30% of their time maintaining paper medical records.

There is one data management clerk who handles all scanning of outside documents into the PTX system and who manages the laboratory interface. As part of her role, the data management clerk also verifies that information generated "outside" the practice gets entered into the correct patient's record. The billing clerks submit bills to insurance companies and follow-up with collections. Ms. Murphy works closely with the billing clerks in handling complaints and assisting them when they have problems or questions.

CASE DESCRIPTION

The Selection

The practice implemented the three integrated components of the PTX office management and EMR system (billing, scheduling and patient records) in August 1994. Prior to this time, patient scheduling and medical records were kept using manual systems. Although the billing function was already automated, Dr. Barron felt it was important for the applications to be fully integrated, so they opted to convert their old billing system to the new one — and implement all three applications at once. According to Dr. Barron, the practice had gotten to a point in which "one full-time person spent her entire day looking for charts." Records were often lost. Dr. Barron recommended that the practice implement the EMR system as a means to manage patient records more efficiently. His recommendation to implement an EMR was deeply rooted in his belief that "medicine was headed in that direction" and that the timing was right to make a change, especially with the practice growing in size.

Besides Dr. Barron, both Dr. Gunther and Mr. Parsons worked in the practice at the time they implemented the billing, scheduling, and EMR applications. Although both Dr. Gunther and Mr. Parsons had an opportunity to view several different EMR systems and conduct site visits to other practices, they were not as knowledgeable as Dr. Barron about the EMR. Dr. Barron was the most influential in the decision to purchase the PTX system. He had visited a specialist practice in a nearby city — who apparently "loved the system" and showed him "all kinds of neat things you could do with it to increase productivity and improve patient care." Both PTX and the specialist using the system told FMC that they could expect to cut at least one full-time equivalent employee by implementing their EMR. Although few EMR systems were available and affordable to community-based physicians at the time, the PTX system had been in use for nearly 10 years and was viewed as a solid product. The owner and CEO of PTX was a physician with a strong computer background; he combined his skills in both areas to develop the PTX EMR system. Since Dr. Gunther and Mr. Parsons knew little about the EMR and were not computer literate, they trusted Dr. Barron's judgment and agreed with the decision to implement the PTX EMR system.

The Implementation

The integrated office management system was purchased from an inexperienced PTX distributor. This was his first sale. Much of the distributor's time was spent working on conversion problems from the old billing system to the new. Little time was left for training of support staff, nurses, and physicians in any of the components, particularly the EMR. Both physicians who were present at the implementation had less than two hours of formal training on the system before they went live. The nurses had a total of three hours of training. Normally, distributors spend one full week training the staff, but because of the billing conversion problems at this site, training time was cut in half. Most of the staff recall that the extent of their training was two half-days. As one nurse recalled, "The entire staff came in for training on Sunday, the day before we were scheduled to go live. Despite having nearly 20 people in the room including nurses, physicians, billing clerks, and receptionists, the distributor had only one keyboard on which he demonstrated the various features of the system. It was a horrible mess. We thought we would die." Dr. Gunther described the distributor as being neither able to explain nor understand the questions asked by the clinicians, because he did not have a clinical background. "It was as if we were just thrown in the water and expected to swim."

The nurses and Dr. Gunther were not alone in their views. Every person who worked in the practice at the time of the conversion found the process to be problematic. Most people described the process as a "nightmare" or a "disaster." Most staff members felt that "too much was thrown at them" in a very short period of time. Even those who were initially enthusiastic about the system as they entered the project admitted that they lost some of their enthusiasm as the problems caused by the inadequate training mounted.

Current Views of the PTX System

The Support Staff

The support staff today report being quite pleased with the PTX system, although many staff do not routinely use the patient record component. If they need to access patient clinical information, it is generally to obtain diagnostic or procedural information to substantiate a claim or to verify prescription information. Since the billing staff is located in the basement, it is particularly helpful to them to have "patient information at their fingertips" and not to have to run up and down stairs. Such ready access has been a tremendous timesaver. Other advantages of using the EMR cited by the support staff include:

- The system facilitates communication within practice through use of the e-mail messaging system (e.g., eliminates need for handwritten messages),
- The system is relatively easy to learn and use, and
- The quality of documentation is improved (e.g., a legible, organized record, with more complete and accurate information is created).

Several support staff feel that their jobs are more "important" now that they are using the computer. In fact, they express frustration with the amount of paper still kept within the practice.

Most support staff members believe that the patients' impressions of the EMR are positive. They describe patients who seem impressed with legibility and with the fact that the staff members are able to retrieve their information quickly. A few staff members feel that the patients are probably unaware of the use of the EMR in practice.

Since the support staff relies heavily on the integrated system, they are at a "standstill" when the system is down. Downtime is overwhelmingly the support staff's biggest concern in using the integrated system. They do not believe that downtime occurs as frequently as it used to. During the first year with the PTX system, they estimate the system was down for at least 30 minutes every week or two. Now, the system is down for a few minutes every two months.

The Clinicians

The physicians and nurses do not share the same positive views about using the EMR as does the support staff. They are, with a few notable exceptions, quite dissatisfied with the system and have a number of concerns. All physicians describe the system as very time-consuming and cumbersome.

Dr. Starmer, who joined the practice a month after the implementation, finds the EMR system extremely difficult to use. He is also quite frustrated with PTX as a company. "If you look at some of the financial packages that people use in their homes, you see that these vendors actually come into your home and watch you use the system. I bought a product from such a vendor and they came over a week later and saw where I had problems. Their next version had the problems corrected. I realize that this takes tons of money to do, but every encounter we have had with PTX has been in the conflict resolution mode… rather than, what can we learn from you to make our product better?"

Dr. Starmer feels that the PTX system has "all the bells and whistles" but cannot do basic functions such as identify the patient. "Everything takes at least four steps and it shouldn't be that difficult. I knew at the end of my first week using the PTX system that it wouldn't work." The other physicians are frustrated too. They attribute this frustration to the fact that they are not using the EMR to its full potential, that they do not have the time or the energy to invest in learning how to use the system more effectively.

Drs. Starmer and Gunther are the most vocal about their dissatisfaction with the EMR system. They expected to be able to increase their productivity and have more time for their families. They also expected that the practice would save on personnel costs over time. Even though the initial cost seemed high, the physicians felt that it would be worth it in the long run if they could improve their efficiency and effectiveness. Both of these doctors are still swamped with dictation — and paper records are stacked all throughout their offices. They complain about the high cost of the system — we have "invested nearly $400,000 in the product — and it still doesn't work right." They do acknowledge that much of their inability to realize cost savings could be attributed to their current practice of maintaining a dual system of both paper and electronic records.

Other disadvantages identified by the physicians include:

- The EMR's inability to handle graphics effectively,
- Ineffective user manuals,
- Inadequate computer support,
- High direct and indirect costs,
- Instability of the e-mail messaging system,
- Excessive downtime,
- Problems with scanning process,
- Inadequate training, and
- Difficulty in learning to use the system.

Dr. Barron, the lead physician, also acknowledges the problems. He maintains, however, that in the long run, the benefits of the EMR will probably exceed the costs because if the practice had continued to use only paper records, they would have had to add on to the building and that would have been a tremendous expense. Dr. Barron "plugs away" at PTX and continues to learn about how to use the system more effectively. He does not, however, routinely share his knowledge and experience with PTX with other staff. This is a result of his leadership style and quiet demeanor, rather than an intentional withholding of information.

Despite the problems, all the physicians agree that they would not want to return to paper records — and that their challenge is to "figure out how to make the PTX system work or replace it."

The nurses also express a lot of frustration with the EMR system. Their general sentiment is the EMR has increased their workload and the amount of stress in their work lives. One nurse said, "Nursing used to be very simple." Now she is expected to enter vital signs, immunizations, prescriptions, and respond to e-mail messages. Nurses are also expected to handle all referrals. The nurses attribute much of their added work to the multiple steps needed to find and edit information, the fact that there is no single central record, and the difficulty in not being able to pull up insurance information while in the patient's record.

Interestingly, this last concern was identified as one of the biggest problems, yet in reality, need not be. Because the nurses handle all referrals and the practice treats a large number of patients under managed care plans, the nurses frequently need access to patients' insurance information throughout the day. To get this information, they exit the EMR component to enter the billing component, a time-consuming and inefficient process. (Authors' note: Although the nursing staff follows this procedure for obtaining insurance information, the system does, in fact, allow a user to switch between systems but the nurses are unaware of how to do this).

The nurses cite a number of other limitations or frustrations with the EMR including:

- Communication problems between DOS computer and Windows computers (two nurses use Windows version; all others use DOS),
- High cost of system,
- Inadequate support when encountering problems with system,
- Inadequate training — both initial and ongoing,
- Difficulty in learning and using the system,
- Some patient dissatisfaction, and
- Some patient concern about confidentiality.

Two nurses have specific concerns about the impact of the system on communication and their relationships with both the providers and the patients. One nurse, in particular, says she "does not talk to the physician as much" as she used to, because messages are all sent by e-mail. She states that she finds it hard to express in writing what she really thinks about how the patient is doing. By not clearly communicating her views, she believes the physician sometimes responds inappropriately to the patient's situation. Other nurses believe that some patients have found it distracting when the nurse is typing. One nurse reports that she has had patients ask her to "please stop typing." Another nurse has had patients express concern about confidentiality of the system. In general, the nurses did not like the EMR system and, for the most part, feel that their patients' impressions of it are not positive either.

Some of the nurses also believe that the EMR system has contributed to difficulty in getting outside nurses to cover for them when they are on vacation. They feel that is very difficult, if not impossible, to get temporary nurses to agree to fill in because they do not know the EMR system. The EMR "scares people away."

Not all of the nurses' opinions about the EMR and its features are negative. One nurse likes the code link feature that is available in the Windows version. This feature allows the diagnosis to be automatically linked with an appropriate diagnosis code. Others like the prescription writer feature. They like the fact that the prescriptions are legible and readily available. The nurses also like the Windows version of the software better than the DOS version and are frustrated by problems they believe are caused by the interaction of the two versions.

Interestingly, two of the nurses with more negative views of PTX are the nurses assigned to Dr. Gunther and Dr. Starmer. Both physicians openly share their frustrations with their assigned nurses.

The clinicians with the most positive views about the EMR are Mr. Parsons, the physician assistant, and the nurse assigned to work with him. They definitely do not want to return to paper records. Mr. Parsons likes entering his notes directly into the EMR at the point of care and finds that his documentation has improved considerably since the system was implemented. Other advantages he sees to using the EMR include legibility, ease of use, the improved communication with his nurse through e-mail, his ability to view the schedule while speaking with the patient, and the easy access to patient information. However, he does share some of the same concerns regarding the problems with scanning and downtime. He would have liked to see the practice more paperless by this point in time. Mr. Parsons also makes it clear that he sees the physicians as having decision-making power. He prefers to take his lead from them.

Current Challenges/Problems Facing FMC

Although the support staff and office manager's impressions of the EMR are much more positive than those of the nurses and physicians, they also agree that there are problems. With the exception of the PA and his nurse, the clinicians are extremely frustrated with the system. They acknowledge that the rocky start and the inadequate initial training may have led to their lack of trust and confidence in both the vendor and the product. Besides the inadequate initial training, the nursing staff expresses concern that there is no one within the practice who knows the intricacies of the system and can provide additional individualized training. In fact, one year after the EMR implementation, Mrs. Young contacted the manager of technical support at PTX to ask for additional group training. However, several clinicians expressed doubt that this was a solution. They believe that the habits they have now acquired in using the system would be difficult to change.

One of the physicians, Dr. Starmer, has sincerely tried to learn more about the system. He has attended several PTX user group meetings and has come back excited about trying new things. But, after spending hours trying to implement what he had learned, he found himself extremely frustrated — and eventually gave up. With his busy work schedule, he has no time to figure it out himself, let alone train others. Even now, a few years later, Dr. Starmer does not believe that changes are being made in the software to make it more user-friendly, and he has conveyed his frustration to the other physicians and nurses in the practice. He and others seem to have given up hope that things will get better.

Drs. Starmer and Gunther are actively pursuing the replacement of the PTX system with another EMR system. These two physicians in particular are convinced that many of their frustrations with the system are software specific. They have seen other EMR programs that they view as more intuitive and easier to use. They would have pushed to replace the system sooner, but they felt that they had financially invested too much in hardware and software to just "dump it."

Dr. Barron does not see the situation quite as dismally. He feels that many of the problems that FMC has experienced were due to the fact that they are "pioneers" in the use of the EMR. He also feels that until all physicians in the surrounding community adopt an EMR, FMC will not be able to eliminate the duplicate paper record. In the meantime, the clinicians at FMC continue to treat patients using both the PTX system and their paper record system and they continue their discussion of changing to a more effective EMR system.

REFERENCES

Anderson, J. G. (1992). Computerized medical record systems in ambulatory care. *Journal of ambulatory care management, 15*(3), 67-75.

Anderson, J. G., Aydin, C. E., & Kaplan, B. (1995). *An analytical framework for measuring the effectiveness/impacts of computer-based patient records systems*. Paper presented at the Proceedings of the 28th Annual Hawaii International Conference on System Sciences.

Balas, E., Austin, S., Mitchell, J., Ewigman, B., Bopp, K., & Brown, G. (1996, May). The clinical value of computerized information services. A review of 98 randomized clinical trials. *Archives of Family Medicine, 5*, 271-278.

Burch, J. G. (1986, October). Designing information systems for people. *Journal of Systems Management*, 30-34.

Dick, R. S., & Steen, E. B. (Eds.). (1991). *The computer-based patient record: An essential technology for health care.* National Academy Press.

Dick, R. S., Steen, E. B., & Detmer, D. E. (Eds.). (1997). *The computer-based patient record: An essential technology for health care* (rev. ed.). Washington, DC: National Academy Press.

Edelson, J. T. (1995). Physician use of information technology in ambulatory medicine: An overview. *Journal of Ambulatory Care Management, 18*(3), 9-19.

Hammond, W. E., Hales, J. W., Lobach, D. F., & Straube, M. J. (1997). Integration of a computer-based patient record system into the primary care setting. *Computers in Nursing, 15*(2 (Suppl)), s61-8.

Ornstein, S. M., Garr, D. R., Jenkins, R. G., Rust, P. F., & Arnon, A. (1991). Computer-generated physician and patient reminders. Tools to improve population adherence to selected preventive services. *Journal of Family Practice, 32*(1), 82-90.

Rittenhouse, D., & Lincoln, W. (1994, June). The angles of a repository. *Healthcare Informatics*, 74-80.

Safran, C., Rind, D. M., Davis, R. M., Currier, J., Ives, D., Sands, D. Z., Slack, W. V., Makadon, H., & Cotton, D. (1993). An electronic medical record that helps care for patients with HIV infection. *Proceedings of the Annual Symposium on Computer Applications in Medical Care.*

Safran, C., Rind, D. M., Sands, D. Z., Davis, R. B., Wald, J., & Slack, W. V. (1996). Development of a knowledge-based electronic patient record. *Clinical Computing, 13*(1), 46-54, 63.

Spann, S. (1990). Should the complete medical record be computerized in family practice? An affirmative view. *Journal of Family Practice, 30*(4), 457-464.

Waegemann, C. P. (1996). The five levels of electronic health records. *M.D. Computing, 13*(3), 199-203.

Wager, K. A., Lee, F. W., White, A. W., Ward, D. M., & Ornstein, S. M. (2000). The impact of an electronic medical record system on community-based primary care practices. *The Journal of the American Board of Family Practice, 15*, 5.

Wager, K. A., Ornstein, S. M., & Jenkins, R. G. (1997, September). Perceived value of computer-based patient records among clinician users. *M.D. Computing*, 334-340.

Worthley, J. A. (2000). *Managing information in healthcare.* Chicago: Health Administration Press.

Yarnall, K., Michener, J. L., & Hammond, W. E. (1994). The medical record: A comprehensive computer system for the family physician. *Journal of the American Board of Family Practice, 7*(4), 324-334.

FURTHER READING

Anderson, J. G. (1997). Clearing the way for physicians' use of clinical information systems. *Communications of the ACM, 40*(8), 83-90.

Cork, R. D., Detmer, W. M., & Friedman, C. P. (1998). Development and initial validation of an instrument to measure physicians' use of, knowledge about, and attitudes toward computers. *Journal of the American Medical Informatics Association, 5*(2), 164-176.

Dick, R. S., & Steen, E. B. (Eds.). (1991). *The computer-based patient record: An essential technology for health care.* National Academy Press.

Friedman, C. P., & Wyatt, J. C. (1997). *Evaluation methods in medical informatics.* New York: Springer.

Kaplan, B. (1997). Addressing organizational issues into the evaluation of medical systems. *Journal of the American Medical Informatics Association, 4*(2), 94-101.

Lorenzi, N. M., Riley, R. T., Blyth, A. J. C., Southon, G., & Dixon, B. J. (1997). Antecedents of the people and organizational aspects of medical informatics: Review of the literature. *Journal of the American Medical Informatics Association, 4*(2), 79-93.

McDonald, C. J. (1997). The barriers to electronic medical record systems and how to overcome them. *Journal of the American Medical Informatics Association, 4*(3), 213-221.

Karen A. Wager, D.B.A., is an associate professor and director of the graduate program in health information administration at the Medical University of South Carolina. She has more than 18 years of experience in the health information management profession and currently teaches courses in systems analysis and design, microcomputer applications, and health information systems. Her primary research interest is in the use and acceptance of electronic medical records in practice.

Frances Wickham Lee, D.B.A., is an associate professor in health information administration at the Medical University of South Carolina. She has more than 25 years of teaching and practice experience in the fields of health information systems and health information administration. Her primary research interest is the use of electronic medical record systems. She was awarded a Doctor of Business Administration from the University of Sarasota (1999).

Andrea W. White, PhD, is an associate professor and director of the Master in Health Administration educational program at the Medical University of South Carolina. She has taught undergraduate, graduate and doctoral level courses in quality improvement and health information management. Her research interests are in quality improvement, electronic medical records and health professions education.

This case was previously published in the *Annals of Cases on Information Technology*, Volume 3/ 2001, pp. 153-168, © 2001.

Chapter XXI

Military Applications of Natural Language Processing and Software

James A. Rodger, Indiana University of Pennsylvania, USA

Tamara V. Trank, Naval Health Research Center, USA

Parag C. Pendharkar, Pennsylvania State University at Harrisburg, USA

EXECUTIVE SUMMARY

A preliminary feasibility study aboard U.S. Navy ships utilized voice interactive technology to improve medical readiness. A focus group was surveyed about reporting methods in health and environmental surveillance inspections to develop criteria for designing a lightweight, wearable computing device with voice interactive capability. The voice interactive computing device included automated user prompts, enhanced data analysis, presentation and dissemination tools in support of preventive medicine. The device was capable of storing, processing and forwarding data to a server. The prototype enabled quick, efficient and accurate environmental surveillance. In addition to reducing the time needed to complete inspections, the device supported local reporting requirements and enhanced command-level intelligence. Where possible, existing technologies were utilized in creating the device. Limitations in current voice recognition technologies created challenges for training and user interface.

BACKGROUND

Coupling computer recognition of the human voice with a natural language processing system makes speech recognition by computers possible. By allowing data and commands to be entered into a computer without the need for typing, computer understanding of naturally spoken languages frees human hands for other tasks. Speech recognition by computers can also increase the rate of data entry, improve spelling accuracy, permit remote access to databases utilizing wireless technology and ease access to computer systems by those who lack typing skills.

Variation of Speech-to-Text Engines

Since 1987, the National Institute of Standards and Technology (NIST) has provided standards to evaluate new voice interactive technologies (Pallett, Garofolo, & Fiscus, 2000). In a 1998 broadcast news test, NIST provided participants with a test set consisting of two 1.5-hour subsets obtained from the Linguistic Data Consortium. The task associated with this material was to implement automatic speech recognition technology by determining the lowest word error rate (Herb & Schmidt, 1994; Fiscus, 1997; Greenberg, Chang, & Hollenback, 2000; Pallett, 1999). Excellent performance was achieved at several sites, both domestic and abroad (Przybocki, 1999). For example, IBM-developed systems achieved the lowest overall word error rate of 13.5%. The application of statistical significance tests indicated that the differences in performance between systems designed by IBM, the French National Laboratories' Laboratoire d'Informatique pour la Mechanique et les Sciences de l'Ingenieur and Cambridge University's Hidden Markov Model Toolkit software were not significant (Pallett, Garfolo, & Fiscus, 2000). Lai (2000) also reported that no significant differences existed in the comprehension of synthetic speech among five different speech-to-text engines used. Finally, speaker segmentation has been used to locate all boundaries between speakers in the audio signal. It enables speaker normalization and adaptation techniques to be used effectively to integrate speech recognition (Bikel, Miller, Schwartz, & Weischedel, 1997).

Speech Recognition Applications

The seamless integration of voice recognition technologies creates a human-machine interface that has been applied to consumer electronics, Internet appliances, telephones, automobiles, interactive toys, and industrial, medical, and home electronics and appliances (Soule, 2000). Applications of speech recognition technology are also being developed to improve access to higher education for persons with disabilities (Leitch & Bain, 2000). Although speech recognition systems have existed for two decades, widespread use of this technology is a recent phenomenon. As improvements have been made in accuracy, speed, portability, and operation in high-noise environments, the development of speech recognition applications by the private sector, federal agencies, and armed services has increased.

Some of the most successful applications have been telephone based. Continuous speech recognition has been used to improve customer satisfaction and the quality of service on telephone systems (Charry, Pimentel, & Camargo, 2000; Goodliffe, 2000; Rolandi, 2000). Name-based dialing has become more ubiquitous, with phone control answer, hang-up, and call management (Gaddy, 2000a). These applications use intuitive human communication techniques to interact with electronic devices and systems (Shepard, 2000). BTexact Technologies, the Advanced Communications Technology Centre for British Telecommunications (Adastral Park, Suffolk, England), uses the technology to provide automated directory assistance for 700 million calls each year at its UK bureau (Gorham & Graham, 2000). Studies in such call centers have utilized live customer trials to demonstrate the technical realization of full speech automation of directory inquiries (McCarty, 2000; Miller, 2000). Service performance, a study of customer behavior and an analysis of service following call-back interviews suggest user satisfaction with the application of speech automation to this industry (Gorham & Graham, 2000).

Speech recognition technologies could expand e-commerce into v-commerce with the refinement of mobile interactive voice technologies (McGlashan, 2000; Gaddy, 2000b; Pearce, 2000). As an enabler of talking characters in the digital world, speech recognition promises many opportunities for rich media applications and communications with the Internet (Zapata, 2000). Amid growing interest in voice access to the Internet, a new Voice-extensible Markup Language (VoiceXML™, VoiceXML Forum) has surfaced as an interface for providing Web hosting services (Karam & Ramming, 2000). VoiceXML promises to speed the development and expand the markets of Web-based, speech recognition/synthesis services as well as spawning a new industry of "voice hosting." This model will allow developers to build new telephone-based services rapidly (Thompson & Hibel, 2000). The voice-hosting service provider will lease telephone lines to the client and voice-enable a specific URL, programmed in VoiceXML by the client. This model will make it possible to build speech and telephony services for a fraction of the time and cost of traditional methods (Larson, 2000).

Haynes (2000) deployed a conversational Interactive Voice Response system to demonstrate site-specific examples of how companies are leveraging their infrastructure investments, improving customer satisfaction and receiving quick return on investments. Such applications demonstrate the use of speech recognition by business. The investigation of current customer needs and individual design options for accessing information utilizing speech recognition is key to gaining unique business advantages (Prizer, Thomas, & Suhm, 2000; Schalk, 2000).

A long-awaited application of speech recognition, the automatic transcription of free-form dictation from professionals such as doctors and lawyers, lags behind other commercial applications (Stromberg, 2000). Due to major developments in the Internet, speech recognition, bandwidth and wireless technology, this situation is changing (Bourgeois, 2000; Pan, 2000).

Internationalizing speech recognition applications has its own set of problems (Krause, 2000). One such problem is that over-the-phone speech applications are more difficult to translate to other languages than Web applications or traditional desktop graphic user interface applications (Head, 2000). Despite the problems, internationalizing speech applications brings with it many benefits. Internationalization of an application helps to reveal some of the peculiarities of a language, such as differences in dialects, while providing insight on the voice user interface design process (Scholz, 2000; Yan, 2000). Furthermore, speech comprehension can work effectively with different languages; studies have documented both English and Mandarin word error rates of 19.3% (Fiscus, Fisher, Martin, Przybocki, & Pallett, 2000).

Speech technology has been applied to medical applications, particularly emergency medical care that depends on quick and accurate access of patient background information (Kundupoglu, 2000). The U.S. Defense Advance Research Projects Agency organized the Trauma Care Information Management System (TCIMS) Consortium to develop a prototype system for improving the timeliness, accuracy, and completeness of medical documentation. One outcome of TCIMS was the adoption of a speech-audio user interface for the prototype (Holtzman, 2000).

The Federal Aviation Administration conducted a demonstration of how voice technology supports a facilities maintenance task. A voice-activated system proved to be less time-consuming to use than the traditional paper manual approach, and study participants reported that the system was understandable, easy to control, and responsive to voice commands. Participants felt that the speech recognition system made the maintenance task easier to perform, was more efficient and effective than a paper manual, and would be better for handling large amounts of information (Mogford, Rosiles, Wagner, & Allendoerfer, 1997).

Speech recognition technology is expected to play an important role in supporting real-time interactive voice communication over distributed computer data networks. The Interactive Voice Exchange Application developed by the Naval Research Lab, Washington, DC, has been able to maintain a low data rate throughput requirement while permitting the use of voice communication over existing computer networks without causing a significant impact on other data communications, such as e-mail and file transfer (Macker & Adamson, 1996).

Pilots must have good head/eye coordination when they shift their gaze between cockpit instruments and the outside environment. The Naval Aerospace Medical Research Lab, Pensacola, FL, has investigated using speech recognition to support the measurement of these shifts and the type of coordination required to make them (Molina, 1991). Boeing Company, Seattle, WA, has investigated ways to free pilots from certain manual tasks and sharpen their focus on the flight environment. The latest solution includes the use of a rugged, lightweight, continuous-speech device that permits the operation of selected cockpit controls by voice commands alone. This technology is being applied in the noisy cockpit of the Joint Strike Fighter (Bokulich, 2000).

Existing Problems: Limitations of Speech Recognition Technology

Even though applications of speech recognition technology have been developed with increased frequency, the field is still in its infancy, and many limitations have yet to be resolved. For example, the success of speech recognition by desktop computers depends on the integration of speech technologies with the underlying processor and operating system, and the complexity and availability of tools required to deploy a system. This limitation has had an impact on application development (Markowitz, 2000; Woo, 2000).

Use of speech recognition technology in high-noise environments remains a challenge. For speech recognition systems to function properly, clean speech signals are required, with high signal-to-noise ratio and wide frequency response (Albers, 2000; Erten, Paoletti, & Salam, 2000; Sones, 2000; Wickstrom, 2000). The microphone system is critical in providing the required speech signal, and, therefore, has a direct effect on the accuracy of the speech recognition system (Andrea, 2000; Wenger, 2000). However, providing a clean speech signal can be difficult in high-noise environments. Interference, changes in the user's voice, and additive noise — such as car engine noise, background chatter and white noise — can reduce the accuracy of speech recognition systems. In military environments, additive noise and voice changes are common. For example, in military aviation, the stress resulting from low-level flying can cause a speaker's voice to change, reducing recognition accuracy (Christ, 1984).

The control of the speech recognition interface poses its own unique problems (Gunn, 2000; Taylor, 2000). The inability of people to remember verbal commands is even more of a hindrance than their inability to remember keyboard commands (Newman, 2000). The limited quality of machine speech output also affects the speech recognition interface. As human-machine interaction becomes increasingly commonplace, applications that require unlimited vocabulary speech output are demanding text-to-speech systems that produce more human-sounding speech (Hertz, Younes, & Hoskins, 2000).

The accuracy of modeling has also limited the effectiveness of speech recognition. Modeling accuracy can be improved, however, by combining feature streams with neural nets and Gaussian mixtures (Ellis, 2000). The application of knowledge-based speech analysis has also shown promise (Komissarchik & Komissarchik, 2000).

Pallett, Garofolo, and Fiscus (1999) pointed out that potential problems associated with the search and retrieval of relevant information from databases have been addressed by the Spoken Document Retrieval community. Furthermore, standards for the probability of false alarms and miss probabilities are set forth and investigated by the Topic Detection and Tracking Program (Doddington, 1999). Decision error trade-off plots are used to demonstrate the trade-off between the miss probabilities and false alarm probabilities for a topic (Kubala, 1999). Security issues and speech verification are major voids in speech recognition technology (Gagnon, 2000). Technology for the archiving of speech is also undeveloped. It is well recognized that speech is not presently valued as an archival information source because it is impossible to locate information in large audio archives (Kubala, Colbath, Liu, Srivastava, & Makhoul, 2000).

Army Voice Interactive Display

Until recently, few practical continuous speech recognizers were available. Most were difficult to build, resided on large mainframe computers, were speaker dependent, and did not operate in real time. The Voice Interactive Display (VID) developed for the U.S. Army has made progress in eliminating these disadvantages (Hutzell, 2000). VID was intended to reduce the bulk, weight, and setup times of vehicle diagnostic systems while increasing their capacity and capabilities for hands-free troubleshooting. The capabilities of VID were developed to allow communication with the supply and logistics structures within the Army's common operating environment.

This effort demonstrated the use of VID as a tool for providing a paperless method of documentation for diagnostic and prognostic results; it will culminate in the automation of maintenance supply actions. Voice recognition technology and existing diagnostic tools have been integrated into a wireless configuration. The result is a hands-free interface between the operator and the Soldier's On-System Repair Tool (SPORT).

The VID system consists of a microphone, a hand-held display unit and SPORT. With this configuration, a technician can obtain vehicle diagnostic information while navigating through an Interactive Electronic Technical Manual via voice commands. By integrating paperless documentation, human expertise, and connectivity to provide user support for vehicle maintenance, VID maximizes U.S. Army efficiency and effectiveness.

SETTING THE STAGE

In support of Force Health Protection, the U.S. Navy has launched a VID project that leverages existing technologies and automates the business practices of Navy medicine. The goal of the Naval Voice Interactive Device (NVID) project is to create a lightweight, portable computing device that uses speech recognition to enter shipboard environmental survey data into a computer database and to generate reports automatically to fulfill surveillance requirements.

Shipboard Environmental Surveys: The Requirement

To ensure the health and safety of shipboard personnel, naval health professionals — including environmental health officers, industrial hygienists, independent duty corpsmen (IDCs) and preventive medicine technicians — perform clinical activities and preventive medicine surveillance on a daily basis. These inspections include, but are not limited to, water testing, heat stress, pest control, food sanitation, and habitability surveys. *Chief of Naval Operations Instruction 5100.19D*, the *Navy Occupational Safety and Health Program Manual for Forces Afloat*, provides the specific guidelines for maintaining a safe and healthy work environment aboard U.S. Navy ships. Inspections performed by medical personnel ensure that these guidelines are followed.

Typically, inspectors enter data and findings by hand onto paper forms and later transcribe these notes into a word processor to create a finished report. The process of manual note-taking and entering data via keyboard into a computer database is time consuming, inefficient, and prone to error. To remedy these problems, the Naval Shipboard Information Program was developed, allowing data to be entered into portable laptop computers while a survey is conducted (Hermansen & Pugh, 1996). However, the cramped shipboard environment, the need for mobility by inspectors, and the inability to have both hands free to type during an inspection make the use of laptop computers during a walk-around survey difficult. Clearly, a hands-free, space-saving mode of data entry that would also enable examiners to access pertinent information during an inspection was desirable. The NVID project was developed to fill this need.

Strengths of NVID

The NVID project was developed to replace existing, inefficient, repetitive survey procedures with a fully automated, voice interactive system for voice-activated data input. In pursuit of this goal, the NVID team developed a lightweight, wearable, voice-interactive prototype capable of capturing, storing, processing, and forwarding data to a server for easy retrieval by users. The voice interactive data input and output capability of NVID reduces obstacles to accurate and efficient data access and reduces the time required to complete inspections. NVID's voice interactive technology allows a trainee to interact with a computerized system and still have hands and eyes free to manipulate materials and negotiate his or her environment (Ingram, 1991). Once entered, survey and medical encounter data can be used for local reporting requirements and command-level intelligence. Improved data acquisition and transmission capabilities allow connectivity with other systems. Existing printed and computerized surveys are voice activated and reside on the miniaturized computing device. NVID has been designed to allow voice prompting by the survey program, as well as voice-activated, free-text dictation. An enhanced microphone system permits improved signal detection in noisy shipboard environments. All of these capabilities contribute to the improved efficiency and accuracy of the data collection and retrieval process by shipboard personnel.

CASE DESCRIPTION

Shipboard medical department personnel regularly conduct comprehensive surveys to ensure the health and safety of the ship's crew. Currently, surveillance data are collected and stored via manual data entry, a time-consuming process that involves typing handwritten survey findings into a word processor to produce a completed document. The NVID prototype was developed as a portable computer that employs voice interactive technology to automate and improve the environmental surveillance data collection and reporting process.

This prototype system is a compact, mobile computing device that includes voice interactive technology, stylus screen input capability, and an indoor readable display that enables shipboard medical personnel to complete environmental survey checklists, view reference materials related to these checklists, manage tasks, and generate reports using the collected data. The system uses Microsoft Windows NT®, an operating environment that satisfies the requirement of the IT-21 Standard to which Navy ships must conform. The major software components include initialization of the NVID software application, application processing, database management, speech recognition, handwriting recognition, and speech-to-text capabilities. The power source for this portable unit accommodates both DC (battery) and AC (line) power options and includes the ability to recharge or swap batteries to extend the system's operational time.

The limited laboratory and field-testing described for this plan were intended to support feasibility decisions and not rigorous qualification for fielding purposes. The objectives of this plan are to describe how to:

- Validate NVID project objectives and system descriptions.
- Assess the feasibility of voice interactive environmental tools.
- Assess the NVID prototype's ease of use.

Components of Questionnaire

To develop an appropriate voice interactive prototype system, the project team questioned end users to develop the requirement specifications. On July 11, 2000, a focus group of 14 participants (13 hospital corpsmen, one medical officer) completed a survey detailing methods of completing surveys and reporting inspection results. The questionnaire addressed the needs of end users as well as their perspectives on the military utility of NVID. The survey consisted of 117 items ranging from nominal, yes/no answers to frequencies, descriptive statistics, rank ordering, and perceptual Likert scales. These

Table 1. NVID focus group participants

Command	Rank/Rate
Navy Environmental Preventive Medicine Unit-5	HM2
Navy Environmental Preventive Medicine Unit-5	HM1
Navy Environmental Preventive Medicine Unit-5	HM3
Commander Submarine Development Squadron Five	HMCS
Naval School of Health Sciences, San Diego	HMCS
Naval School of Health Sciences, San Diego	HM1
Naval School of Health Sciences, San Diego	HMC
Commander, Amphibious Group-3	HMC
Commander, Amphibious Group-3	HMC
USS CONSTELLATION (CV-64)	HMC
USS CONSTELLATION (CV-64)	HMC
Commander, Naval Surface Force Pacific	HMCS
Commander, Naval Surface Force Pacific	HMCS
Regional Support Office, San Diego	CDR

HM1, Hospital Corpsman First Class   HMCS, Hospital Corpsman Senior Chief
HM2, Hospital Corpsman Second Class   HMC, Hospital Corpsman Chief
HM3, Hospital Corpsman Third Class   CDR, Commander

items were analyzed utilizing a Windows software application, Statistical Package for the Social Sciences (SPSS, 1999). Conclusions were drawn from the statistical analysis and recommendations were suggested for development and implementation of NVID. The following discussion presents the results of the survey and how the information was incorporated into the prototype.

The commands and ranks of these participants are shown in Table 1. These participants possessed varying clinical experience while assigned to deployed units (ships and Fleet Marine Force), including IDCs (independent duty corpsmen), preventive medicine, lab technicians, and aviation medicine.

Section 1: Environmental Health and Preventive Medicine Afloat

In the first section of the questionnaire, inspectors were asked about the methods they used to record findings while conducting an inspection (see Table 2).

Response to this section of the questionnaire was limited. The percentage of missing data ranged from 7.1% for items such as habitability and food sanitation safety to 71.4% for mercury control and 85.7% for polychlorinated biphenyls. An aggregate of the information in Table 2 indicates that the majority of inspectors relied on preprinted checklists. Fewer inspections were conducted utilizing handwritten reports. Only 7.1% of the users recorded their findings on a laptop computer for inspections focusing on radiation protection, workplace monitoring, food sanitation safety and habitability.

Table 2. Methods of recording inspection findings

Inspections	Handwritten %	Preprinted Check Lists %	Laptop Computer	Missing %
Asbestos	14.3	50.0	0	35.7
Heat Stress	14.3	71.4	0	14.3
Hazardous Materials	21.4	50.0	0	28.6
Hearing Conservation	21.4	64.3	0	14.3
Sight Conservation	7.1	71.4	0	21.4
Respiratory Conservation	0	71.4	0	28.6
Electrical Safety	14.3	50.0	0	35.7
Gas-Free Engineering	14.3	28.6	0	57.1
Radiation Protection	7.1	28.6	7.1	57.1
Lead Control	0	64.3	0	35.7
Tag-Out Program	7.1	50.0	0	42.9
Personal Protective Equipment	7.1	42.9	0	50.0
Mercury Control	0	28.6	0	71.4
PCBs	0	14.3	0	85.7
Man-Made Vitreous Fibers	7.1	28.6	0	64.3
Blood-Borne Pathogens	0	50.0	0	50.0
Workplace Monitoring	0	42.9	7.1	50.0
Food Sanitation Safety	14.3	71.4	7.1	7.1
Habitability	28.6	57.1	7.1	7.1
Potable Water, Halogen/Bacterial Testing	35.7	57.1	0	7.1
Wastewater Systems	21.4	50.0	0	28.6
Other	0	0	0	100

PCBs, polychlorinated biphenyls, pentachlorobenzole

    In addition to detailing their methods of recording inspection findings, the focus group participants were asked to describe the extensiveness of their notes during surveys. The results ranged from "one to three words in a short phrase" (35.7%) to "several short phrases, up to a paragraph" (64.3%). No respondents claimed to have used "extensive notes of more than one paragraph." The participants were also asked how beneficial voice dictation would be while conducting an inspection. Those responding that it would be "very beneficial" (71.4%) far outweighed those responding that it would be "somewhat beneficial" (28.6%). No respondents said that voice dictation would be "not beneficial" in conducting an inspection. In another survey question, participants were asked if portions of their inspections were performed in direct sunlight. The "yes" responses (92.9%) were far more prevalent than the "no" responses (7.1%).

    Participants also described the types of reference material needed during inspections. The results are shown in Table 3.

Table 3. Types of reference information needed during inspections

Information	Yes %	No %
Current Checklist in Progress	78.6	21.4
Bureau of Medicine Instructions	71.4	28.6
Naval Operations Instructions	71.4	28.6
Previously Completed Reports for Historical References	71.4	28.6
Exposure Limit Tables	57.1	42.9
Technical Publications	57.1	42.9
Type Commander Instructions	50.0	50.0
Local Instructions	42.9	57.1
Procedure Descriptions	28.6	71.4
Other	21.4	78.6

"Yes" responses ranged from a low of 28.6% for procedure description information to 78.6% for current checklist in progress information. When asked how often they utilized reference materials during inspections, no participants chose the response "never." Other responses included "occasionally" (71.4%), "frequently" (21.4%) and "always" (7.1%). In another survey question, participants were asked to describe their methods of reporting inspection results, which included the following: preparing the report using SAMS (Shipboard Not-tactical ADP Program (SNAP) Automated Medical System) (14.8%), preparing the report using word processing other than SAMS (57.1%), and preparing the report using both SAMS and word processing (28.6%). No respondents reported using handwritten or other methods of reporting inspection results. Participants were also asked how they distributed final reports. The following results were tabulated: hand-carry (21.4%); guard mail (0%); download to disk and mail (7.1%); Internet e-mail (64.3%); upload to server (0%); file transfer protocol (FTP) (0%); and other, not specified (7.1%). When asked if most of the problems or discrepancies encountered during an inspection could be summarized using a standard list of "most frequently occurring" discrepancies, 100% of respondents answered "yes." The average level of physical exertion during inspections was reported as Light by 42.9% of respondents, Moderate by 50.0% of respondents and Heavy by 7.1% of respondents. Survey participants were also asked to describe their level of proficiency at ICD-9-CM (Department of Health and Human Services, 1989). An expert level of proficiency was reported 7.1% of the time. Other responses included "competent" (14.3%), "good" (28.6%), "fair" (28.6%), and "poor" (7.1%). Missing data made up 14.3% of the responses.

Section 2: Shipboard Data Communications Technology

In the second section of the questionnaire, end users addressed characteristics of shipboard medical departments, NVID, medical encounters and SAMS. When asked if their medical departments were connected to a local area network (LAN), respondents answered as follows: "yes" (71.4%), "no" (7.1%), and "uncertain" (14.3%). Missing

Table 4. Ranking of device features

Feature	Average	Rank
Voice-Activated Dictation	2.64	1(tie)
Durability	2.64	1 (tie)
Voice Prompting for Menu Navigation	2.93	3
LAN Connectivity	4.21	4
Belt or Harness Wearability	4.57	5
Wireless Microphone	5.29	6
Touch Pad/Screen	5.93	7
Earphones	6.14	8 (tie)
Wearable in Front or Back	6.14	8 (tie)

responses totaled 7.1%. Participants asked if their medical departments had LANs of their own responded "yes" (14.3%), "no" (57.1%), and "uncertain" (21.4%). Another 7.1% of responses to this question were missing. When asked if their medical departments had access to the Internet, participants responded "yes, in medical department" (85.7%), and "yes, in another department" (7.1%). Another 7.1% of responses were missing.

Various methods for transmitting medical data from ship to shore were also examined in the survey. It was found that 78.6% of those surveyed said they had used Internet e-mail, while 14.3% said that they had downloaded data to a disk and mailed it. No users claimed to have downloaded data to a server or utilized File Transfer Protocol (FTP) for this purpose. Missing responses totaled 7.1%.

Table 4 shows respondents' rankings of the desirable features of the device.

"Voice activation dictation" and "durability" were tied for the top ranking. "Wearable in front or back" and "earphones" were tied for lowest ranking. "Voice prompting for menu navigation" and "LAN connectivity" were the number 3 and 4 choices, respectively.

In another question, participants rated their computer efficiency. Just 14.3% rated their computer efficiency as "expert," while 42.9% chose "competent." "Good" and "fair" were each selected by 21.4% of respondents. Participants reportedly used "name of area" as the most used element (85.7%) to identify an inspected area (Table 5).

Table 5. Elements used in reports to identify inspected areas

Identifying Element	Yes	No
Compartment Number	57.1	42.9
Department	57.1	42.9
Name of Area	85.7	14.3
Other	0	100

Table 6. Surveillance areas benefiting from voice automation

Areas	Average	Rank
Food Sanitation Safety	1.21	1
Heat Stress	3.29	2
Potable Water, Halogen	3.86	3
Habitability	4.14	4
Potable Water, Bacterial	4.21	5
Inventory Tool	4.43	6
Hazard-specific Programs with Checklist	4.86	7

Table 6 provides respondents' rankings of the areas of environmental surveillance in which voice automation would be of the greatest value. According to this rank ordering, "Food Sanitation Safety" would most benefit from voice automation. "Heat Stress" and "Potable Water, Halogen" were also popular choices

Section Three: Professional Opinions
In the third section of the survey, participants were asked which attributes of NVID they would find most desirable (Table 7).

Other survey questions provided insights into the workloads of respondents and their preferences related to NVID training. It was reported that 64.3% of respondents saw zero to 24 patients in sick bay daily. A daily count of 25 to 49 sick bay visits was reported by 28.6% of respondents, while 7.1% reported 50 to 74 visitors per day.

When asked how much time they would be willing to devote to training a software system to recognize their voice, 21.4% of respondents said that a training period of less than 1 hour would be acceptable. According to 57.1% of respondents, a training period of one to four hours would be acceptable, while 21.4% of respondents said that they

Table 7. Frequencies of desirable attributes

Opinion	Strongly Agree %	Agree %	Unsure %	Disagree %	Strongly Disagree %
Care for Patients	71.4	28.6			
Reduce Data Entries	21.4	71.4	7.1		
Reduce Paperwork	14.3	57.1	14.3	14.3	
Conduct Outbreak Analysis	21.4	35.7	21.4	21.4	
On-Line Tutorial	14.3	57.1	21.4	7.1	
Lightweight Device	21.4	71.4	7.1		
See an Overview	28.6	50.0	14.3	7.1	
Automated ICD-9-CM	35.7	42.9	7.1	14.3	
Difficulties Using ICD-9-CM Codes	14.2	28.6	28.6	28.6	

would be willing to spend four to eight hours to train the system. To train themselves to use the NVID hardware and software applications, 42.9% of survey respondents said they would be willing to undergo one to four hours of training, while 57.1% said that they would train for four to eight hours. All respondents agreed that a longer training period would be acceptable if it would guarantee a significant increase in voice recognition accuracy and reliability.

Environmental Surveillance Module

Based on the responses from the surveyed population, we chose to automate the following surveys: (1) Food Sanitation Safety, (2) Habitability, (3) Heat Stress, (4) Potable Water, and (5) Pest Control. While requested for automation by the focus group, choices 3-5 were appealing because they already exist in the SAMS program. By including these surveys in the prototype, the research effort hooked into systems already resident in ship medical departments, increasing the appeal of the prototype.

The surveys indicated that the inspectors utilized preprinted checklists most often. NVID automated these business practices by providing checklists for each survey. While some "free dictation" was incorporated, allowing the inspector to include comments during the inspection, predetermined checklists with a limited necessary vocabulary (command and control) allowed the NVID team to use smaller computer devices with slower processors. Extensive "free dictation" capability requires faster processors that do not yet exist on small, portable computing platforms. From the survey, all respondents agreed that most problems encountered during an inspection can be summarized using a standardized list of frequently occurring discrepancies.

A master tickler, a calendar that tracks the progress of surveys and the dates of their required completion, was included in the module. Navy references and instructions were made resident on the system, allowing inspectors access to regulations during surveys. Compatibility of the NVID system with medical department computer equipment was ensured so that downloads and sharing of information between computing platforms could easily be achieved. Final reports may be printed and delivered or distributed electronically via e-mail.

Improving Study Validity

One of the major limitations of this study is the small sample size (n = 14). In the focus group study, the small, conveniently available sample detracts from the external validity of the results. These results may not be generalizable to other populations and situations. More data must be collected to improve the external validity of the study. In the future:

• Having end users from a more geographically dispersed setting would add to the validity of the research.
• Conducting in-depth interviews with end users may help determine missed variables.

Several valid conclusions were drawn from the sample population, though non-response biases may be an issue in determining the generalizability of the results. These results were used to determine features and surveys included in the NVID prototype. First and foremost, a majority of end users indicated that NVID would be very beneficial.

This validated the need for NVID and convinced us that this is a worthwhile effort for the Navy. The Fujitsu Stylistic ™ 3400 Tablet (Santa Clara, CA) with an Intel Pentium® III processor (Santa Clara, CA) was the chosen computing platform. The commercial software included L&H Dragon NaturallySpeaking® 5.0 (Flanders, Belgium). For most purposes, an SR1 headset microphone (Plantronics Inc., Santa Cruz, CA) focused at the lips was adequate for the system tested under conditions of 70-90 decibels of background noise.

CURRENT CHALLENGES/PROBLEMS

Specific challenges and problems of the NVID prototype system that were examined and tested included:

- Shipboard operation in tight spaces
- Operation in high-noise environments
- Data gathering and checklist navigation
- Report generation
- Access to reference materials
- Comment capture capability
- Access to task schedule and survey data
- User system training
- Prototype effectiveness

Shipboard Operation in Tight Spaces

Space and resource constraints on Navy ships make it necessary to complete surveys in enclosed, tight spaces. Ease of use, portability, and wearability of the NVID unit when maneuvering through these areas were validated based on surveys of military users. A study of the ergonomics associated with the use of an NVID computer was also performed. The human factors evaluated included, but were not limited to, the following parameters:

- Safety equipment compatibility

  - Work clothing, including gloves, glasses, and hard hats
  - Sound suppressors/hearing protection
  - Respirators

- Data input comparison and user acceptance (voice command vs. touchscreen) based on the opinions of Navy personnel aboard ship
- User interface evaluation (ease of use)

  - User comfort
  - User adjustability
  - Subcomponent connection procedure
  - Assessment of mean time to proficiency

Operation in High-Noise Environments

Naval ships are industrial environments that contain the potential for high noise levels. To verify the effectiveness of the NVID prototype under such conditions, the difference in error rates using the unit with and without background noise were determined. Voice recognition software training was first conducted by using a script consisting of a repeatable set of voice commands. The following sets of tests were performed with consistent background noise:

- Lab test in a normal office environment (< 70 decibels)
- Lab test with baseline background noise up to the expected level (90 decibels)
- Field test aboard ship with typical background noise (75-90 decibels)

Error Rates Were Then Recorded for Each Test and Compared Between Tests

- **Data gathering and checklist navigation.** NVID prototype system users were capable of navigating through survey checklists by using voice commands, as well as other computer navigational tools, such as a mouse, touch pad, and stylus. The data collected were then automatically stored in an on-system database. To determine whether the system could successfully open each checklist and allow entry and storage of the required data, a script was developed that thoroughly tested the functionality of the hardware and software.
- **Report generation.** The ability to generate reports and save them as files for downloading or printing was verified. Tests were performed to verify that the data were captured during inspection procedures and properly rendered into a usable working report.
- **Access to reference materials.** Users may require access to survey reference materials, schedules, previous survey results or discrepancies found during the survey process. Tests were performed to verify that the application software enabled access to designated reference material as specified within each checklist.
- **Comment capture capability.** The NVID application provides the ability to document the inspector's notes via handwriting recognition, voice dictation and a touch screen keyboard. Verification of all three methods of data capture was performed using a predefined script of repeatable voice commands.
- **Access to task schedule and survey data.** The NVID application software provides the ability to schedule tasks and review past reports. Verification of the software was performed using both voice command and touchscreen technologies.
- **User system training.** To evaluate the effectiveness of user system training, the amount of training time required to achieve the desired level of voice recognition accuracy was first determined. Minimum training for the voice recognition software was conducted, and the length of time required to complete the training was documented. The system was then operated using a scripted, repeatable set of voice commands, and the number of errors was recorded. This process was repeated with additional training until the desired level of voice recognition accuracy was achieved.
- **Prototype effectiveness.** Throughout the test and evaluation process, the current

manual method of shipboard surveillance was compared with the NVID prototype system. A test script was used to exercise the functionality of all components of the software application. The test parameters included:

- The time it takes to perform tasks
- Ease of reviewing past survey data, task schedules and comments

The NVID prototype power source accommodates AC/DC power options. In a lab environment, the battery power meter was periodically monitored to determine the expected usage range. The functionality of the NVID prototype system's voice prompting and speech-to-text capabilities was verified. The usefulness of the device's ability to read menu selections and checklist questions was measured through user feedback.

Although users reported positive responses to the prototype tested, the device exhibited the limitations of current speech recognition technologies. The processors in lightweight, wearable devices were not fast enough to process speech adequately. Yet, larger processors added unwelcome weight to the device, and inspectors objected to the 3.5 pounds during the walk-around surveys. In addition, throat microphones used in the prototype to limit interference from background noise also limited speech recognition. These microphones pick up primarily guttural utterances, and thus tended to miss those sounds created primarily with the lips, or by women's higher voice ranges. Heavier necks also impeded the accuracy of throat microphones.

Accuracy of speech recognition also depended on the time a user committed to training the device to recognize his or her speech, and changes in voice quality due to environmental or physical conditions. Accuracy rates varied from 85-98% depending on the amount of time users took to train the software. Optimal training time appeared to be one hour for Dragon Naturally Speaking software and one hour for NVID software. In addition, current software interprets utterances in the context of an entire sentence, so users had to form complete utterances mentally before speaking for accurate recognition. As speech recognition technologies evolve, many of these limitations should be addressed.

CONCLUSION

The NVID survey established criteria for developing a lightweight, wearable, voice-interactive computer capable of capturing, storing, processing, and forwarding data to a server for retrieval by users. The prototype met many of these expectations. However, limitations in the current state of voice-recognition technologies create challenges for training and user interface. Integration of existing technologies, rather than development of new technology, was the intent of the design. A state-of the-art review of existing technologies indicated that commercial, off-the-shelf products cannot yet provide simultaneous walk-around capability and accurate speech recognition in the shipboard environment. Adaptations of existing technology involved trade-offs between speech recognition capabilities and wearability.

Despite current limitations in speech recognition technology, the NVID prototype was successful in reducing the time needed to complete inspections, in supporting local reporting requirements, and in enhancing command-level intelligence. Attitudes of the users toward the device were favorable, despite these restrictions. Users believed that the prototype would save time and improve the quality of reports.

ACKNOWLEDGMENTS

The authors would like to thank Dr. Cheryl Reed for her timely appearance and Ms. Susan Paesel for her expert editing. Report No. 01-17 was supported by the Bureau of Medicine and Surgery, Washington, DC, under Work Unit No. 0604771N-60001. The views expressed in this chapter are those of the authors and do not necessarily reflect the official policy or position of the Department of the Navy, Department of Defense or the U.S. Government. Approved for public release; distribution unlimited. Human subjects participated in this study after giving their free and informed consent. This research has been conducted in compliance with all applicable Federal Regulations governing the Protection of Human Subjects in Research.

REFERENCES

Albers, J. (2000). Successful speech applications in high noise environments. *SpeechTEK Proceedings* (pp. 147-154).

Andrea, D. (2000). Improving the user interface: Digital far-field microphone technology. *SpeechTEK Proceedings* (pp. 155-160).

Bikel, D., Miller, S., Schwartz, R., & Weischedel, R. (1997). Nimble: A high performance learning name finder. *Proceedings of the Fifth Conference on Applied Natural Language Processing, Association for Computational Linguistics,* Washington, DC (pp. 194-201).

Bokulich, F. (2000). JSF [Joint Strike Fighter] voice recognition. *Aerospace Engineering Online.* Retrieved from http://www.sae.org/aeromag/techupdate_5-00/03.htm

Bourgeois, S. (2000). Speech-empowered mobile computing. *SpeechTEK Proceedings* (pp. 223-228).

Charry, M., Pimentel, H., & Camargo, R. (2000). User reactions in continuous speech recognition systems. *AVIOS Proceedings of The Speech Technology & Applications Expo* (pp. 113-130).

Christ, K. A. (1984). *Literature review of voice recognition and generation technology for Army helicopter applications* (Army Report No. HEL-TN-11-84). Aberdeen Proving Ground, MD: Human Engineering Lab.

Department of Health and Human Services. (1989). *International classification of diseases* (9th revision, clinical modification, 3rd ed.). Washington, DC: Government Printing Office.

Doddington, G. (1999, February). Topic detection and tracking: TDT2 overview and evaluation results. *Proceedings of DARPA Broadcast News Workshop,* Herndon, VA.

Ellis, D. (2000). Improved recognition by combining different features and different systems. *AVIOS Proceedings of the Speech Technology & Applications Expo* (pp. 236-242).

Erten, G., Paoletti, D., & Salam, F. (2000). Speech recognition accuracy improvement in noisy environments using clear voice capture (CVC) technology. *AVIOS Proceedings of The Speech Technology & Applications Expo* (pp. 193-198).

Fiscus, J. G. (1997) A post-processing system to yield reduced word error rates: Recognizer Output Voting Error Reduction (ROVER). *Proceedings, 1997 IEEE Workshop on Automatic Speech Recognition and Speech.*

Fiscus, J. G., Fisher, W. M., Martin, A. F., Przybocki, M. A., & Pallet, D. S. (2000, February/March). 2000 NIST evaluation of conversational speech recognition over the telephone: English and Mandarin performance results. *Proceedings of DARPA Broadcast News Workshop.*

Gaddy, L. (2000a). The future of speech I/O in mobile phones. *SpeechTEK Proceedings* (pp. 249-260).

Gaddy, L. (2000b). Command and control solutions for automotive applications. *SpeechTEK Proceedings* (pp. 187-192).

Gagnon, L. (2000). Speaker recognition solutions to secure and personalize speech portal applications. *SpeechTEK Proceedings* (pp. 135-142).

Goodliffe, C. (2000). The telephone and the Internet. *AVIOS Proceedings of the Speech Technology & Applications Expo* (pp. 149-151).

Gorham, A., & Graham, J. (2000). Full automation of directory enquiries: A live customer trial in the United Kingdom. *AVIOS Proceedings of the Speech Technology & Applications Expo* (pp. 1-8).

Greenberg, S., Chang, S., & Hollenback, J. (2000, February/March). An introduction to the diagnostic evaluation of switchboard-corpus automatic speech recognition systems. *Proceedings of DARPA Broadcast News Workshop.*

Gunn, R. (2000). 'Voice': The ultimate in user-friendly computing. *SpeechTEK Proceedings* (pp. 161-178).

Haynes, T. (2000) Conversational IVR: The future of speech recognition for the enterprise. *AVIOS Proceedings of the Speech Technology & Applications Expo* (pp. 15-32).

Head, W. (2000). Breaking down the barriers with speech. *SpeechTEK Proceedings* (pp. 93-100).

Herb, G., & Schmidt, M. (1994, October). Text independent speaker identification. *Signal Processing Magazine*, 18-32.

Hermansen, L. A., & Pugh, W. M. (1996). *Conceptual design of an expert system for planning afloat industrial hygiene surveys* (Technical Report No. 96-5E). San Diego, CA: Naval Health Research Center.

Hertz, S., Younes, R., & Hoskins, S. (2000). Space, speed, quality, and flexibility: Advantages of rule-based speech synthesis. *AVIOS Proceedings of the Speech Technology & Applications Expo* (pp. 217-228).

Holtzman, T. (2000). Improving patient care through a speech-controlled emergency medical information system. *AVIOS Proceedings of the Speech Technology & Applications Expo* (pp. 73-81).

Hutzell, K. (2000). *Voice Interactive Display (VID)* (Contract Summary Report: Apr 98-May 2000). Johnstown, PA: MTS Technologies.

Ingram, A. L. (1991). *Report of potential applications of voice technology to armor training* (Final Report: Sep 84-Mar 86). Cambridge, MA: Scientific Systems Inc.

Karam, G., & Ramming, J. (2000). Telephone access to information and services using VoiceXML. *AVIOS Proceedings of the Speech Technology & Applications Expo* (pp. 159-162).

Komissarchik, E., & Komissarchik, J. (2000). Application of knowledge-based speech analysis to suprasegmental pronunciation training. *AVIOS Proceedings of the Speech Technology & Applications Expo* (pp. 243-248).

Krause, B. (2000). Internationalizing speech applications. *AVIOS Proceedings of the Speech Technology & Applications Expo* (pp. 10-14).

Kubala, F. (1999, February). Broadcast news is good news. *Proceedings of DARPA Broadcast News Workshop,* Herndon, VA.

Kubala, F. Colbath, S., Liu, D., Srivastava, A., & Makhoul, J. (2000) Integrated technologies for indexing spoken language. *Communications of the ACM, 43*(2), 48-56.

Kundupoglu, Y. (2000). Fundamentals for building a successful patent portfolio in the new millennium. *AVIOS Proceedings of the Speech Technology & Applications Expo* (pp. 229-234).

Lai, J. (2000). Comprehension of longer messages with synthetic speech. *AVIOS Proceedings of the Speech Technology & Applications Expo* (pp. 207-216).

Larson, J. (2000). W3C voice browser working group activities. *AVIOS Proceedings of the Speech Technology & Applications Expo* (pp. 163-174).

Leitch, D., & Bain, K. (2000). Improving access for persons with disabilities in higher education using speech recognition technology. *AVIOS Proceedings of The Speech Technology & Applications Expo* (pp. 83-86).

Macker J. P., & Adamson R. B. (1996). *IVOX — The Interactive VOice eXchange application* (Report No. NRL/FR/5520-96-980). Washington, DC: Naval Research Lab.

Markowitz, J. (2000). The value of combining technologies. *AVIOS Proceedings of the Speech Technology & Applications Expo* (pp. 199-206).

McCarty, D. (2000). Building the business case for speech in call centers: Balancing customer experience and cost. *SpeechTEK Proceedings* (pp. 15-26).

McGlashan, S. (2000). Business opportunities enabled by integrating speech and the Web. *SpeechTEK Proceedings* (pp. 281-292).

Miller, R. (2000). The speech-enabled call center. *SpeechTEK Proceedings* (pp. 41-52).

Mogford, R. M., Rosiles, A., Wagner, D., & Allendoerfer, K. R. (1997). *Voice technology study report* (Report No. DOT/FAA/CT-TN97/2). Atlantic City, NJ: FAA Technical Center.

Molina, E. A. (1991). *Continued Performance Assessment Methodology (PAM) research (VORPET). Refinement and implementation of the JWGD3 MILPERF-NAMRL Multidisciplinary Performance Test Battery* (NMPTB) (Final Report: 1 Oct 89 - 30 Sep 91). Pensacola, FL: Naval Aerospace Medical Research Laboratory.

Newman, D. (2000). Speech interfaces that require less human memory. *AVIOS Proceedings of the Speech Technology & Applications Expo* (pp. 65-69).

Pallett, D. S., Garofolo, J. S., & Fiscus, J. G. (1999, February). 1998 broadcast news benchmark test results: English and non-English word error rate performance measure. *Proceedings of DARPA Broadcast News Workshop,* Herndon, VA.

Pallett, D. S., Garofolo, J. S., & Fiscus, J. G. (2000). Measurements in support of research accomplishments. *Communications of the ACM, 43*(2), 75-79.

Pan, J. (2000). Speech recognition and the wireless Web. *SpeechTEK Proceedings* (pp. 229-232).

Pearce, D. (2000). Enabling new speech-driven services for mobile devices: An overview of the ETSI standards activities for distributed speech recognition front-ends. *AVIOS Proceedings of the Speech Technology & Applications Expo* (pp. 175-186).

Prizer, B., Thomas, D., & Suhm, B. (2000). The business case for speech. *SpeechTEK Proceedings* (pp. 15-26).

Przybocki, M. (1999, February). 1998 broadcast news evaluation information extraction named entities. *Proceedings of DARPA Broadcast News Workshop*, Herndon, VA.

Rolandi, W. (2000). Speech recognition applications and user satisfaction in the imperfect world. *AVIOS Proceedings of the Speech Technology & Applications Expo* (pp. 153-158).

Schalk, T. (2000). Design considerations for ASR telephony applications. *AVIOS Proceedings of the Speech Technology & Applications Expo* (pp. 103-112).

Scholz, K. (2000). Localization of spoken language applications. *AVIOS Proceedings of the Speech Technology & Applications Expo* (pp. 87-102).

Shepard, D. (2000). Human user interface: HUI. *SpeechTEK Proceedings* (pp. 262-270).

Sones, R. (2000). Improving voice application performance in real-world environments. *SpeechTEK Proceedings* (pp. 179-210).

Soule, E. (2000). Selecting the best embedded speech recognition solution. *SpeechTEK Proceedings* (pp. 239-248).

SPSS for Windows, Rel. 10.0.5. (1999). Chicago: SPSS Inc.

Stromberg, A. (2000). Professional markets for speech recognition. *SpeechTEK Proceedings* (pp. 101-124).

Taylor, S. (2000). Voice enabling the Web-modification or new construction. *SpeechTEK Proceedings* (pp. 69-76).

Thompson, D., & Hibel, J. (2000). The business of voice hosting with VoiceXML. *AVIOS Proceedings of the Speech Technology & Applications Expo* (pp. 142-148).

Wenger, M. (2000). Noise rejection: The essence of good speech recognition. *AVIOS Proceedings of the Speech Technology & Applications Expo* (pp. 51-63).

Wickstrom, T. (2000). Microphone voice recognition performance in noise: A proposed testing standard. *SpeechTEK Proceedings* (pp. 211-219).

Woo, D. (2000). Desktop speech technology: A MacOS perspective. *AVIOS Proceedings of the Speech Technology & Applications Expo* (pp. 39-50).

Yan, Y. (2000). An introduction to speech activities at Intel China Research Center. *SpeechTEK Proceedings* (pp. 79-82).

Zapata, M. (2000). LIPSinc: We have ways of making you talk. *SpeechTEK Proceedings* (pp. 273-280).

James A. Rodger is an associate professor of management information systems at Indiana University of Pennsylvania (IUP). He received his doctorate in MIS, from Southern Illinois University at Carbondale (1997). Dr. Rodger teaches network administration, system architecture, microcomputer applications and intro to MIS, at IUP. He has worked as an installation coordinator for Sunquest Information Systems, and presently does consulting work on telemedicine connectivity for the Department of Defense and Management Technology Services Inc. Dr. Rodger has published several journal articles related to these subjects. His most recent article, "Telemedicine and the Department of Defense," was published in Communications of the ACM.

Lieutenant Tamara V. Trank is deputy manager of field medical technologies at the Naval Health Research Center in San Diego, CA. She received a BS in kinesiology and a PhD in physiological science from UCLA. LT Trank has published her research in neuroscience and the neural control of locomotion in major scientific journals, and has presented numerous papers at national and international research conferences. In 2001, she received the Naval Achievement Medal. LT Trank is a member of the Society for Neuroscience, the American Association for the Advancement of Science, and the American College of Sports Medicine.

Parag C. Pendharkar is assistant professor of information systems at Penn State Harrisburg. His research interests are in artificial intelligence and expert systems. His work has appeared in Annals of Operations Research, Communications of ACM, Decision Sciences, *and several others. He is a member of INFORMS, APUBEF, and DSI.*

This case was previously published in the *Annals of Cases on Information Technology*, Volume 4/ 2002, pp. 12-28, © 2002.

Chapter XXII

Geochemia:
Information Systems to
Support Chemical Analysis in
Geological Research

Dimitar Christozov, American University in Bulgaria, Bulgaria

EXECUTIVE SUMMARY

In Bulgaria, mineral resources are the property of the State. The State's Committee of Geology (SCG) executes control and supervision and supports geological research. The latter includes access to information and laboratory services. SCG has established a system, with the following functions:

- *Distribution of soil samples among the laboratories;*
- *Execution of quality control on the chemical analytical methods used in laboratories;*
- *Maintenance of the information fund (archive).*

The enterprise, responsible to the performance of the system is "Geochemia." The case specifies the functions of "Geochemia," information systems and technologies used to support execution of its activities in the time of transition to market economy. It provides basis for development of course projects in three areas:

- *Design of database and TPS;*
- *Design of a MIS for quality control;*
- *Development of DSS: Statistical Inference Procedures for usability and calibration of chemical analytical methods.*

BACKGROUND

Until the 1990s, the government had complete monopoly on all activities related to research and exploration of mineral resources in Bulgaria. In 1973 the State Committee of Geology (SCG) was established to represent the government in these areas. Its objectives include:

- Conducting geological research,
- Establishing a network of enterprises to support researchers, and
- Assisting the other entities in making decisions regarding exploration of mineral resources.

Among other responsibilities of SCG is the correctness of inferences and decisions made according to the results of geological researches. Therefore, the quality of the analysis is a highly important issue.

In late 1970s, SCG had established a network of enterprises to support research activities. Among them was the system of chemical laboratories which specialized in performing chemical analysis on soil samples excavated in geological researches. Organizationally, the laboratories were subsidiaries of SCG. In 1982, all chemical laboratories were united under a specialized entity — "Geochemia," which was responsible for the soil samples processing. The process aims avoid possible bias, mistakes, and inaccuracy of the analyses. "Geochemia" was authorized to represent SCG in the following activities, which defined also its major functions:

1. Distribution of soil samples among the laboratories and control of the results;
2. Control adoption of new chemical methods and quality control of performance of chemical methods in laboratories;
3. Maintenance of sample archive and information fund and providing access to them.

Since 1990, Bulgaria has started a process of transition from completely centralized to market-oriented economy. In 1997, the system of SCG was reorganized to cease the State's monopoly on geological researches, preserving for SCG only the functions of monitoring the research. SCG issues licenses for conducting research and the researchers are obliged to provide to SCG detailed information about findings. The new role of SCG is the role of an information agency. It advises both the government on the issues related to giving concessions of exploring mineral resources, and the geological teams about known findings in the areas of research and about the support they can rely on in the country. In its new role, the position of SCG was lowered to a department in the Ministry of Environment.

The enterprises established to provide support to the research teams have been turned into independent entities. "Geochemia," and also each one of the specialized laboratories, were separated as independent entities of the state and were placed on the free market of laboratory services.

Reorganization of the SCG system requires reevaluation of "Geochemia's" objectives and its interaction with laboratories from one side and with research teams from the other. Also, relations with SCG, as the representative of the state, have to be reconsidered and transform on a contractual basis. In the monopolistic era, prevalence was given on reliable and accurate results and the other aspects of the service were underestimated. Now, "Geochemia" faces the challenges of the free market and has to improve all aspects of its services.

Following we specify the functions of actors on the stage of geological research before reorganization, challenges of the new objectives and constraints, and selected approaches to solving problems.

SETTING THE STAGE

The Role of Geology Research Team

Geological researches are executed in two cases: according to the annual plans for comprehensive screening of the territory of the country and according to a particular project — detail — investigation on a given area to evaluate parameters of the existing mine.

Research starts with screening of the information about the area and planning the research activities. Next is the screening of the field, which includes collection and evaluation of samples. According to the achieved results, the team plans the next steps. This process continues until fulfillment of the stated objectives of the project.

Obtaining a sample is a time-consuming and costly process in many cases. It requires expensive equipment and qualified personnel (e.g., drilling). Therefore excavation of samples must be well planned on the basis of careful analysis of the results obtained in advance and all available information. The period between excavating a sample and receiving the laboratory results is usually an idle time for the team. Its reduction is a major opportunity for increasing the efficiency and cost effectiveness of the research. Reduction of the time for sample processing by "Geochemia" is the most important means of increasing the quality of the service.

The Role of Laboratories

Chemical laboratories, which specialize in performing chemical analysis of geological samples, have adopted a number of analytical methods developed to define the content of a given chemical element (e.g., copper CU) in material extracted from a soil sample. To adopt an analytical method, a laboratory has to posses a set of relevant equipment and chemicals and must train the staff to perform the procedure of the analysis and to interpret results.

We can define analytical method as the set of equipment, adopted procedure (recipe), chemicals, trained staff, and rules for interpretation of the results. According to this definition a method is not only the procedure, but also how the procedure is implemented in a given laboratory. Quality of a method is determined by two characteristics (see ISO 5725, 1994): trueness, which is "the closeness of agreement between the average value obtained from a large series of test results and an accepted reference value," and precision, which is "the closeness of agreement between independent test results obtained under stipulated conditions." It is impossible to observe directly these two characteristics, therefore, the two indicators repeatability and reproducibility are used in evaluation of the quality of a method. Repeatability means that results obtained in several trays on the same sample are close enough or whether the method shows similar results when applied on the same sample many times. It measures the stability of the results. Reproducibility means that the results obtained by the method coincide with the results obtained by other methods applied on the same sample.

Interpretation of the results means that the staff applies existing knowledge about specifics to the method bias, which are evaluated during the period of adoption of the method and its regular control. The process of gathering data, structuring, assessment, and use of such knowledge is called calibration of the method. The formula used to transform obtained measures into final results is called calibration function.

Also, a given method is more precise for certain levels of concentration (content) of the element in the sample and less precise for other concentrations. Therefore the whole diapason of contents of the given element can be split into several intervals (levels of concentration) with different calibration functions and precision ensured by the method.

The laboratory is responsible to keep a method active by periodical calibration. "Geochemia" also performs control on whether a method is using correctly by the laboratory. The statistical procedure for quality control is given in Appendix A.

The Role of "Geochemia"

In 1982, SCG had established the enterprise "Geochemia" as the distribution center for laboratory investigations. In 1985 the three functions of "Geochemia" were specified:

1. **Logistics:** to execute stated procedures of performing chemical analysis on soil samples;
2. **Quality control:** to plan, develop, and execute activities related to both control and calibration on the chemical analytic methods used in the laboratories (repeatability) and control on the results (reproducibility);
3. **Information service:** to maintain an archive holding both soil samples and results achieved by the chemical analysis.

"Geochemia" was given also the authority to serve as the methodological supervisor of the laboratories.

Between 1984 and 1987, a set of documents specifying operational procedures was developed, approved by SCG, and implemented in the SCG laboratory network:

*   "Instruction to guide all activities in organizing and performing the chemical analysis on geological samples," 1984; and
*   "Instruction to specify statistical methods to ensure trueness and precision of chemical analysis," 1987.

The two documents are the basis on which "Geochemia" and associated laboratories operate.

The operational procedures are described below.

The new role of "Geochemia" does not differ from the initially stated one. It was recognized by the laboratories and geologists that the existing system ensures reliable results and it has to be preserved. The only considered issue is to improve the efficiency of the service, which means reducing the time for processing samples.

Procedures

The Procedure of Processing Soil Samples

Every sample of soil, taken in geological research from a certain place (location of the sample) is delivered to "Geochemia". Here, the sample receives a secrete code (this is done to hide information about location of the sample and to hide researchers' expectation from laboratory analysts to avoid bias). The soil is homogenized (usually by mechanical methods) and split into three portions:

- **A portion for primary chemical analysis:** It is sent to laboratories for analysis of the concentration of a given (specified) chemical element, by specified analytical methods (materials from one sample are analyzed by at least two methods in two laboratories);
- **A portion for repeating chemical analysis:** It allows the repeating of the analysis in case of unsatisfactory primary results (lack of reproducibility); and,
- **A portion dedicated to the archive:** It allows the execution of later analysis of the samples excavated in research projects in the past.

A laboratory executes a number of (two to eight) independent analyses on the material and returns results to "Geochemia". The results from all different methods are decoded and all results for the given sample are compared (see Appendix A1). If results coincide (reproducibility), the final result is calculated as a combination (average) of single results. If results do not coincide, materials from the same sample are sent to a third laboratory for additional analysis. This procedure is executed until results coincide or it became possible to exclude outliers with a given level of confidence and the remaining results coincide. The final result is provided to the team of geologists who had excavated it; and it is recorded into the archive.

The activities listed in the operational procedure for soil samples analysis are:

1. Delivery of the soil samples to "Geochemia";
2. Assigning secret code to the sample;
3. Homogenizing and partitioning the sample;
4. Delivering samples to laboratories;
5. Execution of chemical analysis by the laboratories;
6. Delivering results to "Geochemia";
7. Decoding;

8.   Combination and statistically evaluation of the results;
9.   If results do coincide: delivering the final result to researchers and recording it into the information fund; and
10. If results do not coincide: Repeating the process from Item 4.

Annually, over 100,000 soil samples are excavated and processed. Thus, over 1 million chemical analyses are executed, their results are statistically analyzed; and the combined results are recorded.

The Procedure of Quality Control

Among regular samples, "Geochemia" can also send to the laboratory materials or certified samples to be used as a reference value for the concentration of given element with a priori-known content of a given chemical element. "Geochemia" uses results from these samples for quality control of the laboratories and for calibration of analytical methods, during the time when the laboratory adopts them. The goal of quality control is to ensure repeatability and reproducibility of the chemical analysis executed by the laboratories, which guarantees correctness of the results. The statistical inference procedure is described in Appendix A2.

Information Technologies

Initially, all activities of "Geochemia" were executed with paper-based technology. Beginning in 1987, calculations required by statistical tests, were made on computers. Since 1989, all results returned from the laboratories have been recorded into a database, which allows easy search and retrieval in initial screening. The applications were not connected, and they were used independently. The state of IT before the project can be described by three elements:

a.   **Organization and procedures:** The "Geochemia" and associated laboratory network were built up organizationally. Operational procedures for material and data processing were established and proved themselves. Because geological research is conducted basically during the summer, the workload (chemical analyses) is not uniformly distributed annually. Winter is used for adoption of new methods, calibration, training, etc.
b.   **Information technology:** Dominated information technology was paper based. There is a clear understanding of the need to improve the way of managing information resources by implementing computer-based information systems and integration between different elements.
c.   **Computerization:** Computers were used for improving secondary activities and not as a primary tool for management of the information resources. The computer systems used for statistical analysis and for the database were not compatible and were designed for different platforms.

CASE DESCRIPTION

Problems, Challenges, and Constraints

The challenges "Geochemia" faces in its new status as an independent entity, require approaching the following problems:

a.   To improve the geological teams' service in soil samples processing;
b.   To open its database to external users and to ensure easier access to available information (the information fund was recognized as the major owned resource of "Geochemia");
c.   To guarantee international comparability of the results, which will allow:

   •   Serving international teams and participating in international projects, and
   •   That the created information will be useful to international investors.

Recognized also was the need for integration of information resources to be based on CIS. Application of information technologies was recognized as the way to solve the first two problems. Internet technology is intended to help in speeding up the data transfer from laboratories to "Geochemia" and from "Geochemia" to the researchers. It will also be used to provide online access to the information fund.

The difficulties came from:

a.   The new system has to be implemented in parallel with operation of the existing one. The schedule of geological researchers allows wintertime to be used for implementation and training staff and the next summer for testing the new system.
b.   Underdeveloped IT infrastructure in the country, especially in places where research is conducted, does not permit reliance on only computerized information technologies. It requires the preservation work with paper-based and phone-based communication technologies as well as keeping extra staff to work with them. Computerized technologies include both e-mail communication and Internet online access. This complicated the applications but allows smoother implementation of the entire system. During the period of transition everybody can use the most suitable technology.
c.   Integration of existing hardware (equipment for performing chemical analysis) with the information system will reduce typing mistakes in data transfer and will eliminate delays. But, because this equipment is highly heterogeneous, the cost of developing the needed drivers was evaluated as too high to be implemented and was postponed to the next stages of the project.
d.   Training the staff. There are fewer problems in training the staff of "Geochemia" and laboratories, but training the members of geological teams requires permanent efforts and establishment of a training facility. The members of geological teams are appointed for every project.
e.   The project to apply IT has to preserve procedures for sample processing and statistical analysis: evaluation of repeatability and reproducibility of results obtained in different laboratories, quality control and calibration of the chemical

analytical methods. (The researchers and end users recognized them as useful). The project must also enrich the current procedures with internationally recognized procedures (ISO 5725, 1994).

Objectives of the project are to establish an integrated information system and will facilitate interaction between research teams and laboratories, and which will be managed by "Geochemia."

System Specification

The new functionality of the system can be specified as:

a.   Reduction of the time for information transfer between geologists and laboratories;
b.   Information support in decision making:

- **By laboratories:** calibration of methods, measurement instruments, etc.;
- **By "Geochemia":** selection of laboratory and method to analyze certain sample — according to information about the quality of methods used by laboratories, and whether additional analysis are needed;
- **By SCG:** to evaluate the quantity of existing mineral resources in certain area;
- **By research team:** significance of existing minerals.

c.   Involvement of researchers in decision making about repeating analysis.

Development of the system was spilt into three steps:

Step 1. Development of the three information systems:
- SP (sample processing): TPS to facilitate data transfer in sample processing;
- QC (quality control): MIS to support quality control and adoption of new methods;
- GDB (geological database): Database to hold information about findings.

Step 2. Development of interfaces between:
- SP and QC: DSS to assist in selection of the most appropriate method for analyzing a given sample, and for optimal distribution of samples among the laboratories;
- SP and GDB: for automation in updating the information fund.

Step 3. Extending functionality of the GDB to become geographical information system open for external access:

Specification of the systems, included in the first step:

Sample processing: The major objective is to speed up processing of soil samples. Reducing the time for delivery of the physical objects (which the soil samples are them selves) and for performing chemical analysis (which required specific technological time) are not considered here.

Speeding up the following activities is addressed:

- Transfer results from a laboratory to "Geochemia";
- Transfer results from "Geochemia" to research team; and
- Participation of researchers in decision about necessity of repeating analysis.

The last is of special importance because, according to existing standards for reproducibility, results for given element, concentration level and method must satisfy the statistical t-test with a priori given probability. But often researchers could be satisfied with worse results (e.g., if the defined content of the element is far from the expected value in all tries) which do not satisfy the test, but it is obvious (for the researchers) that repeating analysis will not add value.

Quality Control

Notes: This system is intended to provide valuable information about the methods used in the laboratories and to perform required statistical computations. It can serve also in the establishment of the initial parameters of every newly adopted method. It is expected that this system will be accessible only from "Geochemia". Because "Geochemia" has to provide control samples in the same way as regular samples, it is expected that a laboratory also will place results in the SP database. The further development of the system to serve in optimal distribution of samples among laboratories must be considered in the design.

GEOLOGICAL DATABASE

The data includes location where the sample is excavated and the content of chemical elements. On the second stage of the project, the database has to be updated automatically according to the data collected in SP.

Implementation

In 1999-2000, the three systems of the first step was developed and implemented. The systems can be specified as following:

- **Information technology used for SP:** The approach considered in the design of the system is to use a Web database in "Geochemia," which is updated online. The database provides different views for the laboratories and for the research teams (see Appendix B). The laboratory accesses data according to the secret code of the sample, and for the research team the data are visible according to the code of the project. Also, it allows exchanging the data via e-mail or entering data manually (for phone or mail communication), which increases the reliability of transactions. Usually, researchers work far from big cities.

- **Information technology used for QC:** An MS Access database holds information about the available methods: laboratory, equipment, staff, chemicals used, and history of the control: date, control sample (concentration of the element) and results obtained by the method. Interface, between database and the program, which calculates the required statistics, retrieves numerical data from database and form data samples, which are processing by the computational programs.
- **Information technology used for GDB:** An MS Access database holds information about findings: location, research team, findings (tested elements and their content), and comments.

The next two steps will be planned after successful completion of the first step of the project.

REFERENCES

ISO 5725-94. (1994). *Accuracy (trueness and precision) of measurement methods and results.*

Kamburova, E., & Christozov, D. (1997). Assessment of and comparison between Bulgarian and international standards for quality control of evaluation of the elements' composition of geological samples. *Analytical Laboratory, 6*(4), 205-221.

Pollard, J. H. (1977). *A handbook of numerical and statistical techniques.* Cambridge University Press.

State's Committee of Geology. (1984). *Instruction for internal, external, and arbitrary geological control on quality of the analyses of soil samples executed in chemical laboratories.* Sofia: SCG, 1984 (in Bulgarian).

State's Committee of Geology. (1987). *Instruction to ensure coincidence of quantitative measures of chemical content of mineral resources in the laboratories included in the SCG system.* Sofia: SCG, 1987 (in Bulgarian).

APPENDIX A. STATISTICAL PROCEDURES

A1. Repeatability and Reproducibility

Portions of the material extracted from one sample are sent to two different laboratories to be analyzed for the content of a given chemical element, using specified method. A laboratory performs a series of repeating analyses and reports results.

Lab $\{x_1, x_2, ..., x_n\}$

Procedure to Evaluate Repeatability

Repeatability is represented by within laboratory variance, which must be under specified limit. The limit depends on chemical element, method, and concentration level.

1.   Calculation of the mean x.
2.   Calculation of the variance s.
3.   Compare s with table value. Table values are defined in calibration process.
4.   If s is greater than the table values, laboratory is asked to repeat the analysis and procedure continues from 1.

Procedure to Evaluate Reproducibility

Reproducibility is represented by between laboratory variance. The results obtained in the two laboratories are repeatable.

1.   The two means are compared according to t-test.
2.   If we cannot reject the hypothesis of equality of the two means with given probability. (According to Bulgarian standards it must be taken from the series 0.75, 0.9, 0.95, or 0.99). The results are considered also reproducible and the overall mean is considered as a content (concentration) of the given element into the given sample. (When researchers are involved in decision making about reproducibility of the results, it is useful to provide also the p-value, calculated for the t-statistics).
3.   Otherwise, material from the same sample is send to a third laboratory, which performs an evaluation of repeatability.
4.   The t-test is performed for the two couples of results. In the case that we cannot reject the two hypotheses — the mean, calculated according to the three series of results — is considered as the final result. In the case that only one hypothesis is rejected, the final result is calculated as the mean of the two other samples. If the t-test rejects the two hypotheses, a fourth laboratory is involved or the experiment is started from the beginning with use of other chemical methods.

A2. Quality Control and Calibration

To perform calibration on a newly adopted method and also to control the methods in use, "Geochemia" sends to laboratories certified materials with known contents of the given element (in terms of ISO 5725 it is a reference value). An experiment includes sending materials for all concentration levels of the given element, analyzed by the method. A laboratory performs a series of analyses (equal size) for every concentration level:

Level	Reference Value	Results
1	3.00	2.98
	3.02	
	2.95	
	2.99	
2	1.10	1.11
	1.05	
	...	...

The main idea of the method is to build up the linear regression.

$$y = b_0 + b_1 x$$

where x represents the reference values and y — the observed values (results).

The interpretation of the results, according to the model is rather simple and obvious:

- If $b_0 = 0$ and $b_1 =$, the chemical method discovers correctly the content of the element into the whole range of concentration levels, without bias.
- If $b_0 <> 0$ and $b_1 = 1$ — the chemical method discovers correctly the content of the element, but there is a bias b_0, specific to the method. b_0 is used as a calibration constant in results interpretation.
- If $b_1 <> 1$ — the method is not accurate on the entire interval of concentration levels. Additional analysis is necessary, which can highlight where the problems are in the chemical method itself, in its implementation, or it requires more precise calibration.

Activities could be:

- To exclude certain levels from the range of application of the method;
- To develop more complex calibration function, e.g., if data satisfies given non-linear model, e.g., $\ln(y) = b_0 + b_1 \ln(x)$;
- To analyze, whether implementation of the method is correct and all activities are done correctly.

Of course, such clear results are very rare and respective statistical tests to check the above hypothesis are used (see Pollard, 1977). Also, the graphical representation of the regression line can display if the whole range of levels is divided into sub ranges where the method produces correct results and it can be used and the other set of levels where the use of the method is not appropriate.

Note: For application and for numerical methods to apply statistical tests, see Pollard (1977).

APPENDIX B. MAJOR ATTRIBUTES OF THE BASIC DATA ENTITIES

B.1. Soil Sample

Entity: Sample:

Sample ID#
Project code: given by SCG, when the license is issued
Date: of sending the sample
Location: the place where the sample is excavated
Elements: list of elements, content of which has to be evaluated, and evaluated contents (after analysis)
Entity: Analytical results:
Sample ID# (invisible for the laboratories)

    Secret code
    Method ID#
    Element
    Measurement unit
    Results: 4-8 numbers

Note: For every element in a "Sample" must exist at least two related records in "Analytical results."

B.2. Method

Entity: Method:
    ID#
    Laboratory
    Equipment
    Chemicals
    Staff trained to apply method (multiple)
    Source (description of the method)
    Elements, concentration levels and calibration rules (constant, description)

Entity: Control Tests:
    Method ID#
    Date
    Staff member, executed analyses
    Chemical element
    Reference Value
    Results (up to 8)

B.3. Mineral Resources

Entity: Findings
    Location
    Date of research
    ID# of researcher project (according to the license given by SCG)
    Element, content (multiple)

Note: Only chemical elements with content significantly greater than its average content in the soil are recorded.

Dimitar Christozov has more than 20 years of experience in the cross section of the areas of information systems, computer science, quality management, and applied statistics. He graduated in mathematics in 1979 from Sofia University "St. Kliment Ohridski." With the Central Institute of Mechanical Engineering - Sofia (1979-1986), he was involved in the development of information systems to support the research in mechanical engineering. His system "Relia-Soft," built on results of his doctoral thesis (defended in 1986) and addressing the machine reliability evaluation, was widely implemented in Bulgaria. At the Information Center for Technology Transfer "Informa" (1986-1993), Dr. Christozov was involved in the establishment of the national information network for technology transfer. His research includes different aspects of building information systems in the areas of technology assessment, integral quality measures and quality management. He was recognized in these areas as one of the leading experts in Bulgaria. The DSS "Biser" for integral quality evaluation of multidimensional objects was implemented widely in Bulgaria, former USSR, and Hungary. Since 1993, Dr. Christozov has been an associate professor of computer science with the American University in Bulgaria and an adjunct professor at the newly established Faculty of Economics and Business Administration of the Sofia University "St. Kliment Ohridski." He was involved in establishing and implementing the curriculum for the major of computer science at the AUBG and the IS stream of courses at FEBA. Professor Christozov has published three separate volumes, more than 20 articles in refereed journals and conference proceedings, and has over 40 presentations at conferences, seminars, etc.

This case was previously published in the *Annals of Cases on Information Technology Applications and Management in Organizations*, Volume 3/2001, pp. 115-126, © 2001.

Chapter XXIII

ACEnet:
Facilitating Economic Development Through Small Business Electronic Commerce

Craig Van Slyke, University of Central Florida, USA

France Belanger, Virginia Polytechnic Institute and State University, USA

Marcy Kittner, University of Tampa, USA

EXECUTIVE SUMMARY

With the advent of Web-based electronic commerce (e-commerce), businesses of all sizes rushed to take advantage of the potential of e-commerce technologies. While large organizations often have ready access to the resources necessary to implement e-commerce strategies, smaller organizations may lack some or all of these resources. Conversely, the increased reach facilitated by e-commerce may allow some small businesses to be viable in areas where limited access to customers might otherwise prevent success. This increased reach may be particularly beneficial in economically depressed rural areas, which may stand to gain greatly from the economic development potential of small businesses. Small businesses in general, and rural small businesses in particular, face a number of hurdles that must be overcome if they are to benefit from e-commerce. This case describes how a not-for-profit agency, the Appalachian Center for Economic Networks (ACEnet), facilitates the use of e-commerce by rural small businesses as a part of an overall strategy for spurring economic development through small businesses. ACEnet provides a number of resources that help small businesses take advantage of e-commerce, including computer labs, Web site hosting, consultative services, and technical and business training. This case illustrates how these services help rural small businesses overcome many of the barriers to successful implementation of e-commerce.

BACKGROUND

In 1985, a group of community members concerned with the revitalization of the economy of southeast Ohio joined forces and received funding to form the Appalachian Center for Economic Networks (ACEnet). A basic operating premise of ACEnet is that the formation of networks of small businesses can lead to sustainable economic development. Under-employed or unemployed individuals often have the capability to create and operate small businesses, given some initial assistance. These small businesses not only provide employment for the owner but as they grow also provide employment opportunities for others, thus improving the overall economy of the area.

ACEnet has a number of related goals. First, they seek to develop and put into practice innovative economic development models. ACEnet also strives to enable individuals and organizations to develop continual learning-based strategies that will empower communities and their members. Another ACEnet goal is to facilitate and foster relationships among private and public sector organizations. These relationships should be (1) cooperative, (2) collaborative, and (3) inclusive. Finally, ACEnet works to initiate networks among groups committed to economic development, such as businesses, policy, and economic development groups.

ACEnet sees the formation of networks of small businesses as an emerging paradigm for economic development. Focusing on developing and implementing new processes and systems for facilitating economic growth through small business development allows participants in the resulting economic networks to better respond to the dynamic, global business environment.

In the early part of its existence, ACEnet concentrated on helping low-income residents form small businesses that are worker owned and operated. While ACEnet experienced success using the worker cooperative model, it became clear that there was a need for an additional strategy that would allow the continuous creation of a greater number of well-paying jobs. The goal was to supplement, not to abandon, the worker cooperative model. The search for new strategies led to the discovery of European communities of micro businesses (fewer than 20 employees) in the manufacturing sector. These communities use the concept of flexible manufacturing (also known as agile manufacturing) to allow various members of the network to temporarily cooperate in the making of products in emerging market niches. By cooperating, the firms are able to manufacture products that require resources or expertise that exceed those of any single firm. By collaborating, rather than competing, the firms are able to reach markets that otherwise might be closed to them.

To support the network concept, ACEnet established the Cooperative Business Center (CBC), an incubator for small businesses. The CBC provides a number of shared resources including equipment such as computers, laser printers, fax machines, and reception services. All of these are provided at lower prices than those normally available from the local market.

The establishment of the center was ACEnet's first activity aimed at promoting collaboration among small business owners. This tradition of promoting networking among small business proprietors now extends into other areas, including e-commerce.

A more recent effort is the Kitchen Incubator, a fully licensed kitchen facility that small businesses can use to develop food-oriented product lines without having to invest in their own kitchen equipment and licensing fees. The latest effort directed towards helping small business owners gain access to resources is the establishment of the computer opportunities program (COP), which provides computer training to local high school juniors and seniors who, in turn, provide computer consulting services to local small businesses.

The COP illustrates the power of ACEnet's community networking concept. The COP creates a network among high schools, students, and small businesses in the community. There is a circle of benefits for all involved. The students gain valuable computer skills and work experience. The small business owners gain affordable computer expertise and possibly future employees. The school is using its own information technology (IT) more effectively — the students are actually teaching teachers and administrators how to better utilize the technology. The student consultants are also helping their peers find new ways to use computers. All of those involved benefit from their participation. ACEnet's history and major milestones and accomplishments are shown in the Appendix.

ACEnet's Services

A variety of other services are also provided, all of which support ACEnet's overall goals, which were stated earlier. Services provided by ACEnet are summarized below, with the exception of those whose development is the focus of this case.

- **Small business incubator:** ACEnet's small business incubator provides facilities for up to 18 budding small businesses. Both operating space and support services, including computer systems, secretarial, and bookkeeping services are provided at below market rates. New businesses are charged very low rates for these services, and as the businesses grow, the fees increase with the eventual goal of the business leaving the incubator and becoming self-sufficient.
- **Flexible manufacturing networks:** Flexible manufacturing networks (FMN) are associations of small manufacturing firms that join together, often on a short-term basis, in order to produce a product that any single firm in the network would be unable to produce on its own. ACEnet concentrates its FMN efforts in three market sectors: (1) specialty foods, (2) wood products, and (3) computer services. In order to support these networks, ACEnet furnishes access to funding, business advice, workforce development, technology, and training.
- **Business skills training:** ACEnet provides entrepreneurs comprehensive training in basic business skills. This training is provided through a network of local business assistance providers. While other organizations provide most of the general business training, ACEnet personnel provide assistance in a number of areas, including industry-specific business plan development, costing, legal and regulatory issues, and financial systems. These services are delivered through a combination of workshops and individual technical assistance.

- **Kitchen incubator:** The kitchen incubator is a large (3,000 square feet), commercial-grade food preparation and packaging facility that is shared among ACEnet members. The kitchen provides the equipment and facilities necessary for commercial operation. In addition, the kitchen allows a common working environment that fosters networking among the various businesses. Over 100 small food-product entrepreneurs have used the facilities to test potential products (Petzinger, 1997).
- **Marketplace:** ACEnet's Marketplace is an 800-square-foot facility designed to provide a retail outlet in which ACEnet clients can test-market and sell their products. The Marketplace currently sells a wide variety of locally produced food products, as well as other specialty products such as skin creams and soaps manufactured by ACEnet clients.
- **Equipment leasing and funding:** ACEnet works in conjunction with banks and loan funds to develop and implement new small business loan programs. In addition, ACEnet makes donated computers available for a nominal lease fee. When a business's needs grow beyond the capabilities of these leased machines, owners may be able to obtain low-cost loans from ACEnet or associated funding sources in order to procure the new equipment.

ACEnet Organizational Structure

ACEnet currently operates with 15 paid staff and a number of volunteers. Three executives, the president, executive director, and associate director, are responsible for the overall operation of ACEnet. However, the ACEnet management style and organizational culture reflect its roots in worker-owned cooperatives. Management style is very participatory, and all staff are encouraged to innovate in their particular areas of responsibility. In fact, this team-oriented view of the organization is reflected in some job titles. Rather than using the term "manager," ACEnet refers to those in positions of supervisory responsibility as "team leaders." There are three of these team leaders, one each for Administration, Food Ventures, and Telecommunications. ACEnet's formal organizational structure is shown in Figure 1.

ACEnet has an annual budget of $840,000, with approximately half going for IT and telecommunications related activities and half devoted to specialty food industry related projects. The majority of ACEnet's operating funds comes from grants, although a substantial portion comes from fees for services provided by ACEnet.

Economic Climate

ACEnet is primarily concerned with an eight-county area in rural southeastern Ohio. All of these counties are classified as economically depressed. All but one of the counties served by ACEnet were classified in a 1994 study by the Appalachian Regional Commission as being severely distressed. The eighth county was classified as distressed. In order to be classified as severely distressed, a county must have both unemployment and poverty rates that are significantly higher than national averages.

Many the region's economic woes can be traced to the closing of a number of large mining operations. In addition, several large manufacturing employers closed their operations in the region. This resulted in the loss of over 25,000 jobs in the mining and manufacturing industries. While the loss of these jobs was offset by the creation of new

Figure 1. ACEnet organizational chart

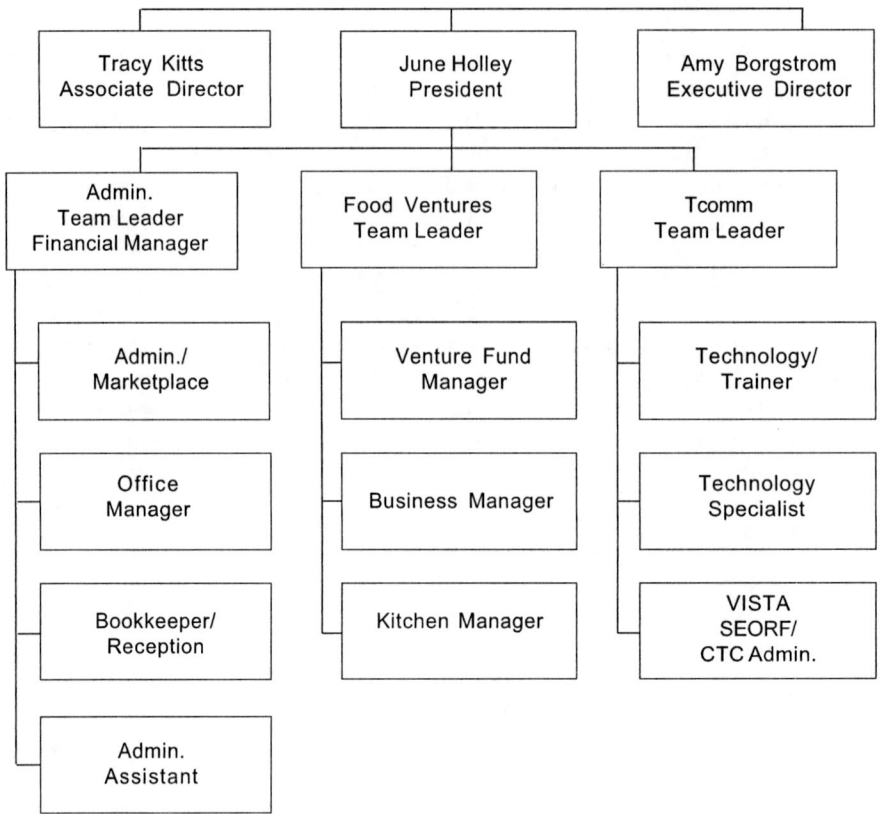

positions in the retail and services industries, these jobs tend to be much lower paying and often do not include basic benefits, such as health insurance and pension plans.

At the beginning of 2000 the unemployment rate for the region was over 12%, compared with 4.1% for the United States (2000). None of the counties in the region had a lower unemployment rate than those for the state and the nation in 1996. The county with the lowest unemployment rate in the region suffers from a high level of poverty. In fact, this county had the highest poverty level in the state in 1993 (34%). Per capita income is quite low in the entire region. In 1993, the per capita income of $13,431 was 32% below that for the state and 35% below that for the United States.

Rural southeastern Ohio has also experienced "outmigration" of both industry and the region's youth. From 1980 to 1990, the counties in the region experienced net migration of -33% for those aged 20 to 24 and -25% for those 25 to 29. The region has also experienced a loss of industry — from 1982 to1992 there was a net loss of businesses. This migration of youth and businesses from the region only serves to degrade the already depressed economic situation.

Interestingly, despite the region's poor economic climate, only 7% of the population received public assistance in 1993. One possible explanation for this is a relatively strong social norm against receiving public assistance.

Understanding Small Businesses

When trying to understand small businesses, the theories and models used for large businesses should not be assumed to apply. Small businesses are not simply smaller versions of large businesses (Welsh & White, 1981). Some of the fundamental differences between small and large businesses, as reported in Thong (1999), include more centralized decision-making structures centering around the chief executive officer (Mintzberg, 1979), the tendency to employ more generalists than specialists (Blili & Raymond, 1993), the tendency to have lower awareness levels about information systems and their benefits (DeLone, 1988; Lees, 1987), and fewer financial resources making them susceptible to more short-term planning (Welsh & White, 1981). Nevertheless, small businesses account for a large portion of today's economy, more than 90% of all businesses in some economies (Longnecker et al., 1994).

A microenterprise is defined as a "very small business with less than five employees ... with the most common case for the business to be a sole-proprietorship" (Johnson, 1998a). Still, these enterprises represent a very large portion of the businesses started in the United States. For example, close to two-thirds of businesses started in this country have less than $10,000 of investment (Johnson, 1998a). One of the reasons the government is encouraging microenterprises through various programs is that it is believed that microenterprises offer the potential for breaking the cycles of social inequality through raising employment and income for minorities and women (Johnson, 1998b).

SETTING THE STAGE

Information technology is an essential component of ACEnet's strategy. In its view, technology amplifies and supports relationships and community development. ACEnet uses technology in a number of ways. These can be divided into four main categories. First, ACEnet employs information technology to enhance its own operations. Second, ACEnet uses IT to build networks of practice among small businesses and resource providers. Third, ACEnet utilizes IT to enhance the economic prospects of individuals in the region through technology skills training. Finally, ACEnet is attempting to help regional small businesses harness the potential of electronic commerce, which is the focus of this case. It should be noted that while it is useful to discuss each of these IT uses separately, they are actually tightly coupled, with all aspects coming together to support ACEnet's overall goals.

Utilizing IT for Internal Operations

Early on, ACEnet staff recognized that they must "walk-the-walk" if they were to successfully champion the use of IT by small businesses. This was necessary for several reasons. First, ACEnet recognized the potential of IT for making its own operations more efficient and effective. In order to remain economically viable and to minimize operating costs, staff needed to apply IT to ACEnet's operations. Second, effectively employing

IT enabled ACEnet to demonstrate "proof of concept" to clients. Finally, by using IT effectively, ACEnet staff gained expertise that allowed them to become better advisors.

One of the first steps ACEnet staff took to learn effective use of IT was to use ACEnet's internal intranet to communicate with each other via electronic mail (e-mail). Staff also joined Foodnet, a network of practitioner groups in low-income communities that provides Web conferencing and an e-mail list service. This allowed ACEnet staff to learn how to harness the power of virtual communities of practice. In addition, the involvement with Foodnet improved staff's knowledge of e-mail and also increased their ability to find useful information on the Web.

ACEnet staff then introduced this new knowledge to clients. This had the effect of increasing interest in IT on the part of small business clients. For example, when a staff member distributed a particularly relevant Web-accessed article to clients, the clients engaged in their own Web searches in order to find additional useful articles.

Building Networks Among Businesses and Resource Providers

The concept of cooperative networks is the cornerstone of the ACEnet philosophy. Therefore, it is natural that the knowledge the ACEnet staff gained be turned to facilitating the emergence of networks among businesses and resource providers. When staff members communicated a use of IT to a client, this information was often quickly shared with other business owners. The shared IT resources available at ACEnet facilitate such sharing. Even before the opening of the Community Technology Centers (discussed below), one room of the Kitchen Incubator was set aside as a public Internet access site, which firms could use at no cost. This sharing of resources encouraged face-to-face interaction among businesses. For example, the owner of a florist shop may help an herb farmer find information about new pesticide-free growing techniques. To date, most of this sharing has occurred serendipitously, with little overt intervention by ACEnet staff members.

Recently, ACEnet opened the first of several planned Community Technology Centers (CTCs), which provide IT equipment in a computer lab environment. While the primary use of the CTC is for training workshops, the equipment is available for community use on a low-cost basis. The CTC has personal computers, high-quality printers, a scanner, and Internet access.

Table 1. Estimated CTC start-up cost

Item	Quantity	Unit Cost	Estimated Cost
Server	1	3,500.00	3,500.00
Workstation PC	15	1,400.00	21,000.00
Scanner	1	400.00	400.00
Laser printer	1	1,500.00	1,500.00
Color inkjet printer	1	350.00	350.00
Misc. network hardware	1	1,000.00	1,000.00
Modem	1	250.00	250.00
Total			$28,000.00

Start-up equipment and installation costs for a CTC run from $25,000 to $35,000 depending on the exact configuration. A typical cost breakdown is provided in Table 1. Note that a CTC is normally located in a public space such as a school or library, so there is no facility cost. Currently, technical support for the CTC comes from existing ACEnet personnel. However, as the number of CTCs increases, additional personnel may be required.

The estimates shown in Table 1 are for typical configurations. For example, the workstation estimates are based on a Dell Dimension L Pentium III system with 128mb of RAM, a 10gb hard drive, a 17" monitor, network card, 100mb ZIP drive, and a CD-RW. Other equipment estimates are based on prices for high-quality devices suitable for a heavy use lab environment. Additional equipment, such as a digital camera, could be added as the budget allows.

A wide variety of technology skills training occurs in the CTC. A series of workshops is available that runs the gamut of basic small business technology skills. Beginning with a basic Windows 95 class, workshop participants progress through word processing, spreadsheet, accounting, database, and Internet-related classes.

The CTC is not simply another place to receive technology training. It also serves as a networking catalyst. Workshop participants routinely network with each other while at the CTC. This allows the businesses to learn from one another, not just regarding the use of technology but also concerning other business issues. For example, the owner of a specialty tea firm may chat with the owner of an herb store and conclude that they can collaborate on the development of a new line of therapeutic herbal teas. Often, mentoring relationships emerge from this networking. Older, more established firms help foster new businesses, guiding them through the many pitfalls of small business ownership.

ACEnet also uses IT to assist its members in gaining access to and building relationships with resource providers. For example, a firm may identify resources through the use of the Internet then use e-mail to build a relationship. An example of this idea in action is an entrepreneur who markets the pulp of a local indigenous crop, the paw-paw. The entrepreneur used the Web to locate a number of research centers that could provide assistance in the processing and storage of the pulp. He then used e-mail to contact and build an ongoing relationship with several staff members of the research centers. A similar example comes from the experiences of a worker-owned cooperative. Members of the local cooperative have used the Web to identify and make contact with other worker-owned businesses. This allows the learning and support network to grow beyond the confines of the local area. Such virtual communities enable interaction that benefits all parties.

Enhancing Economic Prospects Through Technology Skills

ACEnet also uses technology as a vehicle for improving the economic prospects of individuals in the community. This is accomplished by helping community members gain information technology skills, thereby improving their income and employment possibilities. These initiatives have the added benefit of improving the infrastructure of the community since the newly-skilled individuals serve as a valuable source of human resources for community organizations.

Three ACEnet programs are designed to enhance economic prospects of community members. Each of these is described below.

- **Sectoral Training and Empowerment Project (STEP):** The Women's Sectoral Training and Empowerment Project is aimed at allowing women in the region to successfully transition from welfare to work. STEP provides low-income community members with the training necessary to gain the skills that will enable them to secure employment in local businesses. Although STEP involves training in non-IT skills, a major focus of the program is on enabling the participants to successfully navigate the Internet and the Web. For example, the training includes seminars on using the Web to find product ideas, research market trends, and market products.
- **Computer skills seminars and classes:** ACEnet offers computer skills training through seminars and classes. Most of these are held at the CTC. An example is a workshop series that consists of 11 classes. The classes cover a wide range of IT topics, starting with basic Windows 95 operation and continuing through more advanced topics such as financial management and accounting software, database management, and Web searching, marketing and commerce. These classes and seminars are offered at below market prices. For example, the workshop series is only $150 for all 11 classes, substantially below the market rate.

  The seminars and classes are open to ACEnet members and the community at large. Both business owners and individuals can enhance their earning potential through the training. By increasing their computer-related skills, business owners may be able to better employ IT in their business operations and, as a result, increase efficiency and revenues. Individuals may find that gaining computer-related skills significantly enhances their employment opportunities.
- **Computer Opportunities Program (COP):** The Computer Opportunities Program was designed by ACEnet and a group of local businesses, teachers, and students to better meet the need for additional technical training and assistance for community small businesses. Basically, COP trains high school junior and senior students to act as computer consultants to local small businesses. The COP class meets for 80 minutes each school day for an entire academic year. The computer-related learning occurs in the context of entrepreneurship. The students learn how to actually start and operate their own consulting business.

  The COP has had a positive impact on students. Twenty-six students participated during the 1998-1999 school year. Of these, eight now hold relatively high-paying computer-related jobs, two actually own and operate their own computer consulting businesses, and 14 are now in college with five studying computer science.

  In addition to the intended result of training students to be able to meet the demand for small-business IT consulting, there is an added benefit of increasing IT awareness at the high school. COP participants have begun training their teachers in the use of computers, which has led to some classes making much greater classroom use of computers.

  All COP sites require an onsite CTC. In addition, each site requires additional personnel. Each COP site requires two teacher trainers and a technical support person, all of whom are full-time ACEnet employees. In addition, there is a program manager who oversees all COP sites. Of course, as the COP program continues to expand, it may become necessary to add additional managerial positions. In addition to the full-time paid positions, there is currently a full-time VISTA employee. ACEnet provides no funding for this position.

Facilitating E-Commerce

ACEnet also uses information technology to facilitate the use of electronic commerce by client firms, which is the focus of this case study. The next section discusses how ACEnet helps small businesses take advantage of e-commerce.

CASE DESCRIPTION

The explosive growth of the Internet-based World Wide Web (Web) has led to a corresponding growth in the use of the Web for conducting commerce. This form of e-commerce is particularly attractive to small businesses. Organizations of all sizes can benefit from e-commerce, but it seems that small businesses are particularly drawn to doing business over the Web. Small business owners and managers perceive that Web-based e-commerce holds many potential benefits. In a study of Australian small businesses, Poon and Swatman (1999) identified a number of these benefits, which they classified along two dimensions — the directness of the benefits and the time frame for gaining the benefits. Figure 2 shows the resulting grid.

Although comprehensive studies of small business use of e-commerce are just starting to appear, observation indicates that there are at least five different levels of business-related uses for the Web, as illustrated in Figure 3.

Figure 2. Benefits of electronic commerce

	Short-term	**Long-term**
Direct benefits	• save on communication costs • generate short-term revenues	• secure returning customers • long-term business partnerships
Indirect benefits	• potential business opportunities • advertising and marketing	• ongoing business transformation • new business initiatives

Figure 3. Small business uses of e-commerce

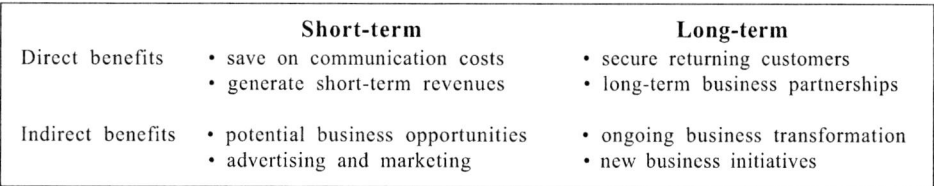

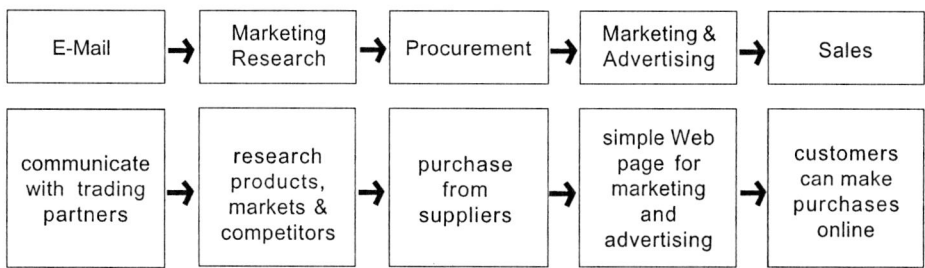

The initial use of the Internet for many small businesses is to use electronic mail to communicate with customers and suppliers. Poon and Swatman (1999) report that most of the firms in their study used e-mail to communicate with customers, although the majority of customer-oriented communications still occurred over traditional media. At the time of their study, however, e-mail was not widely used to communicate with suppliers. Once small business owners and managers become familiar with using e-mail, they may begin to search the Web in order to conduct market research. This research may include investigating new products or markets and gathering information about competitors. While performing the market research over the Web, some small business managers may discover opportunities to improve their procurement activities through interacting electronically with their suppliers. For example, it may be that online procurement offers more convenience, better pricing, and wider choice than more traditional means. In addition, e-commerce may also represent an improved method of obtaining service and support from suppliers. For example, many suppliers' Web sites offer such services as shipment tracking, return authorizations, and technical assistance documents.

The only technology requirements for the first three phases are those that are required to access the Web. The last two phases, however, require the creation of an online presence. The fourth phase consists of the business creating a Web site to advertise its products or services. The sophistication of these sites varies widely, but during this phase, online transactions are not enabled, and the sites are more akin to online advertisements or catalogs. In the final phase, online transactions are enabled, allowing customers to place orders directly via the small business' Web site.

While not all small businesses experience the same progression through these levels, in general it seems that they pass through these phases of e-commerce use in order. However, many small businesses have yet to progress to the final phase of online selling. One reason for many small businesses not taking full advantage of the potential of e-commerce is that these businesses face a number of barriers to the use of e-commerce.

Barriers to E-Commerce

Small businesses face a number of barriers when seeking to effectively employ any information technology — including e-commerce. While the literature points to a number of these, most relate to what Welsh and White (1981) call "resource poverty." This lack of resources refers not only to financial resources but also extends to a lack of expertise in both technical specialties and management capabilities.

Small businesses have a tendency to employ generalists, rather than those with specialized knowledge. This tendency leads to a lack of expertise in specialized areas such as information technology (Thong, 1999). Few studies to date have examined the impact of this lack of IT expertise on small business e-commerce use. However, there is considerable empirical evidence of the impact of lack of expertise on IT adoption and implementation and the level of satisfaction with IT (Cragg & Zinatelli, 1995; Iacovou et al., 1995; Lai, 1994; Thong, 1999).

In addition to the lack of IT expertise, there is sometimes a lack of management expertise (Lin et al., 1993) and awareness of IT in small businesses. Since top management support has consistently been shown to be an important determinant of the success of IT-related projects, the lack of management knowledge and awareness of the capabilities of IT may inhibit the use of e-commerce.

The management of small businesses often suffers from a degree of time poverty (Cragg & Zinatelli, 1995). This is particularly true in microenterprises, in which a manager must serve a number of roles, often trying to manage the organization while at the same time performing a number of other duties. For example, one of the authors is a former small business owner. It was not uncommon for this individual to perform the duties of a salesperson, a technician, an accountant, and a manager in the same day. This time poverty can lead to a short-term management perspective, which can hinder a small business's use of IT (Thong, 1999).

One solution to the lack of expertise is to hire external experts. Unfortunately, most small businesses also suffer from a lack of slack financial resources (Thong, 1999), which may preclude this. The lack of financial resources also prevents many small businesses from purchasing "turn-key" e-commerce solutions, which tend to be expensive (Poon & Swatman, 1999). Scarce financial resources lead to an inability to experiment with e-commerce technologies in order to assess their potential benefits to the business. While large organizations can absorb the costs of such experimentation even when they don't pay off in the short run, small businesses may be unable to survive such losses (Thong, 1999).

The lack of technical skills related to e-commerce may be particularly problematic. Many small businesses may not have any employees with general IT expertise. Those that do may not have employees with e-commerce-specific skills, such as Web page design. Designing, implementing and maintaining an e-commerce Web site requires specialized skills, not just in technology but also in other areas such as graphic design. Few small businesses will be fortunate enough to have employees with the requisite skills.

Some of the lack of expertise can be overcome through networking among small business managers. However, it is still unusual for networking to occur, despite its importance, particularly with respect to e-commerce (Fariselli et al., 1999). Networking can also benefit small businesses seeking to use e-commerce in a number of ways. Perhaps the most important is in increasing the awareness of the capabilities of e-commerce. Research indicates that small business managers often have limited access to external information (Lin et al., 1993). Networking with other organizations may help improve such access.

Another area in which small businesses may lack knowledge is how to market their products over the Internet. In particular, they may not know how to promote their online presence. While larger firms may approach this problem by mounting advertising campaigns, small businesses' financial resource poverty may prevent them from follow-ing this avenue. Many small businesses also need assistance in learning how to provide customer service when using e-commerce. This is a key area of concern, as some consider the quality of customer service to be one of the most important determinants of maintaining customer loyalty.

Small businesses in rural areas, such as southeast Ohio, face additional barriers to their use of e-commerce. One is the lack of a high-quality communications infrastructure. Premkumar and Roberts (1999) put it well: "… rural businesses are caught in a viscious cycle — lack of communication infrastructure reduces the demand for communication services, which further constrains future investment in the infrastructure." Without a good communications infrastructure, connection to the Internet may be difficult, making e-commerce less attractive and perhaps limiting its use in rural areas. This lack of

Table 2. Barriers to small business e-commerce

Barrier	Description
Technical expertise	Lack of employees with IT knowledge. Problem may be particularly acute with respect to e-commerce knowledge.
Management knowledge	Management often suffers from a lack of knowledge in a number of areas, including how IT applications such as e-commerce can impact the organization.
Financial resources	Lack of slack financial resources may prevent experimentation with e-commerce and may also increase the impact of any failed attempts. This also limits the practicality of using outside experts.
Access to external information	Reduced access to external information limits awareness of benefits e-commerce and how to implement e-commerce.
Marketing knowledge	May lack expertise in the area of marketing over the Internet. This includes the ability and resources required to promote an online presence.
Internet connectivity	Poor-quality Internet connectivity may limit attractiveness of e-commerce. This a problem in rural areas.
Technical infrastructure	Limited access to specialized equipment such as scanners and color printers may be more acute in rural areas.

infrastructure even extends to the choices available in Internet service providers (ISPs). Although there are many free or low-cost ISP's available, these may not be accessible with a local phone call. Long-distance or network access charges may quickly make Internet access prohibitively expensive.

Rural small businesses also must deal with their relative isolation from sources of e-commerce expertise. Unlike urban areas, there may be few sources from which small business managers are able to obtain e-commerce expertise. For example, there may be far fewer e-commerce and IT consultants and training opportunities.

Finally, small businesses in rural areas may find gaining access to specialized equipment difficult. In urban areas there may be a number of facilities that provide for-fee access to high-resolution scanners and color printers. Rural areas may lack these facilities, which forces small businesses to choose among time-consuming travel to an urban area, resource-draining acquisition of expensive equipment, or doing without.

As discussed in this section, small business managers seeking to take advantage of the potential of e-commerce face a number of barriers. Table 2 summarizes these. The next section describes how ACEnet attempts to overcome these barriers through a variety of programs.

Overcoming the Barriers

A number of ACEnet programs are directed at helping small business owners and managers overcome the barriers to e-commerce discussed in the previous section. Some programs were specifically created with e-commerce in mind, while others were originally targeted at other development issues but have evolved to play important roles in e-commerce development.

Technical Expertise

The primary vehicle for addressing the lack of technical expertise in the ACEnet service area is the Computer Opportunities Program (COP). The COP is designed to train high school students in IT including e-commerce-related areas. The hope is that this program will have a two-fold benefit. Not only will students who have completed the program have brighter job prospects because of their IT skills, but they also may serve as skilled human resources for small businesses. In fact, a number of the program's participants have actually begun working with small businesses prior to their graduation from high school while others have gained permanent employment with area small businesses.

The technical expertise barrier is also addressed through seminars. ACEnet offers a wide variety of seminars to small business owners and managers. A number of these address areas related to e-commerce such as Web page design and Web search strategies.

Management Knowledge

ACEnet addresses the management knowledge barrier through its extensive program of management-oriented seminars. As discussed earlier, these seminars cover a wide range of IT and e-commerce-related topics. By completing these seminars, small business managers not only gain specific technology-related skills but also gain a better grasp of the capabilities of IT including those offered by e-commerce. This knowledge can help the managers better identify, evaluate, and implement e-commerce projects.

ACEnet also attempts to build networks consisting of small businesses and resource providers. To a degree, this may also address the management knowledge barrier. Through networking, small business managers may find others from whom they can gain considerable knowledge. This can range from informal conversations to more extensive mentoring arrangements.

Financial Resources

Currently, there are few formal ACEnet programs to address the financial resources barrier. However, there are initiatives underway that may help. ACEnet Ventures, Inc. is a newly formed non-profit ACEnet subsidiary that will serve as a community development venture capital fund. ACEnet Ventures will provide risk capital to regional start-up and expanding businesses. ACEnet Ventures has the potential to have a significant impact.

Access to External Information

The formation of networks among small businesses and resource providers is the primary vehicle by which ACEnet attempts to overcome the access to external information barrier. Not only does this networking help small business owners learn from one another, but it also puts them in touch with resources that may provide information to

Figure 4. Public Webmarket home page

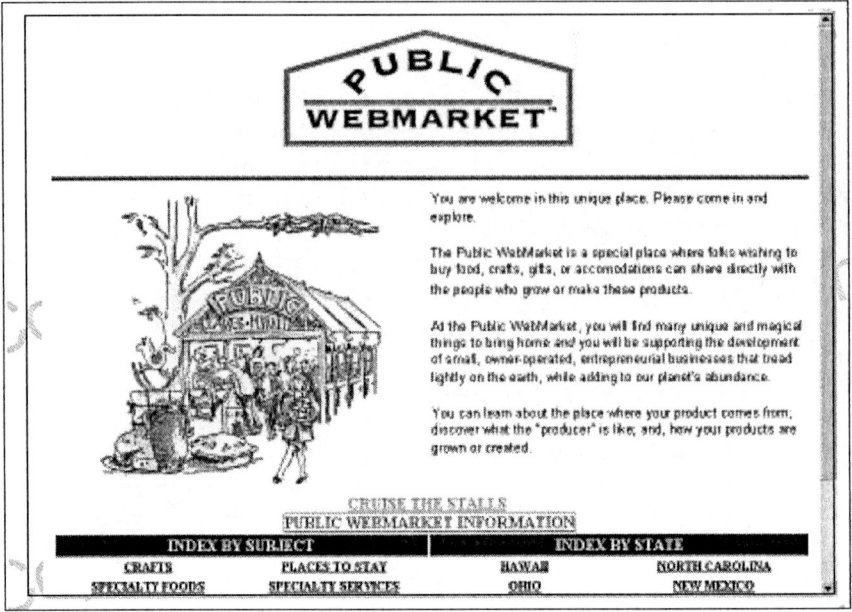

help the small businesses with their e-commerce efforts. For example, one small business owner was recently put in touch with an e-commerce expert who helped the owner identify design problems in her firm's Web site. This led to further, informal conversations about how the owner could use e-commerce to address other aspects of her business.

Marketing Support

The Public Webmarket is a cooperative effort between ACEnet and other public service agencies — primarily the Center for Civic Networking — that is designed to support small businesses' e-commerce marketing efforts. The Public Webmarket is one of the only ACEnet programs that is directed solely at developing small business e-commerce. It is a virtual small business incubator designed to assist small businesses in their e-commerce efforts.

The Public Webmarket provides temporary, low cost Web hosting that enables small businesses to determine whether marketing their products or services through a Web site is feasible. The program also includes assistance in designing Web pages that properly market the businesses' products or services. Another area addressed by the program is how to provide quality customer service over the Web, which is a key marketing consideration. In addition, the Public Webmarket is promoted by the organizing agencies, thus freeing the business from some of this responsibility. Figure 4 shows a screen shot of the Public Webmarket's home page.

Figure 5. Overcoming the barriers

Internet Connectivity

As discussed in the last section, small businesses in the region may have difficulty in gaining low-cost, reliable access to the Internet. ACEnet, in cooperation with other non-profit organizations, sponsors the Southeast Ohio Regional Freenet (SEORF). SEORF is particularly important during the early stages of e-commerce (see Figure 2), when small business managers are gaining knowledge of e-commerce through their use of e-mail and Web searching. All citizens of the region, including potential and current small business owners, can gain access to the Internet through SEORF. While SEORF does not offer commercial Web hosting, small businesses can link to their commercial pages through SEORF. This helps the small businesses in marketing their products and services. SEORF has also helped raise the awareness of the potential of the Internet and Web in the region, which has in turn helped build demand for commercial ISPs. This has led to an increased demand for high-quality Internet connectivity, leading to the availability of digital subscriber line (DSL) connections in the area.

Technical Infrastructure

The primary means for improving the technical infrastructure for area small businesses is through the CTC, which was described in detail in a previous section. The CTC provides low-cost access to specialized equipment such as scanners and high-quality printers. In addition, members can use the CTC to access the Web on a limited basis. Recently, ACEnet has experimented with providing videoconferencing services through the CTC. This will further increase the technical infrastructure available to support small business e-commerce.

ACEnet is currently addressing most of the barriers to e-commerce faced by rural small businesses. Figure 5 illustrates the barriers to e-commerce success and the ACEnet programs that help overcome them.

CURRENT CHALLENGES/PROBLEMS
FACING THE ORGANIZATION

Several challenges still face ACEnet in their efforts to facilitate e-commerce by rural small businesses. One area of continuing concern is the relatively weak IT and communications infrastructure in the area. While SEORF and the CTC have helped improve the situation, more still needs to be done. The key question is how to develop affordable platforms that facilitate communication, collaboration, and mutual learning among businesses and resource providers. As discussed earlier, such activities can help small businesses in their quest for e-commerce success. Free-Nets, such as SEORF, while serving their intended purpose, face increasing pressure to maintain long-term sustainability. These mechanisms may not be able to meet the increasing demands of e-commerce. In order to meet increasing demands, ACEnet now needs to explore new designs that create even more sophisticated partnerships between public and private entities.

Another challenge facing ACEnet is the problem of helping growing firms gain access to capital needed for expansion. While the Public Webmarket has met with considerable success, it is not intended to be a permanent home for a small business's e-commerce presence. Once firms have matured, they may want to expand their e-commerce efforts, taking advantage of emerging technologies. For example, Public Webmarket does not presently have shopping cart technology, nor does it employ cookies. When businesses grow beyond the capabilities of Public Webmarket, they must strike out on their own, which may require considerable investment. ACEnet is currently trying to develop funding sources, such as ACEnet Ventures, that can assist area small businesses in meeting these and other funding requirements.

Another related challenge is to devise ways to help growing firms control their e-commerce operations. As the businesses learn to better employ e-commerce, they may meet with more success than they can handle in their present situation. For example, they may have trouble producing sufficient product to meet demand, or their order fulfillment processes may not be up to the task of filling a large number of orders. Proper use of IT, including e-commerce, may be able to help client businesses overcome these obstacles. However, ACEnet must be able to expand its training and consulting offerings in order to provide assistance in these, and other, areas.

ACEnet is also investigating ways in which to expand on the success of the COP. Although the program has exceeded expectations, more schools and students need to be involved, requiring additional funding and personnel. Program expansion is necessary because, as more small businesses recognize the potential of e-commerce, more skilled human resources are necessary. ACEnet would like to expand the program to several new schools in the next two to five years. They face at least three challenges associated with this expansion. First, agreements need to be secured with additional schools. Second, funding and personnel must be found. Finally, more sophisticated program management processes must be put into place in order to effectively oversee the expanded program.

Perhaps ACEnet's greatest current challenge is rationalizing some of the informal mentoring and communication structures in order to allow greater reach. The informal structures have worked well so far. However, as ACEnet expands, rationalization is required. It is questionable whether the informal structures will allow ACEnet to serve an expanded client base. This problem is not uncommon to expanding organizations in general. Many informal structures and processes that work well in the early history of an organization may fail when the organization expands. Rationalizing these is typically difficult; therefore, ACEnet faces a significant challenge.

CONCLUSION

Small business can serve as a catalyst for economic development in depressed areas. One example of this is southeast Ohio, an area that suffers from a number of economic problems. Recognizing this, the Appalachian Center for Economic Networks (ACEnet) has fostered the growth of microenterprises since 1984. More recently, ACEnet has recognized the potential of Web-based e-commerce to help these small businesses effectively research, market, and deliver their products. As a result, ACEnet has a number of programs that either directly or indirectly support small business e-commerce. So far, these programs have met with success, but several significant challenges face ACEnet.

What can ACEnet do to meet these challenges? Although it is only speculation at this point, there are several steps that ACEnet leaders may wish to investigate. Successful organizations learn from history. While it is critical that ACEnet continues to develop innovative programs, it is just as important to take the time to reflect on and evaluate its current programs. Making such analyses a part of every program may not only identify problems and improvement opportunities in current programs but may also help avoid similar difficulties in future programs. In addition, identifying particularly successful aspects of current programs can help ensure that these elements are included in future programs.

ACEnet should also be sure to work very closely with existing clients to anticipate problems. Close communication with clients is crucial to identifying potential problem areas. By having open, active lines of communication between ACEnet personnel and clients, problems can be anticipated and addressed before they have any detrimental impact. In addition, client input may prove to be invaluable in identifying areas to be addressed by new programs. One way that ACEnet can keep communication flowing is to form a "client council." Many successful organizations have similar structures. For example, Sun Microsystems has a Dealer Council, which is comprised of representatives of Sun resellers. This organization meets periodically to provide input to Sun executives. ACEnet may wish to take advantage of technology to help maintain contact with clients. Online discussions, teleconferencing, and other communication technologies can provide cost effective means for communicating with clients.

ACEnet should also continue to work on innovative partnerships with public and private agencies. ACEnet can not be all things to all clients. By partnering with Small Business Development Centers, angel investors, local Internet service providers, and others, ACEnet can effectively expand the range of services offered to entrepreneurs. For example, ACEnet may be able to match fast-growing clients with angel investors and other funding sources in order to help clients fund expansion.

Finally, ACEnet must focus on innovative ways to assist developing businesses to take advantage of e-commerce. Currently, ACEnet focuses on helping clients use e-commerce technologies for market expansion. E-commerce also holds promise for lowering operating costs. For example, many organizations large and small are seeing significant savings from using e-commerce for procurement. In addition, several application service providers devoted to serving small businesses have recently emerged. ACEnet may wish to help clients explore the potential of e-commerce for improving operating efficiency.

ACEnet must meet the challenges it faces if its success is to continue as its client base expands. If these challenges can be overcome, then ACEnet may reach its goal of facilitating sustained economic growth through small business e-commerce.

ACKNOWLEDGMENT

The authors would like to express their gratitude to Amy Borgstrom for her invaluable assistance in the development of this article.

REFERENCES

Blili, S., & Raymond, L. (1993). Information technology: Threats and opportunities for small and medium-sized enterprises. *International Journal of Information Management, 13*(6), 439-448.

Cragg, P., & Zinatelli, N. (1995). The evolution of information systems in small firms. *Information & Management, 29*(1), 1-8.

DeLone, W. (1988). Determinants of success for computer usage in small business. *MIS Quarterly, 12*(1), 51-61.

Fariselli, P., Oughton, C., Picory, C., & Sugden, R. (1999). Electronic commerce and the future for SMEs in a global market-place: Networking and public policies. *Small Business Economics, 12*(3), 261-275.

Iacovou, C., Benbasat, I., & Dexter, A. (1995). Electronic data interchange in small organizations: Adoption and impact of technology. *MIS Quarterly, 19*(4), 465-485.

Johnson, M. (1998a). Developing a typology of nonprofit microenterprise programs in the United States. *Journal of Developmental Entrepreneurship, 3*(2), 165-184.

Johnson, M. (1998b). An overview of basic issues facing microenterprise practice in the United States. *Journal of Developmental Entrepreneurship, 3*(1), 5-21.

Lai, V. (1994). A survey of rural small business computer use: Success factors and decision support. *Information & Management, 26*(6), 297-304.

Lees, J. (1987). Successful development of small business information systems. *Journal of Systems Management, 38*(9), 32-39.

Lin, B., Vassar, J., & Clard, L. (1993). Information technology for small businesses. *Journal of Applied Business Research, 21*(7), 20-27.

Longnecker, J., Moore, W., & Petty, J. (1994). *Small business management: An entrepreneurship emphasis.* Cincinnati: Southwestern Publishing.

Mintzberg, H. (1979). *The structuring of organizations.* Englewood Cliffs, NJ: Prentice-Hall.

Petzinger, T. (1997, October 24). The front lines: June Holley brings a touch of Italy to Appalachian effort. *The Wall Street Journal,* p. B1.

Poon, S., & Swatman, P. (1999). An exploratory study of small business Internet commerce issues. *Information & Management, 35*(1), 9-18.

Premkumar, G, & Roberts, M. (1999). Adoption of new information technologies in rural small businesses. *Omega: The International Journal of Management Science, 27*(4), 457-484.

Thong, J. Y. L. (1999). An integrated model of information systems adoption in small businesses. *Journal of Management Information Systems, 15*(4), 187-214.

United States: The great American jobs machine. (2000, January 15). *The Economist.*

Welsh, J., & White, J. (1981). A small business is not a little big business. *Harvard Business Review, 59*(4), 18-32.

APPENDIX. ACEnet HISTORY

Milestone	Year	Event
1	1984	Group of community members begins research into effective models for community economic development. Discover system of worker-owned businesses and support institutions in Spain. Decide to adapt model to Southeast Ohio.
2	1985	Worker Owned Network is incorporated and rents first office space. Receives first grant. First projects begin, including a home health care business, a worker-owned restaurant, house cleaning business, and a bakery.
3	1989	Tenth worker-owned business started. Total employment of client businesses reaches 100. Need for new approaches becomes evident, which leads to focus on flexible manufacturing networks.
4	1991	Worker Owned Networks changes name to Appalachian Center for Economic Networks (ACEnet). Moves into larger facility, which includes the business incubator. Begins work with first flexible manufacturing network.
5	1991	Receives major grant from Appalachian Regional Commission. Begins focus on using emerging technologies such as the Internet to network businesses with markets, resources, and each other.
6	1992	Begins focus on specialty food sector. Opens discussions with community collaborators on opening a shared-use kitchen.
7	1994	Provides leadership in the creation of the Southeast Ohio Regional Freenet (SEORF).
8	1995	Participates in the Center for Civic Networking's Public Webmarket project.
9	1996	Grand opening of ACEnet kitchen incubator.
10	1997	Begins research development of product development fund.
11	1997	Starts Computer Opportunity Program (COP) at local school.
12	1999	Continues expansion of COP. Receives support for COP from U.S. Department of Education.
13	1999	Creates ACEnet Ventures, Inc. Shifts focus to working with firms with the capacity to expand very rapidly.

Craig Van Slyke has a PhD in information systems from the University of South Florida, and is currently on the faculty of the University of Central Florida. Dr. Van Slyke has written or co-written a number of papers in journals such as Annals of Cases on Information Technology, Journal of Information Systems Education, Industrial Management and Data Systems, *and* Information Technology, Learning and Performance Journal. *He has also made many presentations at professional conferences, including the Hawaii International Conference on Systems Sciences, and the Information Resource Management Association International Conference. His current research interests are in electronic commerce, information technology skills, and education.*

France Belanger is an assistant professor of information systems and a member of the Center for Global Electronic Commerce at Virginia Polytechnic Institute and State University. She holds a PhD in information systems from the University of South Florida. Her research interests focus on the use of telecommunication technologies in organizations. She has presented her work at numerous national and international conferences, and has published in Information and Management, IEEE Transactions on Professional Communications, The Information Society: An International Journal, Information Technology, Learning and Performance Journal, *and the* Journal of Information Systems Education. *She is co-author of the book* Evaluation and Implementation of Distance Learning: Technologies, Tools and Techniques.

Marcy Kittner is the associate dean for the John H. Sykes College of Business at The University of Tampa and is an associate professor in the Information and Technology Management Department. She holds a PhD from the University of South Florida and has published in the Journal of Information Systems Education, Journal of Information Technology Management, *and* Annals of Cases on Information Technology. *Dr. Kittner has co-authored several texts in information technology management and has made numerous conference presentations at national and international conferences*

This case was previously published in the *Annals of Cases on Information Technology Applications and Management in Organizations*, Volume 3/2001, pp. 1-20, © 2001.

Chapter XXIV

Selecting and Implementing an ERP System at Alimentos Peru[1]

J. Martin Santana, ESAN, Peru

Jaime Serida-Nishimura, ESAN, Peru

Eddie Morris-Abarca, ESAN, Peru

Ricardo Diaz-Baron, ESAN, Peru

EXECUTIVE SUMMARY

The case describes the implementation process of an ERP (enterprise resource planning) system at Alimentos Peru, one of the largest foods manufacturing companies in Peru. It discusses the organization's major concerns during the mid-1990s, including increasing competition, inefficiency of business processes, and lack of timely and accurate information. To address these concerns Alimentos Peru launched several projects, one of which involved the implementation of an ERP system. The case explains the criteria used to evaluate and select the system, as well as the main issues and problems that arose during the implementation process. More specifically, the case focuses upon a set of implementation factors, such as top management support, user participation, and project management. Finally, the case concludes with a discussion of the benefits obtained from the introduction of the system as well as the new organizational challenges.

BACKGROUND

Alimentos Peru manufactures and sells food products for direct or indirect human consumption including cookies, nonalcoholic beverages, bakery products, and sweets, yeast and other ingredients for bread making. It is a subsidiary of *International Food Group (IFG)*, one of the world's largest food products manufacturers and sellers. In Peru, its leading brands are *Turtora, Real, Tako* and *Remo.*

IFG has been present in the Peruvian market since 1939 with the opening of its subsidiary Real Peruana Inc. In 1993, as part of a number of mergers and acquisitions of food producers in Latin America, *IFG* bought the Estrella S.A. cookie maker, the then leader in the Peruvian market. The merger of the Peruvian subsidiaries started operating as *Alimentos Peru.*

Alimentos Peru has two production plants. The first one is located in Lima and concentrates on cookie and candy manufacturing. The other is located in Callao and is devoted to producing inputs for bread making and powder drinks.

Alimentos Peru has faced a long fall in demand as well as intense local and foreign competition. Its executives were aware that their success hinged on introducing a comprehensive strategy that would comprise satisfying the consumers' expectations and needs, as well as reducing operating costs. By the mid-1990s, the company introduced new manufacturing techniques and launched a number of projects to formalize, restructure and standardize its processes.

Its leading production line is cookie making, the source of the company's largest (45%) share of profits. Until 1994, the cookie market was rather dull and led by local brands. However, in 1995, substantial changes started to occur. The acquisition of *Molinos* by the *Atlantic* consortium and the arrival of a new competitor, Chilex — a Chilean company — introduced a new dynamic to the market. Furthermore, imported cookies started to arrive from abroad including those distributed by *Alimentos Peru, Orval, Rose* and *Crasp.* Imported cookies increased their share of the local market from 2% to 10%.

To face the new competitive environment, *Alimentos Peru* changed the packaging in most of its *Estrella* products to make them more attractive and improve their preservation. It introduced new products, including *Chocosonrisa* and *Marquinos*, as well as a new line of imported cookies that are leaders in the international market. Blanca Quino, head of product lines at *Alimentos Peru*, told a local publication: "*We have introduced innovations in our line products at least once a year. Now the winners will be those who can introduce more innovations in a market where the consumer makes the final decision.*"

In 1996, *Alimentos Peru's* share in the cookie market exceeded 30% and its *Estrella* brand name remained as the local leader with 23% market share. However, the consortium *Atlantic* reached the same market share after buying *Molinos*, manufacturer of *Gloria, Zas* and *Ducal*. Moreover, although aimed at a different market segment, *Empresa Galletera*, through its *Grano* and *Pepis* brand names, covered over 21% of the market. This firm's strategy was to sell cookies at a lower price than its competitors.

The beverages and desserts market, a line that creates 29% of *Alimentos Peru's* revenues, also suffered changes due to international competition. The company upgraded the packaging of its *Tako* line of drink powders to meet consumer preferences and introduced a larger variety of flavors. *Remo* beverages went through a number of innovations including a new range of flavors and a ready-to-drink line of products launched at the end of 1995.

In 1996, *ASPA Alimentos* was the market leader for powder beverages with a 53% share at company level, followed by *Alimentos Peru* with 37.1%. *Tako* — *Alimentos Peru's* brand name — was the leader in the sugarless beverages market segment with 19% of the market. It further benefited from the growth in the sugarless market segment. Growing consumer preference for sugarless products sold at a lower price strengthened *Tako's* position against the semisweet *Kino* and *Bingo* products from *ASPA Alimentos*. In the sweet products segments, *Alimentos Peru's Remo* held strong to its 62% market share but faced strong competition from other brands, including locally produced *Dinang* and *Fructal*, a Chilean import.

In its other product lines — bread-making ingredients, candies and chocolates — *Alimentos Peru* rose to the challenges in a similar manner by introducing new products and changing its packaging, and by improving on its distribution system.

Appendix A shows the financial statements of the company for the period between 1994 and 1998.

SETTING THE STAGE

Information Systems

After the merger of the Peruvian subsidiaries, the information technologies divisions at *Alimentos Peru's* Lima and Callao factories merged under one single manager. Carlos Montero became the systems manager in 1994 and was the second person in charge in this area after the merger of *IFG's* Peruvian subsidiaries. Montero's main challenge was to put in place a new information system for the company to replace old systems that were typically fragmented, duplicate and inconsistent. At that time, the new IT division employed 19 persons, mostly programmers, and reported to the local financial manager: "*This was a typical data processing division in the 1960s style,*" says Montero.

The company had two AS/400 IBM servers — one at each facility — that operated at about 80% capacity. Each was connected to about 50 terminals and PCs of varying age and brand names, all operating independently. Each factory ran its own IT systems to suit its peculiar needs. There were more than 20 independent systems, including two parallel systems to process purchases, another two to keep warehouse records; another two for manufacturing; another two for cost and finished product control, and two more for marketing. Furthermore, there were five systems for payrolls: two for laborers, two for clerical workers and one for staff. Lotus 123 and WordPerfect were the standard office software products. The main task at the IT division was maintenance.

Likewise, each plant imposed its own criteria when giving code numbers to their ingredients and finished products. Furthermore, some areas within the same plant, for instance, warehousing or manufacturing, would use different code for the same finished product. In Lima, costing was based on the number of work hours while in Callao it depended on product weight.

Little integration of operations also had an impact on the financial system. Closing of accounts at the end of each month would take more than one week, despite long working hours (overtime) put in by employees. Cost controls were hard to implement as was determining which products were actually yielding a profit. Management reports were put together manually and then sent to *IFG's* offices in the U.S. The first integrated accounting system was introduced in January 1995 thanks to an in-house development. By that time, the IT division had shrunk to seven people after outsourcing programming tasks.

Although both factories were already under one single company, both plants continued to operate separately and their operations showed the same lack of integration that was apparent in their information systems. Not one business process operated under a standardized model, not even account closing at month's end, inventory control or purchasing, or in general any of hundreds of activities in the production process. About this issue, Carlos Montero holds:

Personnel were not used to filling in forms, recording data or examining the manufacturing formulas (or recipes). Inputs requisitions for manufacturing were sent casually: Approximate amounts of ingredients were sent to production and leftovers were returned to warehouse. This led to a large amount of shrinkage and prevented keeping good records on production costs.

Alimentos Peru Strategic Initiatives

By the mid-1990s, the company designed its corporate strategic plan. The general manager, functional managers and the company's main executives met to determine the company's mission and vision, analyze their competitive environment and determine the main strategic actions to take. As a consequence, a number of projects were proposed and assigned to various company executives. The following main strategic actions were introduced:

- To conclude with the corporate merger, led by Tanya Santisteban, the administration and finance manager.
- To develop new products and reorganize the sales force and distribution channels. This task was assigned to Armando Linares, the marketing manager.
- To introduce Total Quality at *Alimentos Peru*, in charge of Mario Neyra, the human resources head.
- To train personnel in MRP II techniques, commissioned to Jorge Figueroa, the logistics manager.
- To improve and integrate information systems, a responsibility assigned to Carlos Montero, the systems manager.

After merging the financial and administration, the systems, and the logistics divisions in 1994, in the years that followed the company continued to merge its other areas. By year-end 1995, plant management in Lima and Callao was placed under a single manufacturing manager office. A few months later, the marketing managers' offices came under a single marketing division.

A Quality Committee was set up headed by the human resources head and including various task forces for each area: Logistics, manufacturing plants, internal and final user physical distribution. The purpose of this set up was to identify and propose ways to improve processes through Quality Circles that remained in operation until the end of 1997.

In 1995, the General Manager hired Oliver Wight LLC, an international consulting firm that had created the MRP II techniques and a training specialist to prepare and put in practice a personnel-training program. Training was mainly directed at manufacturing, logistics and marketing personnel, initially through talks and, in 1996, with video screenings.

Also in 1996, to improve product distribution and response to customer demands, *Alimentos Peru* restructured its sales force. From a geographically based system, it moved to a client-type system. The new system was put in place in coordination with local supermarket chains and allowed to cut operating costs and to increase compliance with purchase orders from these channels.

Jorge Figueroa makes the following comments about this stage in the company's evolution:

IFG put strong outside pressure on the general manager. Alimentos Peru's personnel saw its workload increase substantially when a series of projects were introduced simultaneously. The projects were implemented through work teams but individuals put a priority on the team headed by their immediate boss.

In 1997, *IFG* decided to centralize production, supply and distribution operations at its subsidiaries around the world on a regional basis with a view at establishing "business regions" that would profit from relative advantages in each country. *Alimentos Peru* and the Ecuador, Colombia and Venezuela subsidiaries came under a single production and marketing unit. Venezuela was chosen to become the central management seat for the Andean area.

As a result of the above, the company adopted a new organizational set up. Marketing split into marketing and sales and all other managers' offices, including IT, started to report to the corresponding corporate manager in Venezuela.

CASE DESCRIPTION

A Project to Introduce an ERP System

The aforementioned challenges and problems as well as management's need to get timely and reliable information prompted the project to improve the company's information systems. In this respect, Montero holds:

At meetings between the General Manager and line managers, frequent comments were "We don't have timely information" or "Information is very expensive."

With support from local consulting firm MISPlan, at the beginning of 1995 Carlos Montero[2] prepared an evaluation of the company's information systems. Based on this evaluation, Montero formulated the following recommendations to the General Manager (see Appendix B):

- To introduce client server systems.
- To assess the capability and quality of the central servers.
- To standardize office software.
- To evaluate the IT manager's office structure.
- To introduce an enterprise resource planning (ERP) system for integrated corporate information management.

The question whether *Alimentos Peru* should get its software off-the-shelf or write it in-house was quickly answered. Necessary software functional requirements, the capacity to integrate with other *IFG* subsidiaries and the time for introduction warranted getting an off-the-shelf software product. Marcela Burga, IT Development Manager, holds:

We evaluated the option to develop our own software and estimated a two-year period for implementation, slightly longer than would be needed to implement a commercial package. Moreover, an in-house software would not provide the breadth and scope of functions that could be expected from a commercial package.

On this same decision, Jorge Figueroa says:

In-house development would have required an extraordinary amount of attention from our people, both for design and implementation. On the other hand, this would be custom-made software. At that point in time we did not know if what we had actually suited our processes. Furthermore, using off-the-shelf products would make integration with other IFG subsidiaries easier.

The project to implement the new information system started with software selection. Evaluation of the ERP system started in November 1995 by putting together a task force organized as follows:

- Steering Committee made up by the area managers and headed by Jorge Figueroa.
- Manufacturing and Logistic Function Committee.
- Marketing and Financial Function Committee.
- Technical Committee made up by systems division personnel.

Choosing the ERP system and identifying the corresponding implementation strategies was coordinated with *IFG* whose systems development policy gave its subsidiaries freedom to make their own decisions. There was prior experience of systems introductions in other subsidiaries:

- Ecuador: BPCS for the logistics and manufacturing areas.
- Venezuela: BPCS for the logistics, manufacturing and distribution areas.
- Argentina: BPCS for the distribution and financial areas, and PRISM for the manufacturing area.
- Canada: PRISM for the logistics and manufacturing areas.
- Puerto Rico: J.D. Edwards for the financial and distribution areas.

ERP systems evaluation at *Alimentos Peru* went through two stages. In the first stage, four ERP systems were evaluated: BPCS, J.D. Edwards, PRISM and SAP R/3. The evaluation was based on the following criteria:

- To provide a comprehensive solution including modules that could be enforced within all business processes within the company.
- To have a track record at *IFG*.
- To have a local representative in Peru.
- To propose versions for the AS/400 platform.
- To allow work in a client-server architecture.

In this stage, implementation costs and time were almost totally disregarded. A quick decision was made because Peruvian software suppliers were not numerous and had little experience. J.D. Edwards software had no local representative and included only the financial module. SAP R/3 did not have a local representative either nor were there any experiences of using this system at *IFG*. Taking these considerations into account, BPCS and PRISM were prequalified and went on to the next selection stage. Results from the first evaluation stage appear in Table 1.

As a next step, the task force devoted itself to determining whether either BPCS or PRISM met the company's needs.

First, they evaluated the software supplier and its local representative. The Steering Committee studied the organizations, local facilities and technical support both in Peru and outside. The shareholding structure of the local representative, experience in prior implementations, customers and additional products and services offered were other factors taken into consideration. During visits with local representatives, they were asked to make presentations about their ERP systems and their organizations. Finally, references from clients with previous implementations were checked.

Alimentos Peru IT personnel visited other subsidiaries where the selected ERP software had already been installed and examined the contingencies that emerged during the implementation stage, verified the systems' functionality, transaction processing times and the volume they could support.

Local representatives of each ERP system made presentations before the Function Committees, who also reviewed the corresponding handbooks and demonstration versions. The functionality of each system module was compared with the functionality needed for the business processes by assigning a percent score to reflect the matching

Table 1. Results of the first evaluation of an ERP system

ERP	Comprehensive Solution	Previous Implementation at IFG	Local Representative	AS/400 Compatibility	Client-server Architecture
BPCS	ü	ü	ü	ü	ü
SAP R/3	ü	x	x	x	ü
PRISM	ü	ü	ü	ü	ü
J.D. Edwards	x	ü	x	ü	x

degree between the proposed software and the desired business processes. All the committees gave BPCS a higher percent score than to PRISM.

To evaluate the software's stability, IT personnel resorted to version evolution over time. They also evaluated other aspects, including the working platform, programming language, type of database and handbook language. Cost analysis included the initial investment required, implementation costs, and annual fees for support and software updating.

This second stage took three months, most of the time for evaluation. At the end of this time, the Steering Committee chose BPCS. The decision was favored by the system's previous implementations at *IFG* subsidiaries and the longer experience offered by the consultant, locally represented by XSoft, charged with the implementation.

Results were reported to *IFG* U.S. headquarters in April 1996. James Robinson, general information systems manager for *IFG* came to Peru to bring the approval for the BPCS system as the ERP system to be implemented at *Alimentos Peru*. To conclude with the selection process, the following task was to design an implementation strategy and to start negotiations with the consultant.[3]

The BPCS system would be implemented in the purchasing and warehousing, manufacturing, cost control, accounting and treasury, and marketing divisions, thus fully integrating *Alimentos Peru's* operations. Carlos Montero had estimated this process would take 18 months and involve about 80 persons, including consultants (both to redesign the processes and ERP software specialists), IT personnel and users.

The project's total estimated budget reached US$800,000 of which US$450,000 would go to software licenses and US$350,000 to hardware acquisitions, implementation, consultancy and training. Taking into consideration that other subsidiaries had experienced cost overruns due to contingencies during the implementation period, Montero felt that the time and cost estimates were too optimistic and wondered what factors would facilitate successfully implementing the new system.

The initial discrepancies between software functionality and the business processes then in place had already surfaced in the evaluation stage. In this regard, Marcela Burga, a member of the Steering Committee, holds:

The accounts payable module involved the accounting and treasury divisions. The BPCS software suppressed two functions in accounting and added one to treasury. The head of the treasury division was not willing to take up that function. Despite the fact that the process as a whole was more simple, we saw things as divisions rather than processes.

Towards the end of June 1996, right after the implementation had been launched, the most experienced consultant assigned by the firm left the project. A much less experienced replacement came in and Montero thought IT personnel involvement would become critical in understanding the new project's functionality and ultimate success.

The implementation strategy and methodology were determined in coordination with XSoft, the implementation consulting company that was also the local representative of SSA Inc., the vendor of BPCS.

Three implementation stages or subprojects were devised to be introduced sequentially. The initial subproject would comprise the logistics, purchasing and warehousing, and manufacturing and cost control modules. The second subproject included accounting, and the third one was for marketing.

Montero thought the members on the task force should exhibit a range of qualities, most importantly their capacity to manage a project, experience in information systems implementation, knowledge of business processes, and capacity to lead change.

To lead the project, there would be a Steering Committee comprising the respective area managers charged with identifying business processes and the implementation strategy. To line up the ERP system implementation and the MRP II training program, Jorge Figueroa was named project leader with Marcela Burga as General Coordinator. Burga was also systems development manager and Montero's deputy.

User personnel were chosen to make up the work teams. Each team would include a maximum of seven or eight members, with a leader chosen among them who would be further supported by a member from the IT division. A total of 75 persons would take part in the project's 10 working teams. Personnel selection and working team configuration took place following recommendations issued by each area's manager. The candidates were expected to meet the following requirements:

- To be outstanding members in their divisions.
- To show a participatory and proactive attitude.
- To be capable of using the system and possess a research-oriented attitude.
- To know the process well.
- To be innovative (although in many cases innovations would be the responsibility of IT personnel).
- To have decision-making capacity within their own divisions.
- To be open to communication and have direct contact with their immediate superiors.

Table 2 shows the implementation methodology recommended by the system's supplier.

Implementing the ERP

As mentioned previously, the implementation stage started with the logistics and manufacturing modules in June 1996.

With the ERP system implementation underway, *Alimentos Peru* started to enhance its hardware and software platforms. Both IBM AS/400 servers were upgraded, increasing their speed and storage capacity. The two factories were connected through a client-server network. One of the AS/400 servers was used as a production server and the other as development server. New personal computers were installed while some old ones were upgraded. The company installed a Windows operating system to be used as the computer network software platform. MS-Office was used as office software. Lastly, MS-Exchange provided electronic mail capabilities for both internal interconnection and connection with other *IFG* subsidiaries. All of these tasks, including user training, took about six months.

The ERP system implementation teams were configured at head and supervisor levels. Jorge Figueroa's participation as project leader allowed the logistics division to make timely decisions because lack of decision-making capacity was slowing down the

Table 2. Implementation methodology for an ERP system

Activity	Participants	Description
1. Documenting original processes	• Implementation teams • Organization and methods specialist (one person) • Implementation consultant • IT division	• Define business processes. • Evaluate each process scenario. For instance: in Purchasing: inputs, spare parts, fixed assets, sundries; in Inventory Flow: purchases, transfers, and loans. • Process formalization: process, procedures, rules and policy documentation.
2. Training in process reengineering	• Oliver Wight consulting firm • Implementation consultant • Implementation team • IT division	• MRPII training program with an emphasis on "formula accuracy" and "inventory accuracy" as critical elements to link the implementation of the sales plan, production planning and materials requisitions programming.
3. Training in ERP system use	• Implementation consultant • Implementation team • IT division	• Demonstration versions and handbooks. • Training of implementation team leaders. • Training of implementation teams by their leaders.
4. Process remodeling	• Implementation consultant • Implementation team • IT division	• Identify business processes prototypes with users. • Identify divergences between functionality of the ERP system and business processes.
5. System trial runs	• Implementation consultant • Organization and methods specialist (one person) • Implementation team • IT division	• Selecting real data to test each module. • Stand-alone module testing. • Interconnected module testing. • Parallel trials using original systems and new modules.

project in some processes. During implementation of the manufacturing modules, for instance, there was a step back when the area manager did not directly approve a process change. Regarding the involvement of user division personnel:

Corporate changes led to high personnel turn over and rightsizing. Some key elements in the ERP system implementation project were replaced by others who had to get new training.

Implementing the system required appropriate documenting and recording of each and every purchasing process and stock movement. Montero and Burga realized that best business practices and formalizing people's work would attract division manager's interest as well as attention from the general manager.

Jorge Figueroa, the company's logistics head, says:

When we started implementation, nobody respected the time periods. We had no idea how big an ERP system implementation project would be because there were no previous experiences in Peru. As the implementation moved on, management gained a better understanding of the project's scope and size, leading to a change in mindset. So we were able to make better decisions.

Since the very beginning, the implementation team had to face the difficulties stemming from divergences between the ERP system functionality and business processes. Although the ERP system had been designed as a standard application that does not require significant changes for specific users, the system needed configuration so it could be adapted to each process's individual requirements.

Configuring a system requires much attention and experience. A single change in a configuration table has a substantial impact on the way the ERP system will operate. At *Alimentos Peru*, configuring the ERP system followed the process models prepared by each user division.

About the differences between the ERP functionality and *Alimentos Peru's* business processes, Marcela Burga holds:

Together with the accounts payable module rejected by treasury, we also returned the cost control module. According to users, the cost data supplied by the module did not provide the depth of detail required by IFG. Systems must not only be good; users must also accept them. We had to develop these modules independently and design interfaces with BPCS, thus delaying the implementation process.

Jorge Figueroa adds:

The main implementation issues arose when the ERP system functionality failed to match the processes. When we evaluated the ERP system the consultant told us that the system had the capacity to do whatever we required from it but later we found some surprises.

Carlos Montero remarks:

The guiding principle during implementation was not to modify the ERP system.

Implementing the logistics and manufacturing modules took until May 1998. The last three months were devoted to final user training, in particular factory workers, and to trial runs. Their own bosses trained personnel. Bosses would get their people together and prepare an explanation talk with support from IT personnel.

The final trials included a three-week test running the original systems and the new system in parallel. According to project participants, this was the best way to teach future users how to use this tool.

Marcela Burga has the following comments about the final stage:

Immediately before launching our manufacturing and logistics modules, we found out about difficulties in other countries with the system's start up, and general management asked us to take every possible precaution. However, our personnel felt they were ready. When we started the system, we found only very small errors.

Implementation of the accounting modules started in October 1998 and lasted three months. The sales and marketing modules took four months, starting in January 1999. Youth, a proactive attitude and a greater decision-making capacity among personnel in these divisions led to a fast implementation.

By that time, Montero had realized that implementing the ERP system had effectively introduced changes in *Alimentos Peru's* business practices that would have a positive impact on the company's financial position. Some of the changes were the following:

- Availability of consistent information that suppresses the need for manual integration and reviews that was at the source of many human mistakes and was time consuming. With the new system, company managers had access to a consistent and single version of the data.
- Standardization and simplification: The company started to use a single language. Materials could be identified in a single way throughout the company and criteria for the various activities were likewise unified.
- Formalization of operations: Before introducing the system it was usual to ask and use materials without the corresponding purchase order, as was sending raw materials to warehouses without using standardized and updated forms. When the new information system and the MRP II concept were introduced, personnel were obliged to fill in forms and check the data for each operation. *"From the very beginning of the implementation, bosses were called home, even late at night, to ask authorization to close manufacturing orders needed to close accounts. Everybody had to get used to operating formally,"* says Marcela Burga. The period for closing accounts at the end of every month diminished from more than one week to just two days without any need for the people to work overtime.
- Better business processes: Sales and manufacturing programming depended on end-of-month stocks. This practice led to piling up of finished product and raw materials stocks in warehouses so that orders could be filled at the end of the month. New ideas introduced by MRP II and using the ERP system allowed for operations to be spread out homogeneously throughout the month. Two new positions were created thanks to the new information system: a demand manager and the master production planner.

CURRENT CHALLENGES AT ALIMENTOS PERU

The implementation of the ERP system at *Alimentos Peru* ended in April 1999. A few months later, while he was on his way to work, Carlos Montero thinks about new user requirements. He thinks about the most valuable aspects of this experience and the learning process the company went through while implementing the new information system.

Looking back at the process, Montero remembers the tough decisions that were needed, the many sleepless hours needed to implement the system and redesign the business along a road full of switchbacks. He says:

What did we learn? How could we have reduced total implementation time? What can help us in future implementations?

Montero is aware that there is a new role for the IT manager. *"More than programmers and operators, we are now systems analysts and we have to support users to continuously improve their business processes,"* he adds.

Now *Alimentos Peru* has the ERP system as a foundation for its transactions. The company, however, has new requirements. Some of the technologies under evaluation as part of the new technology plan are the following:

- Data warehousing and business intelligence tools to support the marketing user division in sales planning. In this regard, Montero wonders, *"Shall we have the analysis capabilities to use these new types of tools? Will we be able to use them well and benefit from these new tools? Will anyone arrive at any conclusions using the data provided?"*
- Interorganizational information systems that would provide an EDI interconnection with the company's main customers to enhance supply operations. Montero thinks that any EDI change must go hand in hand with a change in mindset among sales people. *"The sales person will no longer need to provide plenty of information or long price catalogs. They will devote themselves to sell."*

ACKNOWLEDGMENTS

The authors would like to express their gratefulness to Antonio Diaz-Andrade for his collaboration in this case and acknowledge the comments provided by the anonymous reviewers.

FURTHER READING

Appleton, E. L. (1997). How to survive ERP. *Datamation,* 50-53.

Bancroft, N. H., Seip, H., & Sprengel, A. (1997). *Implementing SAP R/3* (2nd ed.) Greenwich, CT: Manning Publications.

Davenport, T. H. (1998). Putting the enterprise into the enterprise system. *Harvard Business Review, 76*(4), 121-131.

Davenport, T. H. (2000). *Mission critical: Realizing the promise of enterprise systems.* Boston: Harvard Business School Press.

Hecht, B. (1997). Choose the right ERP software. *Datamation, 43*(3), 56-58.

Langenwalter, G. A. (1999). *Enterprise resources planning and beyond: Integrating your entire organization.* CRC Press - St. Lucie Press.

Norris, G., Dunleavy, J., Hurley, J. R., Balls, J. D., & Hartley, K. M. (2000). *E-business and ERP: Transforming the enterprise.* John Wiley & Sons.

Shtub, A. (1999). *Enterprise resource planning (ERP): The dynamics of operations management.* Kluwer Academic.

ENDNOTES

[1]   All individual names, company names, and brand names have been disguised.
[2]   Carlos Montero was away from *Alimentos Peru* from mid-1995 until the beginning of 1996.
[3]   After finishing the ERP system selection process and internal restructuring, the company's IT manager started reporting directly to the local general manager.

APPENDIX A. FINANCIAL STATEMENTS

BALANCE SHEET (in thousands of U.S. dollars)					
	1.994	1.995	1.996	1.997	1.998
Current assets	10.241	14.133	16.136	19.273	17.023
Non-current assets	13.064	13.724	15.172	21.197	20.200
TOTAL ASSETS	**23.305**	**27.857**	**31.308**	**40.470**	**37.223**
Current liabilities	9.615	14.648	14.673	21.647	20.119
Non-current liabilities	2.770	1.532	767	0	
TOTAL LIABILITIES	**12.385**	**16.180**	**15.440**	**21.647**	**20.119**
Shareholders' equity	10.920	11.677	15.868	18.823	17.104
TOTAL ASSETS AND LIABILITIES	**23.305**	**27.857**	**31.308**	**40.470**	**37.223**

P/L STATEMENT (in thousands of U.S. dollars)					
	1.994	1.995	1.996	1.997	1.998
NET SALES	48.439	54.318	60.658	63.056	61.865
COSTS AND EXPENSES					
Cost of sales	32.208	37.162	38.823	40.114	41451
Sales espenses	6.869	9.472	11.724	13.584	14.086
Overhead	5.370	4.990	4.160	3.443	3.142
Total	44.447	51.624	54.707	57.141	58.679
RESULTS FROM OPERATIONS	**3.992**	**2.694**	**5.951**	**5.951**	**3186**
OTHER REVENUES (EXPENSES)					
Financial revenues	119	109	138	162	161
Financial expenses	-1.505	-1.376	-1.216	-1076	-1.062
Other, net	-25	47	-170	-353	8
Results from exposure to inflation	714	154	54	-30	-1.708
Total	-615	-1.066	-1.194	-1.298	-2.601
PROFITS BEFORE PARTICIPATIONS	**3.377**	**1.628**	**4.757**	**4.617**	**585**
Participations				-574	-154
Income tax	-372	-436	-841	-1.550	-415
Total	-372	-436	-841	-2.124	-569
NET PROFIT	**3.005**	**1.192**	**3.916**	**2.493**	**16**

APPENDIX B. INFORMATION SYSTEMS EVALUATION AND RECOMMENDATIONS

Executive Summary

Current Situation

The current information systems include both those originally developed for Real Peruana Inc. and those for Estrella S.A. These systems are still being used in Alimentos Peru.

Since the merger in 1993, there has not been any major development/update effort that would improve the systems. As a result, they do not effectively support the current organizational business processes. The major problems with the information systems include:

1. Lack of timely and reliable information.
2. Lack of integration among existing systems.
3. Duplicate systems for a number of functions.
4. Lack of flexibility. Most of the current systems were designed following rigid structures that do not allow the IT staff to easily update the systems. When they were designed no appropriate programming tools were available.
5. Lack of system documentation.
6. The IT division is mainly concerned with maintaining existing systems; there is not enough time for new developments.
7. In accounting, finance and sales divisions information has to be handled or consolidated using special programs.
8. Systems are not user friendly.
9. Lack of standard IT policies, rules, and procedures.

Recommendations

1. The areas needing information systems improvements are the following, by order of priority:

- Logistics
- Employee Payrolls
- Accounting and Financial
- Commercialization
- Manufacturing
- Human Resources

2. Improve or replace basic systems using integrated systems including interconnected modules and applications for these areas, preferably with preprogrammed packages.

3.   Client-server hardware architecture should be adopted by installing a local area network (LAN). Also, evaluate and follow up the use of AS/400 server capacity and standardize PCs.

4.   Organization-level recommendations:

- Setting up an IT Steering Committee under the Management Committee to ensure basically that systems development would be aligned with business goals.
- Enhance expertise in the IT division by hiring new personnel and training for present employees.
- Change the IT division's organizational structure and create project-oriented teams. Also establish a systems career path.

J. Martin Santana is an associate professor of information technology at the Escuela de Administración de Negocios para Graduados (ESAN) in Lima, Peru. He holds a PhD from International University and an MS in information systems from the École des Hautes Études Commerciales, Montreal. His research interests include electronic business, systems development approaches, and conflict management in the development process. He has published in the areas of the use of global applications of information technology, the management of the systems development process, and the consequences of information technology in organizations.

Jaime Serida-Nishimura is an associate professor of information systems at the Escuela de Administración de Negocios para Graduados (ESAN) in Lima, Peru. He received his PhD in management information systems from the University of Minnesota. His research interests include electronic business, strategic impact of information technology, group support systems, and the adoption and diffusion of information technology in organizations.

Eddie Morris-Abarca is a senior lecturer of information technology at the Escuela de Administración de Negocios para Graduados (ESAN) in Lima, Peru. He holds a BS in information systems from Universidad Nacional de Ingeniería (Peru). He is CEO of InfoPlanning, a local consultant firm specializing in IS planning and business process reengineering. He is currently vice president of the Peruvian Association for Computing and Information Systems.

Ricardo Diaz-Baron holds an MBA from Escuela de Administración de Negocios para Graduados (ESAN) in Lima, Peru. His areas of interest are the use of information systems for supporting organizational planning and control. He worked in operational research and control of power systems at Universidad de Piura and was a consultant in the electrical sector in Peru. At ESAN, he worked as a research assistant in the information technology area. Currently, he is a consultant of the Office of Public Investment at the Ministry of Economics and Finance in Peru.

This case was previously published in the *Annals of Cases on Information Technology Applications and Management in Organizations*, Volume 3/2001, pp. 244-258, © 2001.

About the Editor

Mehdi Khosrow-Pour, D.B.A, is executive director of the Information Resources Management Association (IRMA) and senior academic technology editor for Idea Group Inc. Previously, he served on the faculty of the Pennsylvania State University as a professor of information systems for 20 years. He has written or edited more than 30 books in information technology management. Dr. Khosrow-Pour is also editor-in-chief of the *Information Resources Management Journal*, *Journal of Electronic Commerce in Organizations*, *Journal of Cases on Information Technology*, and *International Journal of Cases on Electronic Commerce*.

Index

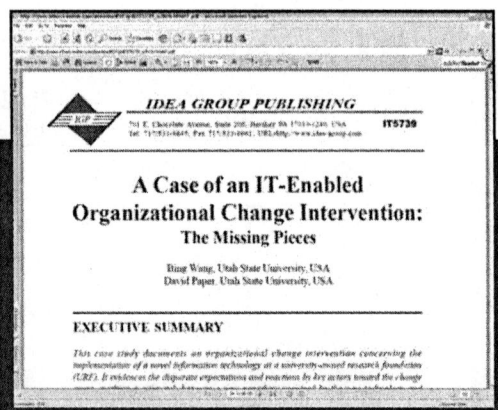

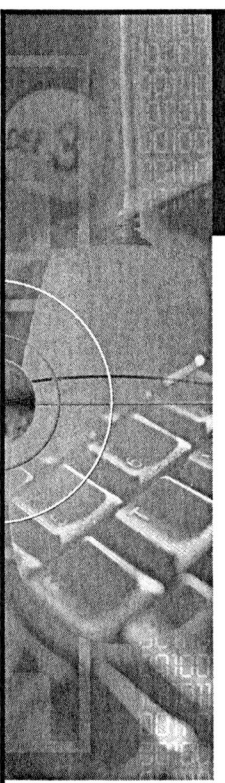